Weather, Climate
& Human Affairs

Weather, Climate & Human Affairs

A book of essays and other papers

HUBERT H. LAMB

ROUTLEDGE
London and New York

First published in 1988 by
Routledge
11 New Fetter Lane, London EC4P 4EE

Published in the USA by
Routledge
in association with Routledge, Chapman &
Hall, Inc.
29 West 35th Street, New York NY10001

Typeset by
Scarborough Typesetting Services
Printed in Great Britain at the
University Press, Cambridge

*British Library Cataloguing in Publication
Data*

Lamb, H. H.
 Weather, climate and human affairs: a
 book of essays and other papers.
 1, Climatic changes – History   2. Man –
 Influence of environment – History
 I. Title
 551.6   GF71

 ISBN 0-415-00674-0

*Library of Congress Cataloging in
Publication Data*

Lamb, H. H.
 Weather, climate, and human affairs.
 Includes index.
 1. Climatology.   2. Man – Influence of
 climate.
 I. Title.
 QC981.4.L36   1988   551.6   87-15347

 ISBN 0-415-00674-0

To my dear wife Moira
who has sustained me
through the writing of this book
and all my work

CONTENTS

PREFACE

I have been asked to arrange this collection of lectures and other papers because of their relevance to any basic understanding of climate, and planning problems of the present day that are affected by climate.

Now that the study of climate and its development has emerged from long years of neglect, this relevance is likely to be more widely appreciated than it was when the earlier papers in this book first appeared. Not all the items included here have been published before, and those that have appeared before were found in an inconveniently wide variety of sources.

My career in meteorology began with research aimed at understanding the haars, or North Sea fogs, which are liable to come in suddenly over the east coast of Scotland. Almost from the outset I also took part in the preparations for the first regular passenger flying on the transatlantic air route, which opened in July 1939. Owing to the outbreak of war, I soon found myself charged with the awesome responsibility of training meteorological forecasters for the then new Irish Meteorological Service at a time when I was still learning many fundamentals of the subject myself. The route at that time was operated by flying boats on the shortest crossings, between Ireland and Newfoundland in summer, and via Lisbon, the Azores and Bermuda (or occasionally farther south) in winter. I stayed on in Ireland to supervise the forecasting service. Despite there being no weather observation reports from the ocean available to a neutral country during the war, the service was operated throughout the war years with only one minor accident. For this I must pay tribute to the conscientious professional standards of my mostly Irish colleagues of those days. The scientific basis of the forecasting was routine analysis of the weather situation on maps covering a larger proportion of the northern hemisphere than had been practised anywhere until that time. Diagnosis of the situation over the Atlantic Ocean was aided by careful study of the weather reported by the aircraft on the routes, thus improving the fixing of the positions of the 'weather fronts' and analysis of their cloud structure and development with time.

Shortly after the war, I was sent by the UK Meteorological Office on a whaling factory ship to Antarctic waters, charged with forecasting advice for two amphibious (Walrus) aircraft carried on the ship for spotting whales and to guide the catcher vessels. With once again no weather observation reports from the ocean, and this time most reporting stations on land at least a thousand miles away, the technique used was to analyse the atmospheric flow patterns and weather over as nearly as possible the whole (southern) hemisphere, once a day, based on the reasonably good coverage of the lower latitudes from South America round to Australia and New Zealand. And to fill in the detail of weather systems approaching the ships' operating region, again the aircraft were used: an aircraft was sent out to explore how far clear weather extended in the direction from which the next disturbance was expected. This procedure again gave a safe and incident-free record. And much was learnt about the development of fronts and cyclones over the Southern Ocean. These experiences convinced me of the importance of paying most careful attention to the structure, or anatomy, of the weather systems and their movements and life history. It seems that the value of this work is still not widely enough known or understood. The lessons to be learnt from it are encapsulated in Chapter 16 of this book.

Later on, when I came to study climatic variability and changes (which up to then had been a largely neglected subject), I used the same methods from the beginning to understand what was producing the changes. One must know the observable facts of how climatic patterns evolve before attempting to develop theory very far. The first tasks must be to establish an actual record of what changes of climate have taken place in the past. Methods of analysis must then be devised to present the picture in simple enough form to make it possible to visualize the evolutions taking place in the course of each and every recurring climatic process, on all the time-scales which appear to be worthy of note in the past record. These include: (i) the roughly 23-year cycle in the frequency of blocking anticyclones occupying the more usually low-pressure zone between Greenland, Iceland and northern Europe, a cycle identified by Flohn and Rex about 1950; (ii) the roughly 80–100 and 200-year cycles in the general wind circulation, which pass through 'high and low index' phases of strength, and through stages with higher or lower latitude 'preferred positions' of the main westerly winds in middle latitudes, as first described by Willett about the same time; and (iii) the change from early mediaeval warmth to later mediaeval and seventeenth- to nineteenth-century cold climate on which the present writer has worked.

Similar efforts to reconstruct the prevailing wind patterns of the last ice age and the successive stages of postglacial climate have taught us a great deal. The patterns of the wind circulation are intimately related to the weather at each stage (as explained in Chapter 14 and elsewhere in this book).

From these beginnings one can proceed to consider the causation of changes, including any intrusions of external influences, and trace the entry and growth of new developments. Through such arguments much more is understood now than was 20 years ago, for instance, about how the changes in the distribution of the sun's heat-supply resulting from cyclic variations in the Earth's orbital arrangements control the timing of glacial (ice-age) and interglacial development and the main stages within each. And much about how veils of volcanic matter in the atmosphere affect the development of the wind circulation and climate has been learnt by similar analysis.

Establishing the facts of the past climatic record involves the use of data and techniques from many branches of learning. We need as much detail, and over as much of the Earth, as possible. And we need to test it. To an important extent, information from independent sources and independent lines of argument can provide tests and corroboration of the facts deduced.

Once we have a firm climatic record, which – as it turns out – reveals some sharp changes that have occurred, we find evidence of events which surely must have affected human history, just as the African droughts and harvest shortfalls in various countries do today.

In the presentation of the record of climate, I have consistently striven for the simplest forms, and have, as far as possible, avoided such fashionable perversities as naming decades such as 1971–80 the 'seventies', rather than 1970–9 which the word implies. Similarly, the chapter on fronts and frontal analysis is a plea for realism. Fronts drawn like the spokes of a wheel, radiating rod-like from the centre of a vigorous cyclone, are unrealistic, when the strongest winds must have swept part of the front farther forward (often much farther forward) than the parts near the calmer regions in the centre and on the outskirts of the depression. Simplification of the picture must not be achieved by distortion, which is bound to obscure what is going on (and results in wrong forecasts). Simplicity is always to be aimed at, but not at that price.

In the essays which make up the chapters in Part I of this book, the history of climate and any noticeable impacts on human history supply the unifying theme. These chapters have been placed first, because they will be the easiest for interested people who are not scientists to read. Part II is devoted to probing the nature and evolution of the processes by which climate and weather develop. Chapters 6–9 already cover some of the processes of change in the environment, in the ocean and in the landscape but these chapters will not present any difficulty to most readers and some may well find these the most interesting parts of the book. Although the last sections of the book are largely devoted to processes and mechanisms, with a view to diagnosing current and future developments, they were written for varied (and sometimes lay) audiences. Many interested people without a specifically scientific education will find these chapters also quite easy to read and will be led to a better insight into important problems and issues of the present day.

Hubert H. Lamb
Norfolk
May 1987

ACKNOWLEDGEMENTS

My thanks are due to the Meteorological Office in the United Kingdom under the leadership of the late Sir Graham Sutton FRS as Director-General. It was there, thirty years ago, that I was first given the chance to work on the history and development of climate and, before that, gained my experience of the Antarctic and some years of work on Antarctic meteorology. This gave me an insight into the global working of the 'climate machine', the general wind circulation over both hemispheres. I also wish to express my gratitude to the University of East Anglia, which made it possible to broaden my work by inviting me to set up the Climatic Research Unit in 1971–2. I also owe thanks to Shell and to the Nuffield Foundation for the funding which established the Unit at that time. And by 1975 I was also greatly indebted to the Rockefeller Foundation of New York and to the Wolfson Foundation in London, through the good offices of Dr Ralph Richardson and Lord Zuckermann OM respectively, for further generous and disinterested funding which rescued the Unit from premature collapse and gave it the lease of life needed to become fully viable. Further support from the Wolfson Foundation has in 1986 given the Climatic Research Unit its first permanent, purpose-built building.

There are many individual friends and colleagues in various countries whom I thank for valuable exchanges of ideas and gifts of data. They cannot all be mentioned here, but their collaboration is much in my mind. Those who must be mentioned include Mr C. K. Folland of the Meteorological Office, Bracknell (England); Dr Knud Frydendahl of the Danish Meteorological Institute, Copenhagen; Dr Christian Pfister of the Historical Institute of the University, Bern, Switzerland; and my colleagues in the Climatic Research Unit at Norwich, Drs T. M. L. Wigley, Trevor Davies, Graham Farmer, Philip Jones and Astrid Ogilvie. Professor André Berger of Louvain la Neuve, Professor H. Flohn of Bonn, the late Professor H. E.

Landsberg, University of Maryland, Dr J. Maley of the Université des Sciences et Techniques de Languedoc at Montpellier, France, Dr Jerome Namias of the Scripps Institution of Oceanography at La Jolla, California, Professor J. Neumann of Helsinki, Professor R. E. Newell of the Massachusetts Institute of Technology, Cambridge, Mass., and Dr S. H. Schneider at the National Center for Atmospheric Research, Boulder, Colorado, and others, have contributed valuably, but sometimes unknowingly, by keeping me in touch with their work and thinking. Some valued colleagues are thanked in the particular chapters to which their work contributed. I hope that none of these will be too much dismayed by my limited acceptance of the fashionable carbon dioxide warming thesis, which looks so straightforward in the physics department but about which I believe there may be reasons for reservation in the wide world of atmosphere, oceans, soils and biosphere.

I wish to thank all those authors and publishers who have allowed me to use the diagrams, maps, etc. in the publications which are here reproduced again either as a whole or in part. Acknowledgments are also due to the editor and publishers of *Contributions to Atmospheric Physics/Beiträge zur Physik der Atmosphäre* for allowing me to keep my copyright on the maps here printed as figs 4.7, 4.8 and 4.9; also to my friend Dr A. Bourke of Dublin for fig. 10.1; similarly to Prof. R. A. Bryson of Madison, Wisconsin for fig. 12.9; and to C. K. Folland for figs 12.11, 12.13 and 12.14; and to Dr J. Murray Mitchell for fig. 13.2; as well as to Prof H. C. Willett of The Massachusetts Institute of Technology for permission to use the figures here reproduced as figs 18.3 and 18.4; as well as to all concerned in providing me with the satellite pictures in fig. 16.19.

Humanity is beset by many problems, among which the threats of failure of the energy supply, of poisoning of the atmosphere and the waters by our own activities, the risks of nuclear disaster (even from nuclear wastes) and the new plague known as AIDS may seem more pressing than any threat from climatic fluctuation and change. But many of them are interwoven with climate. Moreover, the previously unexpected scale of the long droughts in Africa has manifestly provoked disaster on a new scale of starvation and deaths, which has shocked the world in the last few years. We cannot ignore any of these impacts on modern life.

Finally, I express my sincere acknowledgement to the previous publishers for their permission to use in this book so much material from the papers which they first produced in print. And, once again, my thanks are due to Methuen & Co., who have promoted my work in several volumes and over many years, for encouraging me to produce this book. I wish them luck as they become part of the new Routledge.

1
Introduction

Attitudes to climate and awareness of its fickleness have changed down the ages, but probably never more than during the present century. Primitive societies, or at least the wiser heads among their folk, have always known that the vagaries of climate and the longer-lasting changes of its moods were an ever-lurking source of danger.

When the great glaciers of the last ice age melted, replenishing the depth of the oceans, the sea flooded back over the great plains where the North Sea now is, and the peoples living there had to go. No doubt the memory of that, and of the corresponding invasions of the sea over populous low-lying coastlands round the Mediterranean and the coasts of Asia and elsewhere lingered on for many generations; probably it was the origin of most of the legends of the great flood (e.g. Noah's Flood). Erosion of the coasts of the southern North Sea has continued in times of storminess until today, partly because that area of the Earth's crust is slowly downwarping, and therefore sinking, and so allowing the sea-level to rise along those coasts, and partly because of some variation of the world-wide sea-level between periods of warmer climate with glaciers melting and colder times when the glaciers grow. In prehistory, and in early historical times, the coasts of many lands, including those around the North Sea and the Mediterranean, were the sites of the vital salt-making industry, based on the evaporation of sea-water. Salt was the most prized commodity, as it filled the food-preserving role of the refrigerator in modern society. And there must have been great losses of life when the sea overran the populations concentrated near those industrial sites.

The fertile soils of the Ganges delta and Bangladesh are needed to feed a teeming population; and years may pass before the next time a typhoon sweeps the sea in to submerge a huge area. Or again, in times of plenty and expanding population in the past, cultivation in northern Europe spread up the valleys and towards the heights. But when cold, wet summers came, the harvests failed. And in many recorded cases from the Middle Ages to later centuries – notably in the 1690s and around 1740 and 1816 – the sequel was starvation, disease and death or, in the later cases, emigration.

By contrast, the urban and suburban populations of modern wealthy industrialized countries are not often exposed to the weather and only rarely concern themselves about its vagaries, save when a holiday or sporting events are spoilt or the fuel bills significantly increased. And in the earlier part of this century it was widely believed that scientific observations had shown that whatever the short-term variations and disappointments, the climate would always return to 'normal' after a few months or, at most, a few years.

This impression was formed because the first 100-year runs of meteorological observations officially organized in the mid to late eighteenth century, in the leading cities of Europe and northern America, showed that the climate of the 1870s to 1890s was very similar to that of the 1770s to 1790s. There had, in fact, been some noteworthy variations in between, including major advances and retreats of the Alpine glaciers. In the light of twentieth-century observations and the longer records now available (through knowledge of glacier variations, lake and river levels, the yearly growth rings of trees, and so on), the return around 1880 to conditions much as they had been a hundred years earlier might be regarded as more or less accidental, except that there are some signs of a cycle of close to 100 years in length in the longer runs of climatic data. But changes also occur on many other time-scales, down to the daily and yearly changes of weather, and up to the changes over thousands and tens of thousands of years between full glacial (ice-age) climates and the warmest interglacial climates. The over-hasty conclusion that, in the long run, climate does not vary significantly certainly led to complacency about the matter in many quarters and was readily adopted as a convenient working assumption by meteorologists, historians, water-engineers and others.

This is the basis on which the so-called 'return periods' of rare events, such as an extreme gust of wind or an outstandingly heavy rainfall, are calculated, using the observation statistics of a limited period to assess the probable average separation in time between successive occurrences of any particular extreme or unusual conditions. The period of observations used for these calculations is commonly 30 years, rarely very much longer. And yet the assessment of a 'once-in-two-hundred years' gust of wind, flood or frost (or whatever) may be quite misleading if the general condition of the climate does not persist that long. Since 1950 there have been cases in many parts of the world where such assessments, based upon the climatic statistics of the first half of this century, have failed to provide a satisfactory guide for planning. An awkward example occurred in the late 1950s, when, during the building of the Kariba Dam in southern Africa, the estimated once-in-50-

years flood of the River Zambezi was exceeded in each of three successive years. A still more extreme example was provided by heavy snowfalls in parts of South Wales in early 1978, and again in 1979 and 1982, which all exceeded the supposed once-in-200-years expectation.

Continual efforts and education are needed to guard against mistaken actions and policies, which are liable to arise from modern urban societies' increasing remoteness from and unacquantance with the natural world. Happily, the exploration of nature in all its branches is becoming an increasingly popular subject. In Denmark, observation of the conjunction of rapid advances in biological technology and human techniques of intervention in natural processes with the reduced contact of the population with the natural world has led in recent years to reconsideration of biology teaching in schools. It is proposed that the subject should be taught to the youngest grades and should become a compulsory one also in the senior classes, with particular attention to the environmental needs of living organisms. It is surely time for similar thinking about climate and all the other aspects of the natural world which are influenced by climate.

Through most of the first half of the twentieth century the climate was becoming generally warmer in northern lands; also more moisture was reaching the interior of the great Eurasian continent and even into parts of Africa close to the Sahara and Kalahari deserts, while the Indian monsoon became more reliable. These benign tendencies meant that there was no pressure, no sense of need, to support research aimed at understanding climatic change and its origins or physical mechanisms.

Most people are always ready to slip into easy ways and take their continuance for granted. But some, at least, must gain a longer perspective, so that there is some degree of preparedness for changes and the shocks that accompany them and to ensure that timely steps are taken to avert or lessen disasters.

The changes of tendency of the climate that have shown themselves since 1950, and especially since the 1960s, have had much to do with the intensified interest in, and support for, research in the subject. And this has been compounded by the rapid development of wide public awareness of what human activities have been doing to the environment. There has been growing concern about the threats presented by our pollution of the atmosphere to the future development of the global climate itself and, more immediately, through acid rain, to the vegetation and to fish in the lakes and rivers. Extreme illustrations of our pollution of the planet have come to light in such examples as the discovery of DDT in the snow over Antarctica, and of radioactive caesium from the nuclear fuels plant at Sellafield (Windscale) in northwest England being found in the sea surface waters of the polar current off east Greenland. The nuclear accidents in 1980 at Three Mile Island in the eastern USA and at Chernobyl in the Ukraine in 1986 should make clear to us all the danger of establishing nuclear plants, particularly in the heart of heavily populated regions, especially when the alternatives of naturally renewed sources of energy (wind and tidal power, etc.) and restraint and economy in the use of energy have been largely ignored.

Sulphur dioxide and perhaps other invisible by-products of burning coal and oil are commonly blamed for damage widely observed in recent years to the forests of Europe and North America, particularly in regions down-wind from the major industrial areas and conurbations. Carbon dioxide, which is the main end-product of fuel combustion, although not poisonous to people and animals, has effects upon the balance between the incoming solar radiation and outgoing radiation from the Earth that much recent research suggests could soon produce a rapid and drastic upset to the whole world's climate. The effect may be worsened by similar responses of solar radiation to other artificial pollutants, such as the nitrogen oxides produced by fertilizers, and overall it could supposedly lead, within about a hundred years or less, to a radical warming of world climate and a great poleward shift of the rainbelts, deserts and available areas for crop production

These matters are returned to in the final chapter of this book, where other causes of climatic variation both natural and resulting from human activity are briefly surveyed, including the likely consequences if humanity commits the ultimate folly of nuclear war. Lord Zuckerman, OM, FRS, formerly Chief Scientific Adviser to the British governments of Edward Heath and Harold Wilson, has recently written (1986) that 'Man's present political problems are miniscule in relation to what could result from major changes in climate, and someone from outer space viewing our globe . . . could well suppose that nations of today behave like people who quarrel violently and murderously over immediate trivialities on the fiftieth floor of some huge modern Tower of Babel, oblivious of the fact that it is blazing away merrily beneath them.'

There has been a great increase of effort in climatic research in recent years because of these fears, even at a time of cutbacks in many other branches of science. A World Climate Programme of research was set up in 1979 by international agreement, setting out a programme for many years' work under the joint guidance of the World Meteorological Organization and the United Nations Environment Programme. Nevertheless, in this realm, as Sir Crispin Tickell, Permanent Secretary of the British Overseas Development Administration, has recently commented (1986), 'Governments have not so far been willing to support their good intentions with hard cash. In 1983 the World Climate Impact Programme [the international organization's body concerned with the impact of climatic changes] enjoyed less than half the recommended minimum budget.'

Most of the funding goes to support theoretical modelling of the atmospheric circulation and climate, particularly the global climate, its responses to carbon dioxide increase and to other human disturbances of the environment. The latter include (i) the surely irresponsible destruction now proceeding of the great equatorial rain forests, (ii) the more understandable

Soviet proposal for diversion of the Siberian Arctic rivers to water central Asia, where population and industry are increasing and the water table is going down and (iii) the ever-mounting output of waste heat, which is the unavoidable end-product of all our industrial processes and domestic fuel combustion. The Soviet rivers proposal is reported to have been abandoned in 1986, after decades of preparatory research, because the risks to the environment are finally deemed too great. It is now thought that an alternative technology for water conservation can fill the need.

It is, and is likely to remain, a difficult matter to verify how the computed climatic results of modelling theoretical circumstances – such as doubling the atmosphere's carbon dioxide content, or a world without forests – would correspond to reality. To have a basis of reality for our thinking about these problems, it is essential to reconstruct – in whatever detail is possible – the past record of climate, world-wide. We need to know the range of climate's observed behaviour since prehistoric times, when the world's surface looked somewhat different, and the various regimes that have occurred. We should then be able to test the ability of the theoretical models to explain at least what has actually occurred. Periods in the past when, as we already know, great climatic changes took place surprisingly quickly deserve particular attention, for the record shows that climate can appear stable over long periods and then undergo rapid change. Reconstruction studies are much less costly than theoretical modelling, with its advanced mathematical formulations and extended computer integrations of what may be likened to numerical forecasts of the global atmospheric circulation, carried forward not just for several days (as in weather forecasting), but day by day over a number of years ahead. The simpler work on reconstruction of the actual past record still tends to be starved of funds in many countries, though it is vitally needed if the more obviously exciting theoretical work on modelling the future is to be at least partly authenticated.

Work in recent years on the astronomical factors that control the recurrence of ice ages and warm interglacial periods, such as the present, seems to have been adequately verified by dating of the traces of past glaciations registered in the material deposited on the ocean bed all over the world. On this basis, there is an expectation that the first phases of the next glaciation will be upon our descendants within about three to seven thousand years from now (Berger 1980).

Despite our growing knowledge of recurring ice ages and evidence of other lesser, but abrupt, changes of climate within the last 300 years, climate is still very widely taken for granted, and assumed to have no awkward trends. This attitude prevails even when – somewhat inconsistently – the climate is acknowledged as having something to do with Soviet grain harvest shortfalls and the serious drought and food crises in Africa in the 1970s and 1980s. Journalists' accounts and public discussions of what to do for the future commonly ignore the climatic element in these difficulties. The commonest assumptions still seem to be that the climate will soon get back to 'normal' or that, if any irreversible change has taken place, it must be somebody's fault, attributable to human actions in some form. Attempts at aid to Third World countries which were afterwards seen as ill-considered, such as the provision of more wells to get water up from an already sinking water-table, are sometimes mentioned in this connection. Some blame the actions of governments in developing countries, with suggestions of corruption or evil-intentioned political motives. There may have been cases of all these things, but it is abundantly clear also that nature produces changes of climate, and until now these have probably been the main changes. Chapter 12 is concerned with the long record of climate changes in Africa and the severe drought problem in the Sahel, Sudan and Ethiopia in the past two decades, as well as in the corresponding zone of southern Africa, from the fringes of the Kalahari to Zimbabwe and the Transvaal.

From the other side, among those most concerned with the problems of the Third World, climatologists have begun to come under criticism for remaining isolated in their scientific institutions and not becoming involved in agricultural and development planning. This point has been made by Dr Asit K. Biswas (1984), sometime associate of the World Bank. Perhaps some initiative is needed here from the agencies active in such planning to co-opt suitably trained climatologists to their councils. But it must be a matter of some doubt whether there are yet enough scientists suitably trained in the various aspects of climatic development and change and in studying the impact of these in various fields. Undoubtedly, the greatest need now is for education – that means educating more physical and biological scientists in the past record of climate, its recent history as well as the many prolonged processes of its development. It also means educating them to be better at identifying the impacts of climatic change. And of course, the general public needs to be informed – from children in schools to experienced decision-makers in government and industry – about the implications of the subject for society and for the world we live in.

It is clear that climatic changes have sometimes had a great impact on human history and prehistory, and several examples are discussed in this book. In the case of some known historical changes, however, not enough is known at present to establish exactly what was happening to the climate – at least not in the requisite detail – to decide what part it may have played in the changes in society and political organization. For instance, in the Mediterranean region in the Middle Ages and after, and in the times before written history in much of Africa, both climatic researchers and historians may have things to learn from each other and sometimes from working in collaboration. Statistical studies of radiocarbon-dated evidence of cultural changes in the prehistory of the United States Middle West, and of the Sahara and elsewhere, e.g. by R. A. Bryson and his colleagues at the University of Wisconsin (1970) and by M. A. Geyl and D. Jäkel in Germany (1974), have repeatedly pointed up the coincidences with times of change in climate and vegetation. Archaeologists seem generally aware that better knowledge of the climatic record may help towards understanding their own data and what was happening in human society at the time they are investigating.

The drawings on rock in caves in the Sahara, and in the Nile valley in Egypt, of the fauna with which human inhabitants there between about 5000 and 3000 BC were familiar, include cattle, hippopotamus, elephants, giraffes and ostriches. Very arid phases followed, culminating around 2500 to 2000 BC and around 1200 BC. But accounts from Carthaginian and Roman times (e.g. from observations on Hanno's voyage south along the Atlantic coast of Africa in the fifth or sixth century BC and Pliny's writings in the first century AD) point to moister landscapes again. Around 338 BC there were Carthaginian settlements on the west coast of Morocco whose inhabitants had milk and meat and produced much wine from their vineyards, some of it for export. And in Pliny's time great herds of elephants wandered in extensive forests at the southern foot of the Atlas mountains and in winter were able to find rich pastures. These elephants died out finally in the third century AD, when aridity was increasing sharply again. Later variations in north Africa are mentioned in Chapter 14 in connection with present troubles in the region. There can be little doubt that landscape changes of this order affected the human populations as well as the animals.

Other arid regions, and especially their borderlands, yield evidence of similar changes. In Mesopotamia, in the Tigris-Euphrates river valley lowlands, the seat of some of the best-known earlier civilizations, catastrophic declines of population occurred around 3800 and 1000 years ago. What exact combination of drought, political unrest, war and, possibly, disease brought about these disasters is yet to be determined, but it is likely that the climate was involved, since apparently changes were taking place on a global scale about those times. The scale of the disaster around AD 900 to 1000, perhaps the most reliably gauged case of such an event, was horrifying: the population of the region is estimated to have fallen from about one and a half million to between 200,000 and 300,000 (see Adams 1965, 1981; Bowden et. al. 1981, pp. 488–94).

Less momentous events in human history also sometimes suggest the influence of changes in the prevailing tendency of the climate. Thus, in England and northern Europe, the general introduction of glass windows more or less coincided with the sharp cooling of the climate in the sixteenth and seventeenth centuries, as the so-called Little Ice Age set in. And the designing of the elegant terraces of Bath, with grand houses exposed high on a steep hillside, coincided with a time of rapid warming and (it is thought) of greatly increased sunshine, around the 1730s. Similarly, the fashion for flat-roofed houses and office buildings took hold in a warm period in the present century, with the drier summers of the 1930s and 1940s, and has caused difficulties for the owners in recent decades.

Unlike my previous book, *Climate, History and the Modern World* (1982), which this book supplements, amplifies and updates, the chapters in this volume concerned with climate and society in the prehistoric and historical past are grouped together in Part I, as nearly in chronological order as possible. Part II deals with the mechanisms of weather and climate and their changes. The historical chapters provide illustrative case histories of how societies lived in this or that region, in climates that differed from today's and in climates which had become erratic or were changing significantly.

Mechanisms, seen on the widest range of time scales, are discussed in Part II and lead to a final chapter which surveys how far the understanding that has been gained up to now enables us to foresee or, at least, plan more wisely for the future. This arrangement of the book may be more convenient for readers whose main interest is in the problems of human affairs and the intrusion of climatic stresses upon them.

With the enormous increase since 1960 in research activity directed towards better knowledge and understanding of the development of climate, scientific insights have been developing fast. Central to our understanding, and to our ability to reconstruct the past, are the intimate links between climate and the global wind circulation, the subject of a brief review in Chapter 14. Opinions about the effectiveness of various external influences, including of the intrusions of human activity, upon climatic development remain in most cases controversial, since assessments vary. Two cases seem established: (i) the effect of veils of fine volcanic dust and aerosol thrown high into the atmosphere by great eruptions – though there are some differences from case to case and further studies are needed; (ii) in the demonstrated control of the timing of ice-age climate development by the regular variations in the Earth's orbit over thousands and tens of thousands of years. It has been wisely said that the test of any scientific theory is its ability to forecast the future reliably, and it is in the case of these two items that climatology appears to come nearest to the capacity for such prediction.

Recognition of various other influences on climate over various shorter timespans, down to a few days or years, is undoubtedly advancing, as is acquaintance with various mechanisms of climate that regularly link the chain of development of a weather pattern. These links include the main limbs of the world's wind circulation, particularly the prevailing westerlies over middle latitudes, and the importance of eddies of various sizes, particularly the often persistent stationary eddies commonly connected with 'blocking of the westerlies'. Also important – and, in connection with some long-lasting phenomena, even more important – are the interactions between wind and ocean, most dramatically seen in the more or less worldwide pattern changes associated with the so-called 'Southern Oscillation' (the name given to an observed see-saw of prevailing barometric pressure levels between Indonesia and the eastern South Pacific Ocean) and the pattern of the 'El Niño'. This is the name given to the situation which occurs when the usually cold ocean current (with surface temperatures locally below 20°C) which moves west across the Pacific near the Equator, supplied partly by the Humboldt Current from the south along the coast of Chile and partly by up-welling deeper water off Peru and Ecuador, is for some months or longer replaced by warm water from the north. This water has the typical character of the ocean surface near the Equator elsewhere, so that this part of the ocean is then up to 3 degC (and locally more) warmer than in the normal situation when the cold water prevails. This anomaly, extending over a huge extent of the ocean surface in the Earth's main heating zone, must significantly affect the whole heating of the atmosphere. The patterns of change in the world's wind circulation associated with the El Niño extend to the Arctic and Antarctic and include, for example, the probability of cold winters in the eastern United States and in Europe.

Issues which are at present the subject of live, and on the whole inconclusive, debate include the likely effect of: the global increase of carbon dioxide in the atmosphere, resulting from our burning of vegetation and fossil fuels (coal, oil and gas); the dust and other combustion products in the atmosphere which also result; the destruction of forests, particularly the great equatorial forests; and the Soviet proposals, mentioned earlier, to divert the water of the country's Arctic rivers (possibly resulting in loss of some of the Arctic Ocean's sea-ice cover) in order to water central Asia, where the lakes and inland seas are drying up and the water-table is falling. In some of these matters, and in the supposed climatic effects of nuclear war, political pressures show signs of further complicating the position. Money to fund research may be more or less readily forthcoming according to what the results appear (or are expected) to indicate. This irrelevant influence – to which all countries seem liable in only varying degrees – may be backed by powerful interests and threatens to cloud the possibilities of scientific understanding.

Observation further suggests that personalities sometimes have a good deal to do with the incidence and removal of this kind of barrier to scientific advance and that, even within one country, the difficulties may differ from one branch of science to another and from one generation to another. But neither political ulterior motives nor the abuse of power by individuals is the whole story. There are also fashions in scientific work, whereby some theory catches on and gains a wide following, and while that situation reigns, most workers aim their efforts to following the logic of the theory and its applications, and tend to be oblivious to things that do not quite fit.

The swings of fashion among meterological and climatic research leaders over the carbon dioxide effect provide an extreme example. The suggested carbon dioxide effect of an inevitable warming of world climate was more or less strongly held between 1938, when it was clearly argued by Callendar, and 1960, after further contributions by Plass (1956), only to lose ground in the years when it was obvious that the climate in the northern hemisphere was getting colder (despite greater output of synthetic carbon dioxide than ever before) from the late 1950s till about 1974. The theory then rose to renewed dominance around 1980. Between 1975 and 1980 the northern hemisphere cooling levelled off, or possibly reversed, and it became known that there had been a significant warming in New Zealand and the Antarctic without interruption over several decades. This is a very serious matter, because if the claims for the carbon dioxide effect could be substantiated, the commonly estimated warming of the average global temperatures by about 2 degC with a doubling of carbon dioxide in the atmosphere in the next century would be expected to move the climatic belts poleward by three to five degrees of latitude. The effects would doubtless be greater in high latitudes and rather less towards the coasts until the oceans had warmed up; and the situation would be complicated by shifts of the rainfall belts. The economic consequences for agriculture and international trade would be severe and might prove a potent cause of tension between nations.

However, the proposed effect cannot so far be demonstrated, perhaps because it cannot be clearly disentangled from the continual variations of climate due to natural causes. And this is likely to be the position until the carbon dioxide effect is already substantial, if it were actually to come about. Despite many assertions in the literature of recent years, it is not yet possible to estimate convincingly a limit to the range or the rapidity of natural variations in climate. A case might be made out, and has been suggested, for believing that the global climate was warming in the late 1970s because the effect of ever more rapidly increasing carbon dioxide in the atmosphere was at that stage dominant. But since 1980 to 1982 renewed cooling has perhaps set in because of the increased atmospheric burden of volcanic matter following the great, explosive eruptions of Mount St Helens in 1980, and particularly El Chichon in Mexico in 1982, as well as two volcanoes in the Pacific, all between May 1980 and April 1982. The cooling which unexpectedly interrupted the twentieth-century warming of climate between about 1945 and some time in the 1970s may, however, have had other causes, conceivably in the output of the sun itself. No increase of volcanic material in the atmosphere can be adduced to explain the cooling until after 1955.

At least in some respects, our ability to cope with the problems of planning posed by the development of climate has improved. The global situation is now kept under continuous watch and is regularly surveyed in various publications, such as the monthly *Climate Diagnostics Bulletins* of the National Weather Service, Washington, DC and *Die Witterung in Übersee* (published longer in arrears), produced by the Seewetteramt of the Deutscher Wetterdienst in Hamburg, and similarly the *Climate Monitor* issued five times a year by the Climatic Research Unit of the University of East Anglia, Norwich. Also with growing public awareness of the variability of climate's behaviour, mistakes will be less likely to be made through making no allowance for the possibility of change or the continuation of adverse trends when they occur.

References

Adams, R. McC. (1965) *Land behind Baghdad: a History of Settlement on the Diyala Plains*, Chicago, University of Chicago Press.

Adams, R. McC. (1981) *The Heartland of Cities: Survey of Ancient Settlement and Land Use on the Central Flood Plain of the Euphrates*, Chicago, University of Chicago Press.

Berger, A. (1980) 'The Milankovitch astronomical theory of palaeoclimates – a modern review' *Vistas in Astronomy*, 24, 103–122, Oxford, Pergamon.

Biswas, Asit K. (1984) *Climate and Development*, Dun Laoghaire, Co. Dublin, Tycooly International.

Bowden, M. J., Kates, R. W., Kay, P. A., Riebsame, W. E., Warrick, R. A. Johnson, D. L., Gould, H. A. and Weiner, D. (1981) 'The effect of climatic fluctuations on human populations: two hypotheses', *Climate and History*, in T. M. L. Wigley, M. J. Ingram and G. Farmer (eds), Cambridge, Cambridge University Press.

Bryson, R. A., Baerreis, D. A., and Wendland, W. M. (1970), The character of late-glacial and postglacial climatic changes, *Department of Geology, University of Kansas, Special Publication No. 3*, Lawrence, Manhattan, Wichita, University of Kansas Press.

Callendar, G. S., (1938) 'The artificial production of carbon dioxide and its influence on temperature', *Quarterly Journal of the Royal Meteorological Society*, 64, 223–240, London.

Geyh, M. A. and Jäkel, D. (1974) 'Late glacial and Holocene climatic history of the Sahara desert derived from a statistical assay of 14C dates', *Palaeogeography, Palaeoclimatology, Palaeoecology*, 15, 205–8, Amsterdam, Elsevier.

Plass, G. N. (1956) 'The carbon dioxide theory of climatic change', *Tellus*, 8, 140–154, Stockholm.

Tickell, C., (1986) *Climatic Changes and World Affairs*, Lanham, Maryland University Press of America, for Harvard.

Part I

Studies of
climatic history
and its effects
on human affairs
and the environment

Part 1

2

The Earth's restless climate

This wide-ranging essay is the substance of an article in the Encyclopaedia Britannica Book of the Year 1975. Portions have been omitted which would be duplicated in other chapters. It gives an introductory survey of the development of climate over the Earth from the last ice age to today.

Climate is always changing. The fluctuations of weather and climate take place on all time-scales, from the gusts and lulls of the wind, which occur within a fraction of a minute, to the shifts of regime over hundreds of millions of years associated with continental drift and wandering of the poles. Fortunately, however, for the development of life on the Earth and in particular for the evolution of contemporary life forms, the range of the temperature changes has been limited. Since the first appearance of life, there have presumably always been regions of the Earth where the air and water temperature remained generally between 20 deg and 30 degC (70 deg and 86 degF) and extensive regions where the limits of 0 deg and about 35 degC (32 deg and about 95 degF) were never far exceeded nor for more than a few hours at a time. This limited range bears witness that the sun must have been a reasonably constant star and the Earth a planet hospitable to the survival and spread of life. Nevertheless, the changes in the Earth's climate that have occurred have brought innumerable local and regional disasters and have repeatedly challenged the tolerance and adaptability of human and all other living entities.

The greatest and quickest changes of climate, and of the environment dependent on climate, have been connected with the onset and ending of ice ages and with fluctuations near the ice margin at times when the Earth is partly glaciated, as it is at present. The mean temperature in Iceland, for example, rose by about 2 degC during the global warming between AD 1850 and 1950, and the corresponding change averaged over the whole Earth was a rise of around 0.5 degC. These shifts not only moved the limits of the vegetation belts equatorward or poleward, and up and down the mountains, but also affected the rainfall in continental interiors and the position and development of the desert belts. As the ice masses built up on land or melted away, sea-level changed by as much as 100 m between the extremes of ice-age and interglacial climates.

The greatest losses of life directly attributable to weather and climate have resulted from coastal floods, specifically those produced by storm winds and tidal surges at times when the world sea-level had been rising during preceding decades or centuries of warm climate. The histories of China, the Bay of Bengal, and the North Sea coasts of Europe provide many instances in which hundreds of thousands of people have died in such sea floods.

Whether the climatic stress for humanity, animals and plants presented itself in the form of a change in the prevailing temperatures, in aridity, or in a greater incidence of floods and storms, survival often depended on migration. In our more recent history, we have enabled ourselves and our crops and animals to survive and flourish beyond the limits previously imposed by nature, using artificial indoor climates and irrigation; this ability to transcend nature's limitations now depends, however, on the use of a great deal of energy (especially oil) and often on the use of 'fossil' water in underground strata, which may also be limited.

Causes of climatic variation

The causes of climatic change can be classified in four general categories. The first chiefly concerns changes in the Earth's geography. It includes drift of the continents, which change their positions relative to each other and to the poles; uplift and erosion, which change the magnitude and disposition of mountain barriers and thereby affect the flow of the winds; and changes during the Earth's history in the total mass and chemical composition of the atmosphere and oceans. This category also includes variations in the energy output of the sun and changes in the heat flow from the Earth's interior (probably almost always of minor importance). The first three items in this group deal with changes that generally become significant only over tens or hundreds of millions of years.

The second category of causes of climatic change comprises cyclical variations in the Earth's orbital arrangements. These include the angle of tilt of the Earth's rotation axis to the plane of the orbit, which varies by a few degrees over a cycle of 40,000 years and changes the latitudes of the tropics and polar circles and the angle of elevation of the sun. Also in this category is the precession of the equinoxes – the progressive change in the position of the Earth in its elliptical orbit for any given time of the year. This 21,000-year cycle alters the distance of the Earth from the sun at any given season. There is also a 100,000-year cycle in which changes in the ellipticity of the Earth's orbit affect the yearly variance of the Earth's distance from the sun.

The third category consists of changes in the transparency of the Earth's atmosphere to incoming (mostly short-wave) and to outgoing (mostly long-wave) radiation. The best-demonstrated effects under this heading are those that have followed volcanic explosions which throw great quantities of fine dust into the stratosphere, creating a veil there which characteristically spreads over the Earth and lasts for two to three (occasionally as long as seven) years. While the veil lasts, temperatures rise in the stratosphere, due to direct absorption of solar radiation there, and are lowered at the surface of the Earth, due to loss of incoming short-wave radiation. Changes of cloudiness and in the atmosphere's content of water vapour, carbon dioxide, and other substances that are not transparent to radiation on some wavelengths also affect the radiation balance. Some of these vary as a result of the weather itself. Others are increasingly contributed by human activities (though perhaps not yet in sufficient quantities to affect climate).

The fourth category comprises the changes in the amounts of heat absorbed and given off at the surface of the Earth. These are due to variations in the extent of ice and snow, in the distribution of vegetation and of waterlogged or parched ground, and in the amount of anomalously warm or cold water on the ocean surface as a result of variations in the amount of sunshine or up-welling, respectively. These changes, like some mentioned in the third category, are produced by the weather itself and may in some cases increase the likelihood of persistence of the weather pattern that produced them.

Climate during the last billion years

Through the longest stretches of geological time the Earth had only warm climates, with no great polar ice sheets, though it is now thought that there may have been many periods during which any landmasses that were near the poles bore 'permanent' ice. If one accepts this view, the development of the greater ice ages chiefly depended on continental drift to place a continent at one or the other of the Earth's poles or, at least, to put landmasses of continental extent in high latitudes. However, the matter seems to be also affected by astronomical and solar variations. Evidence of the occurrences of greatest extent of ice in the past points to a fairly regular interval of nearly 300 million years, which may be explainable by gravitational effects of the rotating galaxy upon the sun's activity and output. And in each of the times when extensive ice occurred there seem to have been alternations between ice-age conditions and interglacial periods, which, as in the Quaternary (the last one million years approximately), may be attributed to the effects of the Earth's varying orbital arrangements (and tilt of the polar axis) upon the gain and loss of radiation from the sun in summer and winter.

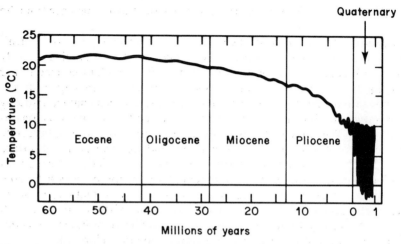

Figure 2.1 Average temperature in the middle latitudes of the Northern Hemisphere over the last 60 million years declined gradually until the Pliocene when it began to drop more sharply. Climate historians have increased the estimated number of ice-age/interglacial fluctuations in the Quaternary to about ten since this diagram was drawn. The time-scale is enlarged for the last one million years.

For reasons probably among those listed but which cannot yet be determined, there were important fluctuations that brought cooler climates at times during the Mesozoic era (between 225 million and 65 million years ago). One of these, about the end of the Cretaceous period and beginning of the Cenozoic era (about 65 million years ago), may have been associated with the extinction of the dinosaurs, which were presumably not adapted to withstand the cold as well as their warm-blooded contemporaries.

During the Tertiary period of the Cenozoic (65 million to 2.5 million years ago) a greater and more persistent cooling set in. In the late Tertiary (Pliocene epoch), superposed fluctuations on a time-scale of about 40,000 years can be traced. These fluctuations were presumably associated with the cyclical variation of the Earth's axial tilt (obliquity), and some of them are now thought to have produced conditions that began to approach the severity of the Quaternary (or Pleistocene) ice ages. The latest evidence suggests that it was during the Pliocene that the first races of humanity emerged as a distinct species from the apes. Perhaps it was a superior ability to adapt to the changes of climate and of the terrestrial environment, even in the warmer regions of the Earth, that enabled the new species to thrive. There is little doubt that it was the emergence during the Tertiary of a geography similar to that of the present, with a south polar continent and an almost complete land ring about the North Pole, that made possible the development of increasingly extensive ice sheets, thus cooling the oceans and therewith the whole Earth (see figure 2.1).

Recent evidence from oxygen-isotope measurements on carbonate-bearing sediments of the Pacific Ocean near the Equator suggests that the end of the Pliocene and beginning of Pleistocene is marked by a sudden change in the pattern of oscillations: from that time onward the temperature history is dominated by oscillations of close to 100,000 years in extent. Whether this means that some astronomical event occurred that made the variations in the ellipticity of the Earth's orbit more important than before is not yet clear, but the periodicity seems to coincide with those variations. The 40,000-year periodicity is still present but seems to play a somewhat subsidiary role. This latest evidence suggests that there have been about ten major developments of glaciation during the last one million years of the Quaternary period, though of varying severity and, as always, with every kind of shorter-term fluctuation superposed.

The last ice age and early human beings

During the last glaciation primitive people made what was probably an easy living by hunting the large grazing animals – reindeer, bison, mammoth – on the open steppe-tundra lands in what is now France and elsewhere on the European plains. They have left a record of their life and of this fauna in cave-wall paintings at Lascaux, France, and Altamira, Spain. During the last glaciation the first people probably entered the Americas, travelling from Asia about 35,000–15,000 years ago over the broad grassy lowland which is now the Bering Strait and Bering Sea, thanks to the drop of world sea-level. The same circumstance probably allowed the first aboriginal people to pass from Asia to Australia approximately 25,000 years ago.

After the ice age

The postglacial climatic regime developed rapidly, particularly in Europe, though there was at least one drastic setback when, for a period of 500 or 600 years, in the ninth millennium BC, glacial conditions returned or re-advanced. With the postglacial warming, there came a time when the rivers were swollen enormously, particularly in summer, by the melting ice. Gravels and sand were rapidly deposited and quickly produced thick deposits in some places; lakes formed and sometimes quickly silted up and completely disappeared. The landscape was changing rapidly, but the greatest change for the human population and for the animals they hunted was the disappearance of the open plains, as the forest advanced northward in Europe and North America.

Species of trees whose pollen and seed are light and are transported far by the winds, or else whose seed is spread by birds, soon replaced the open grassland and tundra. The first immigrant species in middle and northern Europe were generally birch and pine. These arrived very early – about 10,000 years ago – as far north as Denmark. It took thousands of years for the oaks and elms and beeches to spread from their ice-age refuges, beyond the mountains in southern Europe, and first gain an entry into, and then replace, the earlier established types of woodland in central and northern Europe. Forest fires, and later (in the last 6000 years) the clearings produced by man, provided opportunities for change in the composition of the forests. In North America, the ice-age stands of the more warmth-demanding tree types were not remote, as in Europe, but were merely a little farther south on the great plains. Therefore, the arrival of the forest farther north was correspondingly more rapid: probably only a few decades for the first trees, and a later changeover in the course of 100–300 years to the types that eventually dominated the deciduous woodlands.

Humanity seems to have adapted to these changes more successfully than did the animals. The ranges of both moved northward, but the extinction of various species (perhaps including the mammoth) and the disappearance of others from Europe, northern Asia, and North America was probably due to the reduction of their numbers by human activity.

Other great changes in the landscape were brought about by the rapid rise of sea-level, which over some thousands of years proceeded at a rate averaging one metre per century. This ultimately separated Britain and Ireland from continental Europe, created the Baltic Sea and The Sound (Øresund), greatly enlarged the Mediterranean, and caused the loss of vast areas of coastal flatlands in many parts of the world that had previously been easily (and perhaps, therefore, densely) inhabited. It seems likely that there was great loss of life and of the primitive industrial sites for making salt by evaporating seawater. Indeed, it has even been suggested that in this way the end of the ice age probably brought about one of those occasions, rare in history, when the total human population was significantly reduced. This is almost certainly the origin of some of the widespread legends of a flood disaster in the early history of humanity.

The postglacial warming took place so quickly and so extensively that by about 6000 BC, and thereafter for perhaps 5000 years, most of the world was warmer than it is now. Although it took until well after 4000 BC for the last of the great North American ice-sheet to disappear, ultimately the forest spread about 100–200 km (60–125 miles) north of its present limit in northern Canada.

As the warm temperatures moved north, all the climatic belts seem to have moved north too, so that the summer monsoon rains reached farther north into the Sahara than they do now. The levels of Lake Chad and other lakes in tropical Africa rose, and some of them became much larger than they are today. This, and the accompanying rains, may also be the origin of some of the flood legends.

Many thousands of years earlier, during the glaciation, Lake Chad had been an enormous inland sea (as big as the Caspian Sea today) extending from latitude 10° to 18°N and between longitudes 14° and 20°E. This indicates that under ice-age conditions the climatic zones were displaced towards the Equator, especially in the Atlantic, American and European sectors, where the ice sheets were most extensive. Therefore, with rainfall from the fronts of the storm depressions travelling east from the Atlantic – the depression centres probably mostly passed through the Mediterranean and from there north-east into Russia and Siberia – the Sahara and the other desert and nearly desert areas of the Near and Middle East received some regular rainfall. Evaporation would have been much less than now because of the lower temperatures and cloudier skies. For these reasons the ice-age climatic regime left a legacy for thousands of years afterward in the form of a higher level of underground water in the subsoil and rock strata of the present deserts than now exists. Radiocarbon tests have indicated that much of the water in the oases and in the water-bearing strata under those deserts today is about 20,000–25,000 years old.

There is evidence of a much drier phase in the Sahara and the Arabian desert and a lower level of Lake Chad in the earliest postglacial stages. But in the warmest postglacial times there was some renewal of the moisture by means of a climatic pattern that allowed the summer monsoon rains to penetrate farther north.

The development of civilization

The early civilizations in Palestine, in the Nile Valley, in Mesopotamia, and in the Indus Valley area and the vegetation of that era certainly owed something to the great abundance of water stored from an earlier age, as described above. The oases were more extensive and, to judge by the rock drawings from about 5000 to 3000 BC found in the heart of the Sahara, many species of animals were still able to migrate across what is now a desert region. Climatic fluctuations during those times, however, probably caused variations in the frequency of floodwaters in the wadis (desert stream-beds that are normally dry) and in the sizes of the great rivers (as is known to have happened in the case of the lakes). This seems to have been the cause of expansion of population and settlement into areas that are now desert on the fringes of Palestine during two moist phases, around 6000 and 3000 BC. In both cases the maximum expansion was short-lived. Soon after 3000 BC actual records of the time show that the level of the annual floods of the Nile River dropped, and after 2200 BC there came at roughly 200-year intervals some sequences of years when the level was so low as to cause starvation in Upper Egypt. The worst of all these times may have been about 1200 BC, when it is suspected that a widespread increase of aridity provoked migrations of peoples throughout the Near East and an invasion of Egypt by the ancient Libyans and their allies.

The conditions of the postglacial warmest millennia led to an enormous increase in the area of bamboo (*Bambuseae*) growth beyond its present natural limit in China, where it then abounded over most of the great lowland plain of the Yangtze and the Huang Ho. One deduces that the prevailing temperatures in that part of the world were then as much as 2 degC warmer than now over the whole year and 5 degC warmer than now in winter. (A 2 degC excess over modern values seems to have prevailed in many parts of the world, including Sweden.) Since the bamboo plant provides both convenient building and writing material, as well as the use of its shoots for food, it may be that this was important in the early development of a high civilization in China.

The warmest millennia also saw the spread of peoples and highly skilled cultures into northern Europe. About 2000 BC, towards the end of the warmest era, rock drawings in Norway testify to the use of boats (similar in shape to the Viking ships of a later age, but without sails) and skis, and it is known that human settlement reached as far north as latitude 69°N. Agriculture, with wheat and barley, had spread to Denmark and to southern parts of Sweden and Norway about 1000 years earlier. The numerous stone circles, or temples, built about 2000 BC, some of which seem also to have served as astronomical observatories (probably for calendar fixing, to determine the normal round of the seasons), were widely distributed from Brittany in north-west France and in western parts of the British Isles to the Outer Hebrides and Orkney Islands. They indicate a main line of communications by sea, over the fringes of the Atlantic, up to near 60°N and probably on into western Norway.

About 2000 BC the recession of the forests from the coasts of north-west Scotland and Orkney suggests that the climate was beginning to become windier – which probably points to the beginnings of a renewed cooling of the Arctic – and the seas were becoming rougher. Archaeological evidence from the time suggests that it was then that the main population concentrations in Britain and Scandinavia shifted for the first time to the east side. The glaciers in the Alps had receded before 2000 BC, and perhaps only then, after so many prevailingly warm millennia, did they reach their postglacial minimum extent. It seems likely that transcontinental routes of travel became more widely developed than ever before, including routes over the Alpine passes. Between 2000 and 1000 BC salt was mined in the mountains at the Iron Age site near Hallstatt in Austria, and gold was mined high in the Hohe Tauern range, in mines that were later abandoned when the glaciers readvanced.

Neoglacial cooling and subsequent fluctuations

About, or soon after, 1000 BC a sharp cooling of world climates was in progress. By 500 BC the prevailing temperatures in Europe had been lowered by about 2 degC from their previous warmest level. Winter snows and frosts and storminess must have become much more severe in Scandinavia. It has been maintained that it was this that gave rise to the northern legends of the fimbulvinter (an exceptionally long winter heralding the end of the world), and the twilight of the ancient gods and the heroic life of long ago, as well as to similar legends of an icy hell in central Asia (Altai region). Certainly there is historical evidence in the writings of ancient Greek and Roman authors that there were southward movements of peoples from central and northern Europe in the last millennium BC, as well as successive waves of Celtic and Teutonic peoples moving westward over the European plain to France and the British Isles.

During the time of the Roman Empire, and indeed from about 100 BC until around AD 400, the world's climate seems to have again been becoming gradually warmer, drier and more stable. The sea-level was rising, and grapes and olives were cultivated farther north than before (including districts where even in Italy such agriculture had not succeeded in the 2nd century BC). Trajan's stone-piered bridge over the Danube, built between AD 101 and 106 survived for almost 170 years without disturbance by ice. It is reasonable to suppose, as Ellsworth Huntington (1907) suggested, that it was increasing dryness and failure of the pastures used by nomads on the eastern steppes that set in motion from that region the *Völkerwanderungen*,

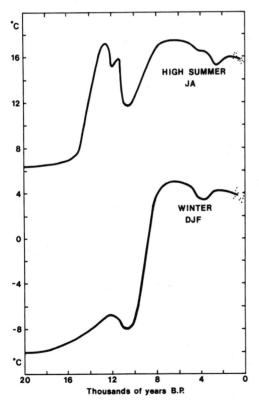

Figure 2.2 Trend of prevailing temperatures in central England during high summer (July and August) and winter (December, January, and February) is shown for the last 20,000 years. a more explicit version of this diagram will be found in figure 6.3 on p. 88.

long known as the barbarian invasions of Europe, which brought about the downfall of the Roman Empire in the West. Indeed, direct evidence of such dryness is the particularly low level of the Caspian Sea at about that time.

A colder period followed, with storms and wetness and more severe winters, particularly between about AD 550 and 800. This was followed in its turn by a remarkable warming, which for several hundred years seems to have restored the temperatures in northern Europe and Greenland to near their warmest postglacial level. Agriculture and human settlement spread rapidly farther north and up the valleys and mountainsides in Scandinavia and northern Britain to levels never before occupied, including some places that became marshy or so exposed after AD 1300 that they have never been cultivated again. (In England these were often the places where the Black Death and subsequent ravages of the plague struck hardest; they were also the places where the population had declined most in the years of bad harvests and famine about AD 1315). It seems likely that the great Viking voyages of discovery and colonization in Iceland, Greenland and beyond, between AD 800 and 1000–1200, were favoured by the absence of sea ice and by calmer seas than prevailed in later centuries.

Evidence of climatic worsening in the late Middle Ages can be traced in the records of an increase in Arctic sea-ice, ice on the Baltic, increased storminess in the North Sea, lowering of the tree-line on the mountains in central Europe and in California and of the limits of cultivation in many places, and advances of the glaciers in the Alps and elsewhere. The first symptoms of the deterioration in Europe seem to have occurred erratically between 1210 and 1320; severe phases followed, particularly around 1430–70 and between 1550 and 1700. A general decline of average temperature level by almost 1.5 degC between the thirteenth and seventeenth centuries can be deduced. Though there were always occasional warm years (sometimes accompanied by severe droughts) and some warmer decades, the general decline and the variability must have borne hard on the relatively primitive economy and the health of the peoples of Europe and elsewhere.

The cooling was worldwide. (Associations with changes in the amount of radioactive carbon in the Earth's atmosphere suggest a solar origin.) Some of its effects included the dying out of the Norse colony in Greenland, unrest in the Highlands of Scotland, retreat of the agricultural peoples who had been spreading up the Mississippi River Valley in North America, and the abandoning of the cultivation of oranges in the Kwangsi Province of China. There was a similar southward shift of the limit of vineyard cultivation in Europe. The Southern Hemisphere was similarly affected.

Climate in modern times and human activity

The cooling trend was reversed about 1700, and apart from the setback that followed the remarkable warmth of the 1730s and some smaller fluctuations in the nineteenth century, the history of the climate from the time when thermometers came into general use until the 1940s was a one-way trend towards ever increasing warmth. The warming became much stronger about 1900. In the Arctic, where it was strongest of all and caused a reduction of about 10 per cent in the extent of sea-ice, the greatest warming took place after 1920. The climatic change in the first half of the twentieth century could well be described as a general improvement of world climates because, in addition to the increasing warmth, significantly more rainfall was reaching the continental interiors (except the US Middle West), the summer rains extended farther north into the southern fringe of the Sahara, and the Indian monsoon rarely failed. These aspects seem to have been linked with general intensification of the global wind circulation. They certainly eased the problem of producing enough food for the world's increasing population.

The coincidence of the latest period of general warming of world climates with the industrial revolution and of the more intense warming with the industrial growth earlier in the present century led to the suggestion that the warming was due to artificial production of carbon dioxide. Since the 1890s there has been a 10–15 per cent increase in the atmosphere of the proportion of this gas, which acts as a trap for the Earth's outgoing radiation. But since 1945–50 the global average temperature has fallen again somewhat, despite an even greater production of carbon dioxide by human activity. This suggests a natural climatic fluctuation that is strong enough to outweigh the effect of the increasing amount of carbon dioxide. With better knowledge of the past record of climate, it may become possible to identify the nature of this most recent fluctuation or the entire series of cyclic changes to which it belongs.

Analysis of the longest available series of weather data, and of data related to weather, such as the yearly layers in the ice of the Greenland ice sheet and the yearly growth rings of trees, indicates the presence of quasi-periodic, or cyclic, oscillations, each of which probably corresponds to the normal time-scale of some process in the atmosphere, the oceans or the cosmic environment. The period lengths that are most commonly found are about 2.2, 5–6, 9, 11, 19, 22–23, 50, 90–100, 170–200 and 400 years. Periods of about 700, 1000–1300 and 2000–2600 years are increasingly suggested by workers who have studied series of exceptional length. Few of these cyclic processes, except perhaps the 2.2-year cycle and those of 200 years and longer, are strong enough to appear dominant. They all interfere with each other. The 9-year and 19-year cycles, and perhaps a cycle of about 2000 years, are probably caused by changes in the range of strength of the combined tidal force exerted by the sun and moon. Many of the others are often supposed to be due to cyclic variations of solar output, but some may be no more than the frequencies of 'beats' (as in musical sounds) between cycles of shorter length.

Some computations of the increase of human carbon dioxide output, including its redistribution between atmosphere and oceans and its effects upon radiation exchanges in the atmosphere, indicate that doubling it might be expected to raise the overall average temperatures prevailing at the surface of the Earth by about 1.9 degC.

The future

The increasing scale of human activity and the variety of ways it pollutes the atmosphere are causing anxiety, not least because of its possible effects on local and, ultimately, on global climate. Chemical pollutants are being injected near the Earth's surface from industry, traffic, and domestic heating, while the stratosphere is being polluted by high-flying aircraft and rockets (and, it could be, by dust and ashes from the use of nuclear bombs in war). Some of the substances introduced into the atmosphere may one day significantly affect the radiation balance. A straightforward projection of continuity in the growth of human population, industrial activity and demands indicates that within about 100–200 years at the longest, the output of heat will come to have a dominating effect on world climate. It has even been suggested by a leading Soviet scientist that, at that time, it may become necessary to spread artificial dust veils in the stratosphere in order to reduce the incoming solar radiation and thereby control the heat and prevent melting of the great ice sheets on land, which would lead to a disastrous rise in world sea-level. Despite these anxieties, however, it appears that at the present time natural fluctuations are still the dominant element in the Earth's climatic changes. It is important, therefore, that research should also be devoted to improving knowledge of the past record of climate and gaining better understanding of these natural changes.

The seriousness of these fluctuations in a world that is heavily populated, and where adquate food reserves cannot be constantly maintained in every country, can hardly be overemphasized. A great international effort is now devoted to monitoring climate and climatic changes as they occur. Satellites and the latest methods of sensing conditions at many levels in the atmosphere and in the oceans are used in this world climate watch. Every year 'climate diagnostics' discussion meetings are held, and efforts are devoted to detecting the causes and mechanisms of whatever changes are observed. Particular emphasis is laid on detecting any possibility of identifying effects attributable to human activities. However, if we are to be able to detect human disturbance of the natural regime of the climate, we must gain the fullest possible knowledge and understanding of the natural behaviour of climate. And that must be seen at its best in times and circumstances before any effects of human actions could be significant – hence the need to reconstruct the past.

Reference

Huntington, E. (1907) *The Pulse of Asia*, Boston, Mass., Houghton Mifflin.

3

Climate and history in northern Europe and elsewhere in the last thousand years

This chapter is an adapted and updated version of a presentation in Stockholm in 1983 at the Second Nordic Symposium on climatic changes and related problems, published as pp. 225–40 in the book of that symposium, Climatic Changes on a Yearly to a Millennial Basis, *N. A. Mörner and W. Karlén (eds), Dordrecht, Holland, Reidel, 1984.*

Although this review concentrates on a general survey of the climate and its effects over the past 1000 years in Europe, there are clearly lessons to be drawn which are applicable to problems in the Third World, and perhaps elsewhere, today. The concentration on Europe is justified because there is more information than for other regions (although the possibilities of retrieving data for China and Japan have probably not yet been fully exploited).

Introduction

Historians in most countries have generally been reluctant to believe that changes of climate affect the subject of their concern and need to be considered as part of the story. The situation is rather different in northern Europe, where Professor Andreas Holmsen and the Nordic Research Project on Deserted Farms and Villages 1300–1600, the Ødegardsprojekt, have been pioneers in the field. In Iceland the late Vilhjalmur Stefansson (e.g. 1943) and Sigurdur Thorarinsson (1956), and others, opened up the subject some decades ago. Yet, even in the case of Greenland, it has been argued by some historians – by historians whose own experience is of life in middle latitudes – that the demise of the Old Norse Colony during the later Middle Ages was not really due to the worsening of the climate and increase of sea ice, so much as to the colonists' failure to adapt themselves to an eskimo way of life.

We shall look at this Greenland question further in what follows. But we may do well to recognize that there seems to be a general and popular resistance since the modern scientific and technological revolution to the idea that our way of life may be drastically affected by natural changes and fluctuations in the physical environment. This attitude may hold dangers for the future. Our forebears had a very different view, based on bitter experience of disasters from famine and disease. But even in our own day climatic fluctuations have imposed similar distresses on the poorer peoples of Africa – particularly in the Sahel zone and in Ethiopia and Somalia – and have led to unfulfilled plans in the agricultural development of no less a country than the Soviet Union. It is important to arrive at a sound and well-informed understanding.

The climatic record

The necessary background to our study is the history of the climate itself since the last ice age. Our first outline knowledge of this came from the work of Scandinavian botanists, from Blytt (1876) and Sernander (1910, 1926) to Faegri (1964) and Iversen (1973). This knowledge has in recent decades been greatly improved by a host of new lines of evidence that provide corroboration and by new techniques, particularly radiocarbon dating (e.g. Olsson 1970) and its refinement by calibration against the American bristlecone-pine tree-ring chronology (which goes back many thousands of years) and against other objects of known age. The Swedish varve chronology, derived by study of the individual year-layers in the deposits of old lakes formed as the ice withdrew after the last glaciation, is also subjected to continual refinement and offers another countercheck on the radiocarbon datings (Strömberg 1985). The contribution of botany to improving the detail of the record has continued to be big, and quite recently Hafsten (1981) has reported finds of tree remains well above the present limits of trees of comparable size on Dovrefjell and Hardangervidda in southern Norway, shown by radiocarbon measurements to have grown in the warmest postglacial times and in the Roman Iron Age of AD 0-400 respectively. Stumps of full-grown trees occur widely in Scotland 30 to 70 m above the present tree-line, generally dating from 2000 to 5000 BC (Pears 1972): in a few cases, the heights are up to 200 m above the present natural forest limit. In central Europe, it is reported that the upper forest limit descended by 200 m between AD 1300 and 1600, after the mediaevel warm period (Firbas and Losert 1949).

Assessments of past temperatures over the last thousand years or rather more have also been derived from statistical studies of the details in historical documents (Lamb 1965, 1977, 1982). And these estimates can be controlled by reference to field evidence of various kinds, from pollen-analytical studies in peat bogs, from cultivation records, from glaciers, and so on.

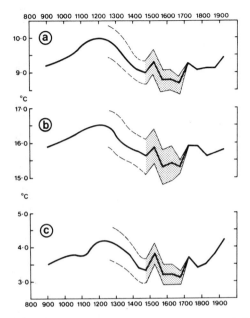

Figure 3.1 Temperatures in central England since AD 800. Probable averages for:
(a) the whole year; (b) July and August; and (c) December, January and February.
The shaded area indicates the range of uncertainty of the values (approximately 3
times the standard error of the estimates).

Here we shall base our considerations on the reconstructed temperature
record for central England and a reconstruction of the variation over recent
centuries in the sea-surface temperatures prevailing between Iceland,
Scotland and south Norway (see Chapter 9). The English temperature
record is chosen because it has been worked out in most detail, covering the
period AD 800 to the present, and the sea area mentioned is so far the only
one for which an estimate of conditions at the climax of the Little Ice Age is
available. Figure 3.1 shows the prevailing temperatures derived for England
from AD 800 to 1950 (as successive 50-year averages from 1100) and gives an
indication of the margins of possible error. The changes in prevailing
temperature level over the last 20,000 years range from the final climax of
the last Ice Age to the warmest postglacial times, some 4000 to 8000 years
ago, and a wavering decline since (see figure 2.2).
    The salient features which may have affected human history seem to be:

(i) the temperature levels established in the warmest and coldest times,
     including the mediaevel warm epoch around AD 900–1300 and the
     Little Ice Age in more recent centuries, and how these may have
     affected the extent of land available for human activities – cultivation,
     settlement, and so on;

(ii) the rapidity of the changes, and the still largely unknown detail of year-to-year variations during the periods of rapid change between one more stable regime and another; the year-to-year variability must, it seems, have been abnormally great at such times and there are indications in some tree-ring sequences that this was so (see Lamb 1977, and data in Fürst 1963); disappointing harvests, sometimes leading to famine, and disasters of quite various kinds (droughts, landslides, great snows and frosts, storm floods and incursions of the sea) are most likely to have occurred at these times;

(iii) changes in the ocean are also implied, which presumably affected fish stocks, spawning grounds, etc. – they are known to have done so within the last 400–500 years and continue to do so today;

(iv) changes in the penetration of polar water southwards from the East Greenland Current into the Atlantic, passing both sides of Iceland (i.e. along the coast of Greenland, also round the east coast of Iceland and at times near to the Faeroe Islands) are observed to occur on all time-scales (Lamb 1977, 1979, 1982; McIntyre et. al. 1972) from days and weeks at the present time up to many thousands of years. In the present epoch, the Faeroe Islands, with a 100-year average sea-surface temperature of 7.7°C, are normally in the midst of the main drift of warm, saline water of Gulf Stream origin flowing into the Norwegian Sea and the Arctic, but at times the polar water approaches from the north and lowers the water surface temperature by several degrees. For 30 years between 1675 and 1704, at the climax of the Little Ice Age development, the ocean surface between Iceland and the Faeroe Islands seems to have been about 5°C colder than the 1870–1970 average.

Over and above these variations, indications have been found in recent years that an oscillation of prevailing temperature level, somewhat parallel-ing the change from the warmth around AD 1100 to the colder regime around 1700, may have occurred in each of the previous four millennia. A parallel series of events in the last mellennium before Christ has long been known in outline (see e.g. Brooks 1926; Gams 1937).

If we could also establish a rainfall history in corresponding detail to the temperature history summarized above, we would again find variations. Over northern and central latitudes of Europe generally, the long-term averages may have been 10–20 per cent greater in the warmest postglacial epochs and 5–15 per cent less than now in the Little Ice Age. Experience within the present century indicates that significant further variations of rainfall may have been superimposed over more localized regions, as the rainfall difference between the western and eastern sides of mountains changed with changes in the prevailing winds. (The frequency of westerly winds covering the British Isles as a whole seems to vary between about 20 and 40 per cent over periods lasting up to a century.)

The dryness or wetness of the ground is, however, a different question, since it depends on the amount of evaporation as well as the rainfall. It seems that the colder periods of our climate and much of the periods of cooling were characterized by generally wetter ground, especially in the higher latitudes of Europe, including northern parts of the British Isles. The greater rainfall in the warmer periods seems rather to have been associated with heavier thunderstorms in summer, and the greater frequency of clear skies, warmth and sunshine meant that the ground dried more readily. The wetness of the Little Ice Age in northern Britain, particularly from the late sixteenth to the eighteenth century, and in some of the earlier stages around AD 1400, is well illustrated by studies of peat bogs (as in the example of figure 3.2) here and elsewhere in northern Europe. This confirms the impressions of many local documentary accounts. Examples are provided by the frequency of landslides and other difficulties on the farms in west Norway (Grove 1972) and accounts of the condition of the lowlands and glens of Scotland in the sixteenth to nineteenth centuries as 'dissected . . . by immense watery tracts in the shape of countless lochs and lochans. In Fife there were twenty water bodies as big as Loch Leven is now which have since shrunk, by no means all as a result of drainage work' (Fraser 1970). By contrast, in the thirteenth century, some of the greatest buildings of the time seem to have suffered subsidences and collapse with shrinkage due to drying out of the soil about their foundations.

In considering the history of soil-wetness in north-west England shown by figure 3.2, it is important to realize that the correlation between rainfall in north-western Britain and either south-eastern parts of the country or central Europe is not strong. There are bound to be differences. In particular, the marked dryness of the second quarter of the eighteenth century in the latter regions, which is well authenticated, and the probably still more marked dryness of the tenth century, as also the periods 1270 to about 1300, and 1370 to 1390 or 1400, seem not to have affected north-west England.

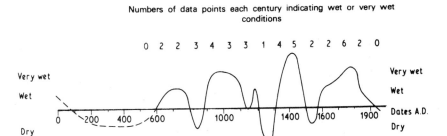

Figure 3.2 Variations of the surface wetness of Bolton Fell Moss, a peat bog north-east of Carlisle near the England–Scotland border, derived by Dr K. Barber from pollen and macrofossil analysis, soil chemistry measurements and records of land-use history.

The main effects of the changes of the overall average temperature level, referred to under (i), above, were probably the changes in the average length of the growing season, accumulated summer warmth, incidence of frost (especially prolonged frosts), and the average number of days with snow cover. In England, the growing season in the warmest decades of the present century, between 1930 and 1949, was about 10–20 days longer than in the late nineteenth century and probably 3–5 weeks longer than in the late seventeenth century. In the high Middle Ages, before about 1300, with a still higher temperature level, the average growing season was probably 5–7 weeks longer than in parts of the fifteenth, sixteenth and seventeenth centuries. Thus, the late mediaevel climatic decline between AD 1300 and 1600 can hardly have failed to have serious effects on European agriculture. In more continental climates, the shortening of the growing season may have been as great as this only within 200-300 m of the upper limit of cultivation and on the upland pastures (saeters and alps), but there the effects seem to have been sufficient to cause widespread abandonments. In the case of the duration of snow cover, however, the change was greatest on the low ground. Even in southern England, there are a few reports in the seventeenth century of winters with 70–102 days with snow lying, and the average was probably about 20-30 days, compared with twentieth-century averages of 5-15 days and an extreme about 60 days (in 1962-3). In the 1960s and between 1979 and 1986, averages have again risen to over 15 days in many places, but were low during much of the 1970s.

As regards accumulated summer warmth, Parry (1975, 1978) has shown by close examination of meteorological data and historical farm records how this must have affected the issue in the Lammermuir Hills in south-east Scotland, not far from Edinburgh. (See also Sandnes 1971 for an example in Norway.) In figure 3.3, Parry has taken the average accumulated day-degrees above a base level of 4.4°C for the season in which oats would grow at the known upper limit of their cultivation in this area over the period

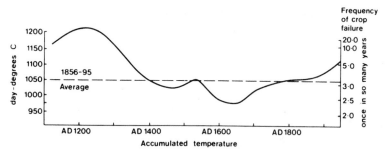

Figure 3.3 Estimated average accumulated warmth of the growing seasons at different times, and probable frequency of failure of the oat harvest on high-level farms in the Lammermuir Hills in south-east Scotland at the height of the late-nineteenth-century upper limit of cultivation. (From Parry 1978.)

1856–1895 and has applied the historical variation shown by our reconstruction of the temperature variations since AD 1100 in central England to show how the accumulated warmth of the growing season presumably varied. The right-hand scale shows the frequency with which the oat crop would be expected to fail at the various levels of summer warmth. This scale was derived by consideration of the best and worst years within the period for which we have full records of both temperature and crop results. It appeared that other aspects of the climate, such as raininess and windiness of the summers, were sufficiently correlated with the temperature that the latter could be treated as the one basic indicator. We see that in the present climate, failures may be expected about one year in seven at the upper limit of cultivation: in the thirteenth century, this may have been reduced to one year in 20, and in fact cultivation then was extended to levels almost 100 m higher. But in the late seventeenth century, the average failure rate approached one year in two, and many examples of failures in successive years would be expected. This would assuredly lead to eating the next year's seed-corn and then to famine. And, of course, the upper limit of cultivation had to come down to lower levels in consequence.

An aspect of climatic behaviour which may be very awkward for the human economy and a potential cause of disaster, particularly famine, e.g. through drought or plant diseases, is the observed – but not yet explained – tendency for several years with the same markedly unusual character to occur in clusters. Examples of such close successions of years – in this case, with anomalous frequency of easterly or northwesterly winds or calms at Copenhagen – are seen in figure 3.4. The succession of cold years in the 1690s, when the harvests failed in all the upland parishes of Scotland in seven years out of eight, is a notorious example of what the effects can be.

Effects on human history

When we come to consider how the various fluctuations and changes of climate may have affected human history, we must treat as separate issues the case of communities directly hit by a climatic event and the more difficult problem of tracing the influence on communities which were much less severely affected or not directly affected at all. It is logical to look for most effect near the margins of the inhabited world, in the North, on the uplands, as in the Alps, and near the deserts of Africa and Asia.

In extreme cases, as when the steady rise of sea-level with the melting of the ice-sheets and glaciers of the last glaciation flooded the great plain in the North Sea basin, the people that were living there had no choice. And there followed a long succession of disasters, involving sometimes hundreds of thousands of deaths, as great storm surges of the North Sea took more and more of the low-lying coastlands (see Lamb 1977, pp. 120–8, 1982, pp.109, 155) until the late Middle Ages and after. A similar compulsion affected the peoples living in North Africa when the Sahara and much of Arabia turned to desert from the Savanna-like terrain where they had herded cattle and hunted wild game before 3000 BC (Lamb 1982, p. 117). In some valleys in southern Arabia, too, cultivation of land which subsequently became desert had been practised.

Was the situation really less compelling for the Old Norse Greenlanders when the sea ice increased over the period between AD 1190 and about 1500, ultimately blocking their coasts? This and the increasing storminess of the northern seas cut them off from Europe. No doubt, they could have tried an eskimo way of life, but the Old Norse Greenlanders – as G. H. Petersen of the Zoologisk Museum, Copenhagen, has pointed out (personal communication, May 1983) were themselves skilled hunters and fishers and may by the fifteenth century have had little to learn from the eskimos. Archaeology indicates that their health and stature was declining, and a point would surely be reached at which a people depressed in mind by their situation, and probably by watching an increased death rate among the frail children and old people, would lack the morale to undertake bold changes. And what of the eskimos that were encroaching more and more into the area as they abandoned their former hunting grounds farther north, presumably because of increased freeze-up in the Arctic north and movement of the fish and seal populations? It may be that the resources of the area would no longer have yielded sustenance for the human populations of both races crowded together in a limited part of south-western Greenland. A likelier suggestion,

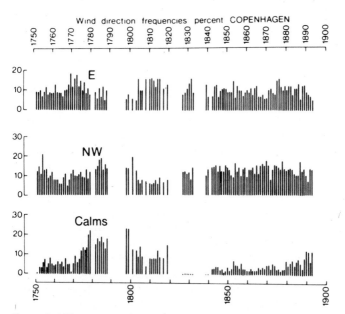

Figure 3.4 Frequency of surface winds from E and from NW, and frequency of no wind, reported at Copenhagen in different years, 1752–1893.

also by Petersen, is that the final blow (the *coup de grâce*) to the Norse population – and probably to many of the eskimos – was the ill-chance that a visiting ship, or shipwrecked fishermen arriving from England or Germany, may have brought an infection (the plague?) that flared up as an epidemic among the Greenland population, whose run-down state and long isolation left them with no resistance to it. This would be no more than a repeat in Greenland of the consequences which afflicted the American Indian populations of North and South America after the arrival of the Europeans about the same time.

In the Sahel zone of western Africa within the last decade, an analogous situation developed. After six years of drought between 1968 and 1973, the rainfall has never returned to its former average level. And in the first extreme phase of the drought, around 1972–3, with hundreds of thousands of cattle dying and multitudes of human deaths besides, the population fled southwards, crossing the national frontiers that were an awkward relic of the old colonial administrations. Some complete national territories may, in the long run, become more or less uninhabitable if the development continues and goes further. The 1980s have seen another extreme phase, with 1984 the worst year so far. We return to examine this more deeply in Chapter 12.

In all these cases, the compelling power of a change of the prevailing climate has virtually dictated the action taken by the peoples affected. But when we look farther afield, there are hints of a much farther reaching influence of some climatic trends and, on the other side, no great difficulty in finding examples of apparently contrary effects.

Students of history can hardly fail to be struck by the apparent coincidence of the high points of cultural achievement in the late Stone Age and Bronze Age development of trade and communications across Europe and of seagoing communications all along the Atlantic coasts and island chains, and again later in Roman times, and thirdly in the high Middle Ages, with the main crests of the temperature curve. And the parallelism seems to go into some closer detail too.

In central Norway, as we have seen, between AD 800 and 1000, the area of settlement, forest clearance and farming, which had been more or less unchanged since early Iron Age times, and had retreated in places since AD 400, rather quickly spread 100–200 m farther up the valleys and on to the heights (Holmsen 1961). It retreated again just as decisively between AD 1300 and 1500, not only due to the Black Death, which did not strike until after the population decline and abandonments had begun. This seems also to be the finding of investigations of the contemporaneous retreat and social unrest in Scotland, England, Denmark (Christensen 1938; Gissel 1973) and central Europe, as well as Holmsen's (1978) studies in Norway, Salvesen (1979) in Jämtland (Sweden) and of the Ødegardsprojekt in general. The parallelism with the temperature curve in the course of the mediaevel warm-up and later decline is close. In the case of abandonments of many villages and smaller settlements, where the dates are known, they appear to be concentrated in the severest decades (see Lamb 1977, 1982). Examples include Hoset in eastern Trøndelag in the 1430s and again in the 1690s, numerous villages in central and eastern England between 1430 and 1480, and one in the Southern Uplands of Scotland in the 1690s. And it was during the long climatic decline that the cultural capital, as one may describe it, of northern Europe moved southward from Trondheim, successively to Bergen and Oslo, and later farther from the ocean to Copenhagen and Stockholm. In 1536, Norway lost its existence as a separate country. So, too, in Scotland in 1436, after the murder of the king, the capital was moved for security to Edinburgh Castle from the former northern seats in Dunfermline and Scone (near Perth); after the famine and other disasters of the 1690s and later, the Parliament and administration in Edinburgh were moved to London in 1707 with a deeply felt loss of the identity of Scotland.

Was the parallelism with the temperature curve just coincidence? Or was there a remorseless pressure exerted by the harvest failures and increasing dislocation of life in the most exposed places, however well some of the more fortunate communities and individuals elsewhere fared? One prominent member of the Old Scottish Parliament in Edinburgh in 1698, Andrew Fletcher of Saltoun, berated the well-to-do people of the prosperous, corn-growing lowland districts near Edinburgh for their unconcern over the disaster in all the upland parishes where the people were reduced to gathering nettles to eat and were dying in huge numbers of starvation and disease.

There seems little doubt that when, in the sixteenth century, the Baltic herring shifted to the North Sea, England and the Netherlands prospered while people in the northern countries suffered. For a long time during the climatic deterioration after AD 1300, the cod fishery seems to have been improving up the Norwegian coasts, as if the increased flow of ice southward in the East Greenland Current was being compensated by an increased flow of Atlantic water northwards off Norway. However that may be, the improvement in the coastal fishery led to more human settlement along the coast, partly an actual movement northward of fisher folk from the districts near Bergen (precisely during the time of general climatic decline) and partly a drift to the coast of those who were abandoning farms in the interior in north Norway. In Denmark and Sweden, too, there are complicating details, with serious effects of the climatic worsening (storm losses of coastlands and abandonment of farms) in Jutland but much less effect, if any, in the more eastern islands. In southern Sweden, there were abandonments also, but much less in the north (Osterberg 1977), perhaps because settlement had been so sparse there that only the best sites were occupied.

Most remarkable of all the apparently contradictory developments is the great age of expansion of Norway's shipping and seagoing trade between 1680 and 1710, just coinciding with what appear to have been the bitterest years of the climatic development. There is no doubt, of course, that that was the time when the fisheries finally went into decline (and at one stage in the 1690s disappeared) along the whole Norwegian coast. It is recorded, however, that it was in the years of poor harvests in the late seventeenth century that the farmers who were near enough to the coast of south Norway and had good timber on their land – some of them had oak – took to shipbuilding and trading their timber abroad in their own ships. The development was also helped by the immigration of experienced mariners and shipwrights from Holland, seeking the advantages of a neutral land while the powerful nations of Europe further south were at war with each other.

At the same epoch in Iceland, life was at a low ebb. The advancing glaciers were overrunning farms at the same time as the fisheries were failing, and there were notable outbreaks of disease probably partly due to under-nourishment. Gunnar Gunnarsson (1980) has interestingly stressed how the situation in that country was made worse by its social conditions and the customs of the time. The landowners used their power in what was still a largely feudal society to restrict the drift of workers from the farms to try their luck with fishing on the coast. And there were regulations restricting fishing, allowing no more than one hook on a line and forbidding the use of worms as bait. Custom also dictated the use of open boats. And even 100 years later, when the government of Copenhagen tried to stimulate recovery in Iceland by introducing decked vessels and the use of many hooks on the lines, it met with prolonged and general opposition. This is surely an example of lack of resilience and more than ever stubborn clinging to old ways in peoples whose morale has been sufficiently lowered by climate-induced adversity.

It seems to the writer undeniable that pressures due to climatic change were present underlying all the historical events we have discussed. The impact is clear in the case of the most devastating climatic events upon the societies that were directly hit. These we may describe as impacts of the first order. But, as we turn our attention to societies which were less and less affected, we find their reactions more and more diverse, depending on the range of choices open to them and, in some cases, there were opportunities to profit at the expense of the communities whose former activities declined. Among those who suffer the greatest restriction of their former activities, depression, undernourishment and ill-health seem liable to undermine their capacity to adapt to different patterns of life; and so they miss such opportunities of self-help as exist. And they doubtless ascribe their troubles more to ill health and bad luck, or everything going against them rather than to the climate. This characteristic has been noted by the Brandt Commission Report (1980) as operating in the Third World today. Perhaps we should also note a lesson from the exodus in recent years of the so-called 'boat people' from south-east Asia, and of Africans from Uganda and the

Ogaden, in years of both climatic (monsoon irregularities) and political stress: that political prejudice may also enter in and dictate which groups shall be the main victims of a natural disaster.

References

Blytt, A. (1876) *Essay on the immigration of the Norwegian flora during alternating rainy and dry periods,* Christiania, A. Cammermeyer.

Brandt, W, and others (1980) *North–South: A Programme for Survival. The report of an Independent Commission on International Development Issues under the Chairmanship of Dr. Willy Brandt,* London and Sydney, Pan Books.

Brooks, C. E. P. (1926) *Climate Through the Ages,* London, Benn.

Christensen, A. E. (1938) 'Danmarks befolkning og bebyggelse i middelalderen, in *Nordisk Kultur II: Befolkning i Middelalderen,* Copenhagen, Adolf Schück.

Dansgaard, W., Johnsen, S. J., Reeh, N., Gundestrup, M., Clausen, H. B. and Hammer, C. U. (1975) 'Climatic changes, Norsemen and modern man', *Nature,* 225, 24–28, London.

Faegri, K. and Iversen, J. (1964) *Textbook of Pollen Analysis,* Copenhagen, Munksgaard; New York, Hafner; and Oxford, Blackwell.

Firbas, F. and Losert, H. (1949) 'Untersuchungen über die Entstehung der heutigen Waldstufen in den Sudeten', *Planta,* 36, 478–506, Berlin.

Fraser, Lady Antonia (1970) *Mary Queen of Scots,* London, Weidenfeld & Nicolson.

Fürst, O. (1963) 'Vergleichende Untersuchungen über räumliche und zeitliche Unterschiede interannueller Jahrringbreitenschwankungen und ihre klimatologischer Auswertung', Flora, 153, 469–508, Jena.

Gams, H. (1937) 'Aus der Geschichte der Alpenwälder', *Zeitschrift des deutschen und Osterreichischen Alpenvereins,* 68, 157–170, Stuttgart.

Gissel, Sv. (1973) 'Nogle kronologiske synspunkter vedrørende agrarkrisen i senmiddelalderen', Arbejdspapir 1972–3 nr. 47 (København 28.3 1973). Copenhagen, Det Nordiske Ødegardsprojekt (Dansk Afdeling).

Grove, J. M. (1972) 'The incidence of landslides, avalanches and floods in western Norway during the Little Ice Age', *Arctic and Alpine Research,* 4(2), 131–138, Boulder, Colorado.

Gunnarsson, G. (1980) 'A study of causal relations in climate and history, with an emphasis on Icelandic experience', *Meddelande från Ekonomisk-Historiska Institutionen,* nr. 17, Lund University, Lund.

Hafsten, U. (1981) 'An 8000 years old pine trunk from Dovre, south Norway, *Norsk Geografisk Tidsskrift,* 35, 161–165.

Hafsten, U. (1981) 'Palaeo-ecological evidence of a climatic shift at the end of the Roman Iron Age', *Striae,* 14, 58–61, Uppsala.

Holmsen, A. (1961) *Norges Historie,* Oslo, Bergen, Universitetsforlaget.

Holmsen, A. (1978) *Hva kan vi vite om agrarkatastrofen i Norge i middelalderen?* Oslo, Bergen and Tromsø Universitetsforlaget.

Iversen, J. (1973), 'The development of Denmark's nature since the last glacial', *Geol. Survey of Denmark, V Series,* No. 7c, Copenhagen, Reitzel.

Lamb, H. H. (1957) 'On the frequency of gales in the Arctic and Antarctic', *Geographical Journal,* 123, 287–297.

Lamb, H. H. (1965) 'The early medieval warm epoch and its sequel', *Palaeogeography, Palaeoclimatology, Palaeoecology,* 1, 13–37, Amsterdam, Elsevier.

Lamb, H. H. (1977) *Climate: Present, Past and Future, Vol. 2, Climatic History and the Future*, London, Methuen.

Lamb H. H. (1979) 'Climatic variation and changes in the wind and ocean circulation: the Little Ice Age in the northeast Atlantic', *Quaternary Research*, 11, 1–20.

Lamb, H. H. (1982) *Climate, History and the Modern World*, London, Methuen.

McIntyre, A., Ruddiman, W. F. and Jantzen, R. (1972) 'Southward penetrations of the North Atlantic polar front: faunal and floristic evidence of large-scale, surface water-mass movements over the last 225,000 years', *Deep-Sea Research*, 19, 64–77, London, Pergamon.

Mörner, N.-A. (1980) 'The northwest European "sea-level Laboratory" and regional Holocene eustasy', *Palaeogeography, Palaeoclimatology, Palaeoecology*, 29, 281–300.

Ogilvie, A. E. J. (1984) 'The past climate and sea-ice record from Iceland, Part 1: data to AD 1780', *Climatic Change*, 6, 131.

Olsson, I. U. (ed.) (1970) *Radiocarbon Variations and Absolute Chronology (Nobel Symposium 12)*, Stockholm, Almqvist & Wiksell, and New York, Wiley.

Østerberg, E. *Kolonisation og kriser: bebyggelse, skattetryck, odling och agrarstruktur i västra Värmland ca. 1300–1600*, Lund. Det Nordiska Ödegardsprojektet. Nr. 3, Gleerup.

Parry, M. L. (1975) 'Secular climatic change and marginal agriculture', Transactions of the Institute of British Geographers, 64, 1–3.

Parry, M. L. (1978), *Climatic Change, Agriculture and Settlement*, Folkestone, Dawson and Hamden, Connecticut Archon Books – Studies in Historical Geography.

Pears, N. V. (1972) 'Interpretation problems in the study of tree-line fluctuations', in *Research Papers in Forest Meteorology* J. A. Taylor (ed.), Aberystwyth, Cambrian News.

Salvesen, H. (1979) *Jord i Jemtland*, Östersund, A/B Wisenska bokhandelens förlag.

Sandnes, J. (1971) *Ødetid og gjenreisning*, Oslo, Bergen and Tromsø, Universitetsforlaget.

Sernander, R. (1910) 'Die schwedischen Torfmoore als Zeugen postglazialer Klimaschwankungen', 11th International Geological Congress, Stockholm.

Sernander, R. (1926) 'Postglaziale Klimaverschlechterung', in Ebert's *Reallixikon der Vorgeschichte*, Berlin.

Stefansson, V. (1943) *Greenland*, London, Harrap.

Strömberg, B. (1985) 'Revision of the late glacial Swedish varve chronology', *Boreas*, 14, 101–105.

Thorarinsson, S. (1956) *The Thousand Years Struggle Against Ice and Fire*, Museum of Natural History, Miscellaneous Papers No. 14, Reykjavik.

Tooley, M. J. (1974) 'Sea level changes during the last 9000 years in northwest England', *Geogr. J.*, 140, 18–42, London.

4

Climate and life during the Middle Ages, studied especially in the mountains of Europe

The analysis reported in this chapter was prompted by an inquiry from a representative of the Walser people living today in northern Italy and compiling a history of their settlement in the region. This analysis sheds new light on how the earlier mediaeval warm climate broke down in the late Middle Ages.

I am indebted to Mr Tony West, of St Vincent, for much information about the history of this area and for introducing me to the local references cited.

West adds that the need for water must have been great when the Canale di St Vincent was built, in view of its length and the difficult construction problems. After the fifteenth century, no other water-supply canals were built in the Val d'Aosta region. Certainly, the economic and political situations changed as well as the climate. The Val d'Aosta region was entangled in the fortunes of the Duchy of Savoy from the twelfth century. Among the disasters which struck the region was the plague of 1630, which killed 70 per cent of the population. Several of the canals were abandoned for ever, and it may be that the population had lost heart and become too sparse to manage the upkeep which all the canals demanded every year.

Europeans are aware of the signs of a high point of culture and achievement in the Middle Ages, after preceding centuries of turmoil, and before the time of plagues and increase of troubles from other diseases in the centuries after AD 1300. The average length of life in England seems to have gone down from 48 years around the 1270s to 38 years around 1400 (Comfort 1966; Lamb 1977, p. 264). The outburst of energy among European peoples in the best of mediaeval times carried Norse settlers to Iceland, Greenland and the coast of North America, Viking voyagers far into the Arctic and to the Mediterranean, Scandinavian warriors and traders into Russia and to

Byzantium, Christian missionaries establishing monasteries across the plains of Europe into Russia; it later saw the conquests of the Teutonic knights in eastern Europe and the Crusades to the Near East, to Syria and Jerusalem. Twentieth-century opinion has regarded the latter items of this history with less approval than formerly, but of their significance as an indicator of overflowing energy and confidence there can be no doubt. There seems to have been a fairly steady build-up to that point from the time of Charles the Great around AD 800, though some would trace the origins earlier: to the checking of Muslim expansion in Europe at Poitiers in 732, to the efforts of the Christianizing monks beginning in the centuries before that, and the gradual cessation of the invasions and marauding by barbarian bands across the whole of Europe that had marked the decline and fall of the Roman Empire in the West. The most obvious surviving monuments to the achievement of European peoples in the Middle Ages are the great cathedrals and other buildings: the abbeys, castles, city walls, bridges, and drainage works, etc.

Historical evidence of warmer climate in the Middle Ages

Evidence in high latitudes

It has been known for a long time that there are indications of a long period of more genial, generally warmer, and in some ways steadier, climate in the high Middle Ages than in the centuries that followed. This particularly applies to the Arctic sea-ice about Greenland, which seems to have been much reduced from the time of the Norse settlement in AD 987 or earlier until about 1200, as was noticed by the Swedish oceanographer Otto Pettersson as long ago as 1914. In Iceland, intrusions of the Arctic sea-ice had rarely been mentioned, and certainly seem to have been of no concern, from as far back as AD 900; and the deterioration after AD 1200 has been documented from the writings of the time, originally by Thoroddsen (1916–17), then further by Koch (1945) and others. Most specifically, we have the report of Ivor Bárðarson (or Baardsøn)* in his *Description of Greenland*, written probably soon after 1364, that the old sailing route from Iceland to Greenland in latitude 65°N, and thence around the southern coasts of Greenland to the colonies on the western side, had been abandoned by about 1342 because of the increase of ice and that a new route farther south had become customary. Deterioration of the ice situation in Iceland waters is confirmed by Bishop Arngrim Brandsson, writing before 1350.

---

* A modern Danish translation was produced by F. Jonsson (1930). I am indebted to my colleague Astrid Ogilvie for the information that Ivar Bárðarson was a Norwegian priest who went to Greenland in 1341 and may have stayed there for 20 years.

A letter of Pope Alexander VI in 1492 lamented the loss of communication with the churches in Greenland, so that 'no bishop or priest . . . has been in residence for eighty years or thereabouts' and 'shipping to that country is very infrequent because of the extensive freezing of the waters – no ship having put into shore, it is believed, for eighty years' (quoted fully in Lamb 1977, p. 6).

Not only on the seas around the coasts, but also inland in Iceland and Greenland, the evidence of a warmer climate in the earlier centuries is clear. In the heyday of the Norse settlements in south-west Greenland in the Middle Ages, the dead were buried and the roots of small trees or bushes grew in ground which afterwards became permanently frozen. In Iceland, grain crops (chiefly barley) were grown until the practice was largely abandoned by the fifteenth century (Thorarinsson 1956) and, finally, given up in the sixteenth century: some resumption from the thirties of the present century has met with less success since 1960. In the best period of the Middle Ages, grand farms were established in southern Iceland at places which were subsequently encroached upon by the glaciers. At Kvisker farm (near 64°N 16½°W) on the southeast coast, which is today close to the Breiðamerk spur of the Vatnajökull ice-cap, and where successive farmhouses have been built at roughly 100-year intervals on the same site, the foundations show that the oldest mediaeval building, from before AD 1090, was the biggest and richest: finds in the midden testify to a standard of living and diet (including imported oysters) never afterwards equalled (G. S. Boulton, reporting on-site investigations at a University of East Anglia colloquium, 18 March 1975). At this farm, oats were grown in addition to barley until about AD 1200, after which date pollen analysis reveals no more trace of oats and the amount of barley seems to have been halved.

Evidence in North America

Archaeological studies indicate a more or less parallel sequence on the other side of the Atlantic in North America during the times discussed in this chapter. Key dates for the turning-points in the climatic trend there have recognizable correspondences with events in Greenland–Iceland and in Europe. The dry western sector of the Great Plains towards the Rocky Mountains, the upper Mississippi Valley, and the arid south-west of what is now the United States, all seem to have enjoyed a moister climate than now between about AD 700 to 800 and 1200. The Indian tribes were able to spread their settlements and their agriculture up the valleys, north and west into Wisconsin, and after 800, on into eastern Minnesota (Griffin 1961; Woodbury 1961). It was a warm period from New Mexico all the way to northern Canada, where forest remnants up to 100 km north of the present limit (Nichols 1970) have been radiocarbon-dated to just these centuries (see also Bryson *et al.* 1965). All this tendency towards extra warmth and moisture in the region went into reverse about AD 1150–1200. Pollen investigations in the middens of the Indian Mill Creek culture on the northern plains in Iowa indicate a rapid decline of the tree pollens, and the middens reveal that the people quickly went over to bison-hunting and a meat diet: first, their agriculture and then the settlements themselves were abandoned in what had evidently become a drier climate. Bryson (1977) reports that when the first Europeans arrived in the region a few centuries later, they found 'vestiges of a thousand small villages' across the plains from Iowa to Colorado which, by that time, had been drifted over by blown soil. The moisture and the former inhabitants had both withdrawn far to the south: in Texas and Oklahoma conditions for agriculture had apparently improved about the same time (Bryson *et al.* 1970). The meteorological explanation of such a marked increase of dryness in the plains regions east of the great Rocky Mountain chain is obvious: an increase in the frequency of westerly winds and the effective 'rain shadow' of the mountain barrier. And that development in the latitude of the northern half of the USA points to an expansion of the cold polar region, spreading over Arctic Canada south-wards, and a corresponding southward shift and intensification of the storm belt along its edge – the zone of eastward-travelling subpolar depressions. These are the very developments which we have already noticed in the Greenland and Iceland region from the late 1100s onwards.

Glacier evidence and related historical reports in Europe

Study of past variations of the glaciers in Iceland, Norway, Sweden and the Alps indicates a similar history, with a minimum extent of the ice in mediaeval times, mostly before AD 1200, followed by various advances. In parts of the Alps, after 250 years or more of glacial retreat and the establishment of tracks over some of the high passes, a first major advance of the glaciers took place about AD 1250–1280 (Schneebeli 1976, and referen-ces in Ladurie 1967–71). Traces remain of ancient routes up the valleys of the Swiss canton Valais, or Wallis, over several of the high passes into the Aosta region of Italy. These included one over the very high Col d'Hérens (3460 m or over 11,300 feet) and others over the Theodulpass (3317 m), both approached from the Zermatt Valley, the Monte Moro pass (2882 m) from Saastal and the Fenêtre Durand (2803 m) at the head of the Val de Bagnes farther west. (A key map of the passes and other places in the Alps most referred to here is given in figure 4.1.) The Col d'Hérens seems to have been used in Roman times by a different track from the Italian side to the Zermatt valley from the route used in the Middle Ages, along which Roman remains have been found (Röthlisberger 1976). The change of route suggests that traffic over the mountains had ceased in the interim. It seems likely that parts of the routes became overrun by advances of the glaciers: this has been effectively demonstrated, and dated by the radiocarbon method, in the case of the Val de Bagnes between AD 400 and 800. There the glaciers advancing from the mountains on either side crossed the floor of the valley and blocked the drainage, so that a temporary lake formed, later broke the ice dam and flooded down the valley. This catastrophe occurred

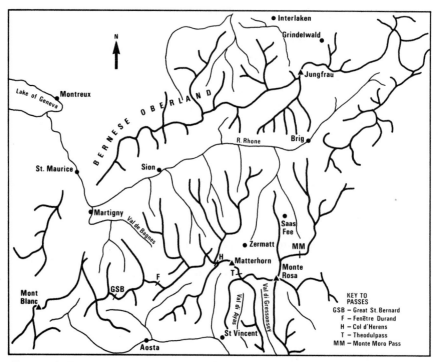

Figure 4.1 Outline map of the Alpine ridges, passes, river valleys and places referred to in this chapter.

about the year 580 and was repeated much later on, in the well-known Little Ice Age period, at least four times, in 1549, 1595, 1640 and 1818 (Schneebeli 1976). Similar disasters occurred at least 12 times in the Saas valley during the Little Ice Age, in 1589, 1633, 1680, 1719, 1724, 1733, 1740, 1752, 1755, 1764, 1766 and 1772; there were repeats on a more minor scale in 1968 and 1970, when part of the village of Saas Balen was destroyed. Major events of this kind may well have been more effective even than the increase of ice on the passes in stopping use of the routes over the mountains. It seems that there was a period, or periods, after 1300 or 1350 when some renewed retreat of the Alpine glaciers took place, but probably the retreat was rather slight and certainly the periods concerned cannot have lasted very long. Pfister (1985) believes that the Aletsch glacier's late mediaeval advance continued until AD 1400 or thereabouts.

The dating of these mediaeval glacier changes, which depends on radiocarbon measurements on wood found in the moraines where trees were pushed over by the advancing glaciers, is unavoidably imprecise because of the awkward margin of error of the C^{14} method in the late Middle Ages. This is related to variations in the production of radioactive carbon in the atmosphere around that time. The error may be more than 100 years.

A shorter late mediaeval interruption in the cooling and the state of the Alpine glaciers

Patzelt's studies (1974) in the Austrian Alps indicate up to 100 years of glacier recession in the late Middle Ages, centred on AD 1500, agreeing reasonably well with much evidence from documentary sources of a warm period in Europe around 1500 or somewhat before to 1540 (which can be seen in figure 4.4). Pfister (1984) has deduced from documentary material of the period 1530–64 in Switzerland that the climate of that time matched the warmth of the warmest decades in the twentieth century, the springs and summers being on average up to 4 degC warmer than in this century.

One date that appears confusing indicates, from C^{14} measurements on birches that grew near the Findelen glacier, near Zermatt, that the main mediaeval recession of that glacier had caused it to melt back to above its twentieth century position as late as AD 1400. This is hard to reconcile with the other evidence cited above, and it is conceivable that the radiocarbon date needs a correction to almost 200 years earlier. There may also have been differences in different parts of the Alps in and about the fifteenth century, between glaciers on the north and south, or east and west, sides of the main massifs. Pfister (1985) believes that the Aletsch glacier's late-mediaeval advance, which probably began in the thirteenth century, continued until AD 1400 or thereabouts. A water-supply duct, which had been built earlier in the Middle Ages to bring water from the Aletsch glacier, in an area just north of the River Rhône, was overrun by the advancing ice and had fallen into disuse by 1385. The Allalin glacier was blocking the upper part of the Saas valley by AD 1300 and had at that time reached a position near where it stood in 1589 and 1850 (Delibrias *et al.* 1975; Grove 1979).

In 1546, a certain Sebastian Münster, travelling on horseback up the upper Rhône Valley, in the same region, to go over the Furka pass to St Gotthard, described how he encountered the Rhône glacier at the valley floor, before the climb up to the Furka: its front was a mass of ice 10–15 m high and at least 200 m wide. His account, translated from the original Latin, is given by Ladurie (1971, p. 130). There were also some blocks of ice, the size of a house, separated from the main ice-front. Presumably the glacier was, or had recently been, advancing. And thereafter it, and the other Alpine glaciers, advanced to reach their most forward positions generally around AD 1600–40 and 1800–50. Schneebeli (1976) and Röthlsberger (1976) report that one of the glaciers in the Val de Bagnes was already in a very advanced position in 1549, comparable to Sebastian Münster's observation of the Rhône glacier.

The picture we derive of glacier variations in the Alps over the last 2000 years can be outlined as in figure 4.2 but hardly in greater detail. There were glacier advances between about AD 400 and 700 and in the later Middle Ages from about AD 1250. There was an interval, probably short, of glacier

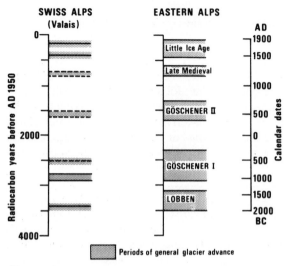

Figure 4.2 Known phases of glacier advance of the last 4000 years in the Swiss and Austrian Alps. (Adapted from Grove 1979.)

recession that has not so far been precisely dated or delimited, around the year 1500, roughly in the span 1470–1530, before the cold Little Ice Age climate and glacier advances of the late 1500s. In the Swiss Alps, the maximum extent of the glaciers in the last-named period came in clearly separate phases in the seventeenth century and around 1820 and 1850. The diagram has been extended to show the indications of other glacial maxima in the two previous millennia.

Late mediaeval cooling and the forest limit in central Europe

Other evidence of the secular temperature fluctuation that must be presumed to underlie these changes comes from studies of the upper tree limit. Firbas and Losert (1949) established that the tree-line in the Vosges, in the Black Forest (the Schwarzwald), and in the Erzgebirge mountains on the borders of Germany, Poland and Czechoslovakia was lowered by 100–200 m between about AD 1300 and 1500 – a height change which, if it were all attributable to a change of prevailing temperatures, would indicate a cooling of 0.7–1.4 degC. Flohn (1967) stressed that the greater height of the upper forest limit in the high Middle Ages on the Alps and other ranges mentioned here is attested by numerous tree remains, and that in some places the glaciers are known to have been smaller than today; an example is provided by the Findelen glacier.

It has been noted that the upper tree-line in the Swiss Alps and elsewhere on the mountains of central Europe has never regained its mediaeval height.

But the writer does not know how far this can be explained by increased use of the upper pastures for grazing.

Mediaeval climatic oscillation, glaciers, and the limits of settlement and agriculture in northern Europe

Holmsen (1961) has recorded that the limits of human settlement and farming in central Norway advanced rather quickly 100–200 m farther up the valleys and hillsides between AD 800 and 1000. Retreat began in the fourteenth century and became wholesale after the Black Death reached the country in 1349.

Some disaster had already befallen the farms in Olden and Loen close to the Jostedalsbre ice-sheet in 1339 (Grove 1985). And climatic difficulties seem to have contributed to keeping the farms on the uplands mostly unoccupied for 200 years thereafter (Holmsen 1975). Holmsen reports that there were deserted farm settlements in Trøndelag (the Trondheim district of Norway) as early as 1260. In another work, Holmsen (1978) has examined the wealth of information from thousands of farms in the land account books of the bishopric of Oslo in 1393 and the archbishopric of Nidaros (Trondheim) in the 1430s (both published in the nineteenth century, see References). These record a massive fall in corn production by around 1400: the tithes then were only a small fraction of what they had been in 1290, although no new norms had been set. The figures for that earlier time were still remembered as the supposed normal. The bishop could only produce enough oats for his own household on his farm in the good district near Oslo, and Norwegian corn farther north was proving to be of unsaleable quality. In parishes far inland where many farms were deserted, the tithes were sometimes as little as one-eighteenth of the thirteenth-century norm, elsewhere they were more generally in the range 21–55 per cent. By the 1430s, new, lower norms were recognized.

Parallel changes affecting the mines and human settlement among the mountains in central Europe

While this was the pattern of events in Norway, these seem to have been the times of most notable agricultural decline also farther south, and therewith of distress, unrest and increasing ill health among the people and animals. English visitors to a royal wedding in Denmark in 1406 reported much sodden uncultivated ground and that wheat was grown nowhere. Christensen (1938) writes that many farms were deserted, especially in Jutland but also elsewhere in Denmark, and that those farmhouses that were still in use were shared by several families. In England, too, there was a great retreat of settlement and of cultivation, especially in the east and north, after the thirteenth century peak: in the midland county of Warwickshire, which was probably fairly typical of the country as a whole, 69 per cent of all the earlier villages that disappeared were lost between about 1400 and 1485 (Beresford 1954, pp. 148–9, 158–9). On Roten Moor, one of the most isolated and desolate areas of central Germany, which was cultivated in Roman times, population grew to a peak around AD 1200, followed by the area being almost abandoned in the course of the fourteenth and fifteenth centuries. Roten Moor was reoccupied in the warmth of the early sixteenth century, but became once more deserted during the cold seventeenth century (Abel 1955, p. 47).

Similarly, ancient goldmine workings in the Hohe Tauern in the Austrian Alps (north of the Dolomites), which had been abandoned long before, because of encroaching glaciers in the last millennium before Christ, were opened up again in the high Middle Ages. They were abandoned once more in the deteriorating conditions starting about 1300. This history applied also to other mines in the hills in northern parts of central Europe. It was reported in AD 1360 that water had been increasing in the mines in the Harz mountains for more than 50 years. In Bohemia, some mines had been abandoned for that reason as early as 1321. And in the Alps some of the mine entrances were once more closed by the glaciers. In various parts of Europe, the desertion of farm and small village sites, which became widespread in the fifteenth century, began about this time: desertions all over England, noted in the *Nonarium Inquisitiones* survey of agricultural production in 1341, were attributed to population shrinkage in the famine years around 1315, soil exhaustion and seed shortages, i.e. in ways more or less directly attributable to adverse weather.

One may reasonably ask how far people living at the time were aware of the changes in conditions. Certainly, those directly affected were aware when rapid or decisive changes for the worse took place, as when the Greenland sailing routes were changed and, later, when any access to that country became difficult and ultimately contact was lost. Ladurie (1967–71) has been able to quote documents dated about 1640 in which the inhabitants of Chamonix, in the French Alps, and other villages near there, complained of 'a third of their good and cultivable land having been lost in about the last ten years . . . through avalanches, falls of snow, and glaciers'. Similarly, from west Norway later in the same century, and in the century that followed, Grove and Battagel (1983) have found in the local archives records of numerous hearings of pleas for tax reductions on account of deterioration of the land through avalanches, floods, landslides and glacier advances. (In these cases at least, the landlords and governing authorities were made aware through the effects on their rents and tax revenues.) Of course, the people of the time lacked a verifiable long record or any documented time-perspective of climate, except where this was enshrined in features of the landscape and its vegetation, or in accessible archives, deeds, titles to land-use, etc. And when disease struck, they certainly attributed all their difficulties from plague or other illnesses, and crop and animal sicknesses, directly to those diseases rather than to any climatic events which may have played a part in bringing about the epidemics.

A recent study by Galloway (1985) has shown how, among people living at or near the bare subsistence level, year-by-year variations in the food supply due to weather and harvests are related to the incidence of all those diseases that are affected by malnutrition and hence to the death rate. As long ago as 1662, Graunt perceived this, arguing that few Londoners died of starvation in poor harvest years, the increase in deaths came from greater susceptibility to diseases through undernourishment.

Investigation of the weather history

The present author has tried (Lamb 1965) to see how far the changes of weather pattern, which accompanied and apparently in some ways underlay the history of Europe between about AD 800 and 1600, can be traced in the surviving reports of the weather of the time in various sorts of documents, despite their writers' lack of a long time-perspective. Descriptions of individual seasons, and sometimes the specific days of great floods, frosts, snowstorms, etc., occur in the audited accounts of ecclesiastical, private and royal estates, in occasional diaries and chronicles, in legal papers, and in the records of (river) bridge masters, of building operations, and so on. Many of these have been extracted by meteorologists and others with a historical turn of mind and published in some famous collections, e.g. Short (1749), Arago (1858), Hennig (1904), Speerschneider (1915), Easton (1928), Britton (1937), Müller (1953), Buchinsky (1957) and Ladurie (1967–71). (A longer list and more information is given in Lamb 1977, esp. pp. 31, 120–6.) Great care is needed in identifying the dates, particularly the exact year, to which the individual reports relate. The collections have been criticized (e.g. Bell and Ogilvie 1978) for inaccuracies and duplicated references to the same event, arising from insufficient attention to authenticity of the sources used and questions of the reliability of the scribes and original reporters, which differ from case to case. For these reasons, historians prefer not to use secondary sources, such as collections of excerpts, and most would insist that checking should go back to the original source. However, textual criticism, itself, can cause distortions and waste of valuable material where strong preconceptions on the part of critics bias their consideration of the material. In the case of the many reports of severe winter weather in various parts of Britain in the 1430s, the compiler, Britton (1937), discusses at some length the problem of which winter in that decade was the severe one. He seemingly never seriously considered the possibility that the reports were trustworthy and that several, or even most, of the winters of that decade were severe. Britton's choice fell on 1433–4. Yet if we put the reports from all over Europe on meteorological maps, it seems that there was indeed a run of severe winters. This has been demonstrated in the case of 1431–2 by the author's detailed map of the reports from many parts of Europe and two ships at sea (Lamb 1982; see also Taylor 1963). No collusion between the different reporters or compilers can be presumed to have contrived this result. The usefulness of the mapping method as a test of consistency between different reports, as well as to indicate the weather pattern that prevailed, seems so far to have been generally overlooked in historical climatology. To put the available weather reports on maps, when possible, might indeed be regarded as an essential part of the testing of their historical authenticity (see also Lamb 1987).

The reports of weather available for study tend to be abundant in seasons of any dramatic character, so that some individual seasons and some individual decades are well covered by reports. Erroneous reports can be expected to stand out on the maps in these cases, and so can be disregarded or eliminated from the analysis.

Such an abundance of weather reports exists, which could be used for mapping the weather season by season over Europe from the Middle Ages to today, that the searching of all possible original sources, translation of the various (often archaic) languages, and thorough critical examination of all reports would take many people many years and demand very great expenditure. A first survey was, however, possible (Lamb 1965), using little more than a large number of the collections referred to and relying on the number of reports incorporated in considering 10- or 50-year groups to suppress the effect of an occasional faulty report. By considering the relative numbers of reports of warm or dry months or seasons, compared with those of cold or wet character in each decade, index numbers could be devised which register the prevailing dryness/wetness of the summers and mildness/severity of the winters of the decade. The counts were based on unambiguous reports of such things as parched ground, persistent rains and flooding, frosts and snows, unseasonable flowering of plants in winter, gales, winter thunderstorms even in the continental interior, and so on. This was done for three different areas of Europe separately, about longitudes 0°, 12° and 35°E, near latitude 50°N (respectively in England, Germany and Russia), where the wealth of reports was greatest. The procedure has been more fully described and the results tabulated in my earlier books (Lamb 1966, 1977). The first attempt to produce a numerical index to show the sequence of changing temperature levels in Europe from the Middle Ages to our own times was a Winter Severity Index for western Europe by Easton (1928). The results were quite similar to mine for longitudes 0° to 12°E.)

My results support the indications of a period of warmer climate in Europe generally, at least between about AD 950 and 1200 or somewhat after, especially as regards the summers. The winters of those times appear not so different from nowadays, in some decades mostly mild, in others colder, and occasionally a severe one. A preponderance of mild winters, however, appears in most European longitudes between 1150 and 1200 and in the west of Europe until 1300. In much of western and central Europe,

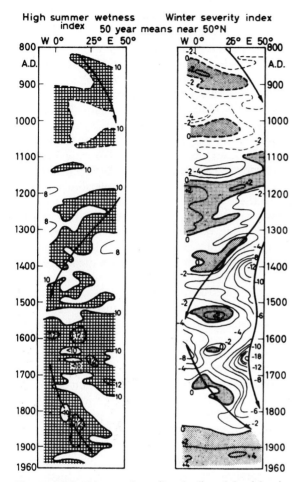

High summer wetness index Winter severity index
50 year means near 50°N

Figure 4.3 Incidence of wet (hatched) and dry (clear) summers, and of mild (stippled) and severe (clear) winters, in different longitudes across Europe near latitude 50°N from AD 800 to 1960. Derived from 50-year averages of Lamb's summer wetness and winter severity index numbers (shown by the numbered lines). The arrows indicate apparent progressive shifts eastward and westward across Europe of features with a recognizable continuity.

there was again some mildening of the winters in the mid-fourteenth century between about 1340 and 1390 and in part of the 1420s. This means that, decade by decade, experiences were not quite alike all across Europe, somewhat as with the more familiar differences across Europe in individual summers and winters.

The investigation was carried further by taking five-decade (50-year) averages for each of the three areas and plotting them out on a diagram (figure 4.3). This procedure meant that no excessive reliance was being placed on the index values for the individual decades.

We find that, apart from the period of summer warmth and relatively stable climate all across Europe, which is seen as most marked between about AD 1050 and 1200, there was a period of more or less opposite character, with often wet summers and cold or severe winters, right across Europe between around 1550 and 1700. There is also an appearance, emphasized by the arrows in figure 4.3, of some features that moved progressively west across Europe between 1200 and 1500, when the climate seems to have been mostly cooling, and returned eastwards during the times of warming of the climate between AD 800 and 1000 and between 1700 and 1900, or 1930. There may be a straightforward meteorological explanation for these movements of any of these regions in the latitude of mid-Europe. The changes of position are what meteorologists would expect if there were progressive changes in the prevailing wave-length of the meanders in the zone of broadly westerly upper winds encircling the Earth in middle latitudes, and specifically affecting the spacing of features in the flow east of the great Rocky Mountains barrier. The implication seems to be a shortening of the wave-length (westward displacement of prevailing features over Europe) during the time of cooling climate after AD 1200 and a lengthening of wave-length (eastward shift over Europe) when the climate was warming after 1700.

Lamb (1965) further used the series of summer and winter index values over England to derive probable values of the prevailing temperatures in England. This became possible when the indexes were continued decade by decade after 1700: the values then could be compared statistically with the known prevailing temperatures, based on instrument measurements, given in Manley's (1974) series. The resulting temperature curves, again based on 50-year averages to avoid excessive reliance on individual year or decade values, are seen in figure 4.4. In arriving at the final estimates of temperature prevailing in England in the Middle Ages, some adjustments were made to take account of various indications from botanical studies and of the significantly higher ocean surface temperatures in the northern Atlantic, which were most obvious near Greenland, where the excess over today's values may have reached 4 degC and even more in the local fjords in summer. The curves do show a mediaeval warm climate period, which in England seems to have reached its height between 1150 and 1300.

It is interesting to notice, from table 4.1, how far the long histories of the occasional freezing over (in exceptional winters) of Bodensee/Lake Constance (47°N, 8° to 9°E) and of the lagoon at Venice (near 45° 25′N, 12° 20′E) parallel the temperature history we have arrived at, despite the haphazard element in the occcurrence statistics of such rare or extreme phenomena.

It seems obvious from what has been written earlier in this chapter, and from other details known, that the peak of mediaeval warmth in Iceland and

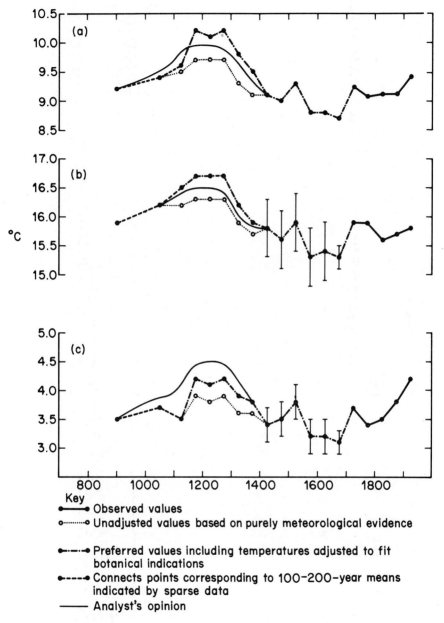

Key
●——● Observed values
○········○ Unadjusted values based on purely meteorological evidence

●—·—● Preferred values including temperatures adjusted to fit
      botanical indications
●----● Connects points corresponding to 100–200-year means
      indicated by sparse data
——— Analyst's opinion

Figure 4.4 Estimated course of the average temperatures at typical sites in central England from AD 800 to 1950: (a) for the whole year; (b) for high summer (July and August); (c) for the winter (December, January and February). (Derived from the index values used in figure 4.3.)

Figure 4.5 Part of the Ru Courtaud or Canale di St Vincent, where it passes along the cliffs below Mt Zerbion, a few kilometres before it reaches the pass, Col de Joux, over which it goes to the pastures above St Vincent.

This 25-km-long channel was constructed between AD 1393 and 1433, to bring irrigation water to the St Vincent pastures.

In 1980 the channel was finally put into pipes, to save the burden of the yearly maintenance and clearing tasks. (Photograph kindly supplied by Mr A. West.)

Table 4.1   Number of winters frozen over per century, in two European lakes

Century	Bodensee	Venice Lagoon
9th		5
10th		0
11th		0
12th		1 (or 2?)
13th	2	3
14th	3	2
15th	4	7 (or 8?)
16th	6	4
17th	5	4
18th	1	9
19th	2	3
20th	1	3

Note: I am indebted to Dr D. Camuffo, of Padua, and Professor J. Neumann, of the Department of Meteorology, University of Helsinki, Finland, for the information about the lagoon at Venice and Bodensee.

Greenland was between about AD 900 and 1180, earlier than in the British Isles and mainland Europe. The survey of index values plotted in figure 4.3 may indicate that the warmest time was before the year 1200 in eastern Europe as well, although there seems to have been a rather marked second peak of summer warmth there in the fourteenth century, at a time when the winters showed signs of quite sharp cooling. It was pointed out in Lamb (1977) that whereas temperature reconstructions from proxy climatic data, such as tree studies and oxygen isotope measurements in the Greenland ice, produce histories with this sort of rough correspondence with England in the cases of Iceland, Greenland and also New Zealand, there are rather greater differences from the European record in the results from studies of Californian trees (at the upper tree-line). Reconstructions from China (and, perhaps, Japan) show a more radical difference, with a cold climate phase there in about the tenth to twelfth centuries AD (Lamb 1977, p. 402).

Diagrams such as figure 4.3, and maps of the prevailing weather reported in individual seasons, decades or other periods, can be used to gain hints of the prevailing wind circulation characteristics that produced the weather observed.

Mediaeval rainfall, drought and water supply problems in the Alps

An aspect of the mediaeval climate that has received much less attention, and no analysis, is the evidence of dryness and water-supply difficulty in central Europe. Also, as in Norway, there may have been some influence of the climate's changes upon folk migrations and the patterns of settlement.

That the Alpine region should have been drier during the main warm summers period in the Middle Ages is understandable. Meteorological reconstructions generally interpret the period as having been marked by frequent anticyclones extending from the subtropical region of the Atlantic near the Azores to central Europe. Moreover, that would imply that the summers were on the whole rather sunnier than in most later times, and so there would be more evaporation. There probably were nevertheless some years with a good many thunderstorms. Reports from European countries north of the Alps seem to confirm this and that, with the warmer times, some of the thunderstorms were notably severe.

Human settlement in and near the Alps was spreading during the Middle Ages. The permanent settlement in part of Switzerland and in Alsace of German-speaking people from Swabia (Schwaben) can be traced back to the fifth century AD. Some of the Alemanni tribe are thought to have arrived in the very first centuries AD. They were already used to living in the high country of south-west Germany. In the eighth and ninth centuries, they spread into the upper regions of the Aar basin, the high midlands of Switzerland about Bern, and later from there over the mountains into upper Valais (Wallis) canton.

At some time, probably before AD 1200, and perhaps from their arrival in upper Valais in the tenth century onwards, these people found it necessary to construct water-supply channels to carry water from the upper reaches of the Alpine streams to water the pastures along the often steep valley-sides. These ducts included the one already mentioned, known as the Ober-riederin, which took water from beside the Aletsch glacier (near $46°\ 23'N\ 8°E$) to the pastures just north of the Rhône valley above Brig; this was overwhelmed by the glacier by the late fourteenth century, as discussed above. Other ducts were constructed in the same early period, for example, in the Saas Valley, south of the Rhône, where two of the ancient 'Heidenkanäle' still exist in a ruinous state near Visperterminen (above Visp). Others, of unknown date, are still in use in Saas. There are over 300 such channels in this canton today; their average length is about 6.7 km. Other channels of the same type were constructed in the Dolomites, farther to the south-east. These small canals, then, have been a normal feature of life in the mountains on either side of the upper Rhône valley in the present century and in the Middle Ages, though at least some of them fell into disuse at times after AD 1300. The need for them to support the present population level is undoubted. The average number of days a year with rain or snow falling is only in the range 40–90 in the valleys today. Construction of the water ducts involved daring feats of engineering in the Middle Ages. It was costly on labour, and they have always demanded much regular effort to keep the channels flowing. The whole local communities are involved in the maintenance work, and the sharing out of the water by branch channels has always been strictly regulated.

Disasters must have made it difficult, and at times impossible, to keep these irrigation channels going during the intervening centuries. And perhaps they were not always so obviously needed. Avalanches, landslips and, particularly, lake-burst floods afflicted some valleys, notably Saastal, repeatedly during the cold-climate period which has come to be known as the Little Ice Age. The Allalin glacier above Saas pushed forward across the valley bottom, blocking the river and causing a lake, the Mattmarksee, to build up, till it overcame the strength of the glacier tongue and poured down the valley. Twelve repeats of this disaster are recorded in Saas between 1589 and 1772, as detailed earlier. The worst cases, which carried away many houses, were in 1633, 1680 and 1772.

A changing régime in the later Middle Ages

What is more surprising is that such channels continued to be built on the Italian side of the main ridge of the Alps during the period 1300–1450, when Europe north of the mountains was undergoing the most obvious change towards a cooler, wetter climate. The most remarkable channel of all, the Canale di St Vincent (figure 4.5) is in the Val d'Ayas, where it takes water coming from the Ventina and other glaciers of Monte Rosa, at a point about 2100 m up, runs along the west side of the valley to the cliffs below Mt Zerbion and, finally, over a pass (Col de Joux) at 1640 m, to water the

Figure 4.6 A fifteenth-century painting of one of the builders of the water-supply ducts in the Val d'Aosta region. (Photograph kindly supplied by Mr A. West.)

pastures above St Vincent in the Val d'Aosta, a total distance of 25 km. It was built as late as 1393 to 1433. Flohn (1967) has stressed the remarkable (statistically demonstrable) incidence of extremes of both winter mildness and severity, and also of summer wetness and droughts, in Europe in two sections of this period, about 1310–50 and 1430–80. The period between approximately 1310 and 1347, in particular, seems to have been extraordinarily rainy over Europe west, north and east of the Alps. And farther east, the Caspian Sea reached a high stand, 15 m above its mid-twentieth century level.

The changeability from year to year in those times surely means that the westerly winds which often dominate Europe's climate, and bring a similar experience of mild oceanic climate to all parts, with more noticeable day-to-day than longer-term changes, were replaced for much of this time by something quite different. To meteorologists, it suggests an uncommon frequency of 'blocking of the westerlies'. There seems to be some confirmation of this in the tree-ring studies of Dr J. Fletcher, done on oak wood panels from paintings and furniture at the Research Laboratory for Archaeology and the History of Art at Oxford some years ago. These showed a high level (70–75 per cent) of correspondence between the successions of yearly growth-rings shown in wood of English and German origin before AD 1250, but only 50–55 per cent agreement in the fifteenth century, as might be expected if occasions with the same wind-stream blowing over both countries had become rarer. Another curious pointer to the same conclusion is that occurrences of locusts (presumably from Africa) in Switzerland were reported in no less than 17 different years between 1280 and 1380 – an unusually high incidence.

Effects of the climatic changes on life, settlement, and travel in the mountains

Many of the irrigation channels in the Aosta region of Italy on the south side of the high Alps between Mont Blanc and Monte Rosa, were built by German-speaking 'Walser' (Walliser) people – the name means people from Wallis or Valais – who had migrated over the high passes, bringing their animals and their customs, including the style of chalet, with them from upper Valais. A documentary record of 1218 exists relating to a massive colonization of the Val de Gressoney, in the eastern part of the Aosta region, supposedly arranged by the bishops of Sion (Sitten) in Valais, whose lands had already for centuries extended without dispute as far south as Issime (near 45½°N) in Gressoney. At that date, there were already some German-speaking families who had arrived in Gressoney some time before: on some evidence of house archaeology, they were probably there by 1100 or soon after. When the migrations of these folk into upper Valais, and later over into the valleys on the Italian side took place, the most fertile lands farther down the valleys were already occupied. The new settlements in the Italian valleys were therefore established high up: some of the hamlets were at over 2000 m above sea-level. This and their German language meant that intercourse tended to be kept up over the passes rather than down the valleys, though later the increase of the glaciers effectively closed some of the routes.

Trading links had been kept going by trains of pack-animals over the passes to the north. In the eleventh and twelfth centuries, the glaciers had probably receded altogether from the Theodulpass. But later, when the glaciers once more covered it, it was still traversable because the slopes covered by the ice were gentle. So although some of the links were lost when the glaciers grew greatly after AD 1500, at least some traffic continued over the Theodulpass until a disaster in 1852, when a whole party disappeared into a crevasse and were not found for many years: the pass is still ice-covered today. Until around 1935 or later, the only way up the Saas valley to reach the villages Balen, Saas Grund, Saas Fee and Saas Almagell, 20 km above where the valley joins the Zermatt Valley at Stalden, near the Rhône, was an ancient mule-track, in places uneven and steep, and at some points not much over a metre-and-a-half wide. This track led on to the Monte Moro pass into Italy at 2882 m above sea-level: it had been regarded for centuries as 'the high road' from Saas to Gressoney, although that involved crossing two more, almost equally high, passes on the Italian side. Access from Zermatt over the Theodulpass required crossing only one more pass, at less than 2700 m, on the Italian side to reach Gressoney. In Gressoney, the German language survives, but in the western valleys the descendants of the Walser have gone over to a French patois akin to Provençal. It is believed that this happened in the eighteenth century, a date which might well indicate that part of the loss of links with Switzerland was due to the increase in the glaciers.

Many irrigation channels (known as *ru* on the Italian side, *bisse* among French speakers in Switzerland, and *Bed* in old German) were constructed in the Aosta region, south of the main Alpine ridge, between AD 1250 and 1400 or after. Twelve are listed by von Fels (1962), the earliest (the Ru de Joux) in 1250, two more by 1300, and the remainder later, four around 1400. Construction of the longest and evidently most difficult one – the Ru Courtaud or Canale di St Vincent (see figures 4.5 and 4.6) – was decided upon in 1393: documents detailing the agreement and parties to it survive. Despite the fact that the driest part of the region, with present average yearly precipitation less than 600 mm (locally only 500 mm), is between St Vincent and Aosta itself, the strange coincidence between the dates of this activity and the time of greatly increased wetness in Europe north of the Alps (registered by growth of the peat bogs and by river sizes, and of climatic variability) is remarkable. The rivers of England, in particular, are believed to have been generally rather bigger and more navigable in the periods about 1300–70 and 1450–90 than in our own day (Trevelyan 1928).

Weather and wind patterns over Europe in the late Middle Ages

Flooding of England's rivers was particularly great in the summers of 1437, 1438 and 1439. In European Russia, there are mentions of great troubles with floods of the main rivers in western and southern parts of the country during this period, but also of a good many drought years (Buchinsky 1957). The biggest reported flood event of the time was in China in 1332, when 7 million lives and countless human habitations are said to have been lost in the flooding of a huge area of the great river plains. It is probably no coincidence that the bubonic plague epidemic, which ultimately swept the world as the 'Black Death', started in 1333, in China. There can be little doubt that the waters had dislocated the habitats of the wild life as well as the human settlements, including those of the plague-carrying rodents (information personally supplied by Dr M. Bloch and based on the works of Dr J. Needham, FRS). Gesnet writes (1982) that 'The Yellow River floods . . . had grown worse since 1327 and were causing deadly famines almost every year. In 1344 the dykes broke downstream from K'ai-feng after continuous rain. The river flooded huge areas and it was not until five years later, after eight months work, that the breaches could be filled.'

From a new, most carefully examined collection of seasonal weather reports from mediaeval Europe (Alexandre 1986), considered together with (and in good overall agreement with) the X-ray examination of tree-ring wood-density structure year by year of the pine trees at Lauenen in the Bernese Oberland at the Swiss Forestry Research Institute, Biemensdorf (Schweingruber *et al.* 1978), we see a great number of years in the fourteenth century with failed summers with which only 1816 in recent centuries can be compared (Pfister 1985). The worst case was probably 1315. In the Alps, there were years with repeated (up to 20 times) renewals of the snow cover all through the summer to heights below 2000 m. The years named by Pfister (1985) in this connection are 1302, 1315, (1330,) 1335, 1345, 1346, 1347, (1350,) 1359, 1366 and 1370. That three such years occurred in succession in the 1340s must have affected the health and stamina of people, cattle and crops: those were the last years before the Black Death, bringing the greatest outbreak of bubonic plague, arrived in Europe. But no more such summers appeared between 1370 and 1400. As late as the 1330s there had been one last occurrence of three summers within a few years of each other (1331, 1333 and 1336) producing late-wood densities corresponding to a level of warmth and drought that had been familiar earlier – four occurrences between 1270 and 1300 (in 1270, 1273, 1287 and 1300) and four more near approaches between 1291 and 1301 – but which has been registered only once since (in 1473) right up to today.

Thus, the changeability characteristic of the late mediaeval breakdown period, as stressed by Flohn (1967), is confirmed. A time of extraordinary wetness in much of the fourteenth century in Europe, at least from the main Alpine ridges northwards, is also confirmed. There was somewhat of a change of character from about 1400, though the high levels of the rivers in England and of the Caspian Sea seem broadly to have continued, as did the frequency of great changes of the weather from one year to the next. The 1430s, 1440s and 1490s, along with 1310–19, produced the most extreme figures for Lamb's (1965) winter severity index in central Europe in the whole series of decade values from AD 1100 to the present; the winters of the 1430s seem unmatched in western Europe as well. In eastern Europe, notable severity appeared in the 1280s, 1370s, early 1400s, 1430s and 1440s, but the severity of all these seems to have been surpassed there by the winters of the mid-seventeenth century. The summers of 1310–19, the 1340s (apart from 1340 itself and 1348), the 1360s (notably in England and central Europe), 1400–9 and the 1450s seem to have been extraordinarily wet, particularly in central Europe. This situation seems to have recurred from the 1560s to the 1590s and on to 1630 or so.

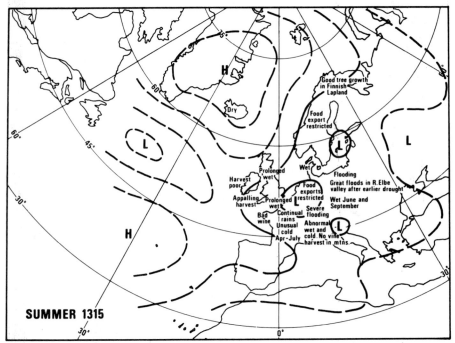

Figure 4.7 Map of the reported weather and harvest results, summer 1315. The suggested average barometric pressure pattern to explain the reported weather is shown by supposed isobars (lines of equal pressure). The winds would blow clockwise around the high pressure regions and anticlockwise around the low pressure – this implies prevailing northerly winds over the Norwegian Sea and much of Europe, though more variable in the regions of very wet weather in central Europe. The summer seems to have been fine only in Iceland, Lapland and probably parts of north Russia and Spain.

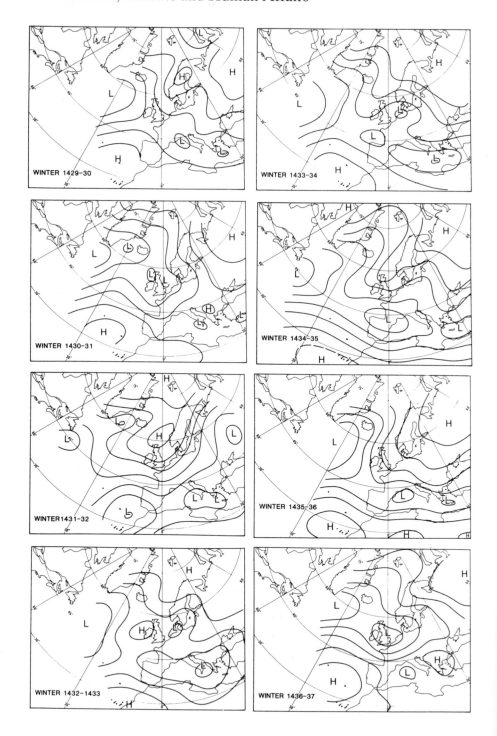

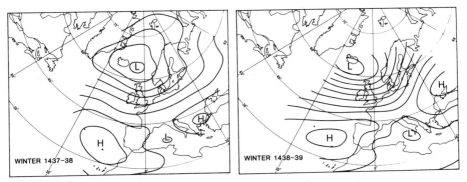

Figure 4.8 Maps of the suggested barometric pressure and implied wind patterns derived for each winter, 1429–30 to 1438–9.

But the main, seventeenth-century stage of the Little Ice Age climate development seems to have differed somewhat from the fifteenth-century stages which we have examined here. It produced a reputation for dry windstorms shifting the loose soils and sands of the Breckland in East Anglia and the Veluwe district in the Netherlands. John Evelyn (Diary, 10 September 1677 (Old Style)) wrote that 'The Travelling Sands . . . have . . . damaged the country . . . like the Sands in the Deserts of Libya' and have 'quite overwhelmed some gentlemen's whole estates'. One such storm in 1668 carried huge quantities of sand 8–10 km from an area known as the Lakenheath Warren to bury the village of Santon Downham in 'a sand flood'. It is recorded (Thomas Wright in the *Philosophical Transactions of the Royal Society, London* (No. 37, p. 722) that 'the sands first broke loose 100 years since', i.e. in about the 1560s, and first reached Downham 'about 30–40 years ago', evidently around the 1630s. That period was, however, another one marked by heightened year to year and decade-to-decade variability: some of the seasons were wet, particularly in the 1620s, 1640s, 1650s and 1690s, just as there were some notably hot summers despite the more general coldness of the time.

There is enough information about the weather across Europe in the appalling summer of AD 1315, which was followed by several years of famine far and wide over the continent and in the British Isles, to make it possible to sketch out the probable pattern of prevailing winds and barometric pressure (figure 4.7). Similarly, the abundance of information resulting from the dramatic character of the European winters between 1429–30 and 1438–9 makes it possible to sketch the patterns which probably prevailed in each of those winters, using the same methods. Some of the individual seasons in the 1420s and 1440s could be equally confidently analysed. About the same coverage exists for them, an example being the severe winter of 1422–23, when it is reported that the whole Baltic froze over and people could ride on the ice all the way from Mecklenburg and even Danzig (now Gdansk) to Slesvig (now Schleswig) and the Danish islands, and which was preceded

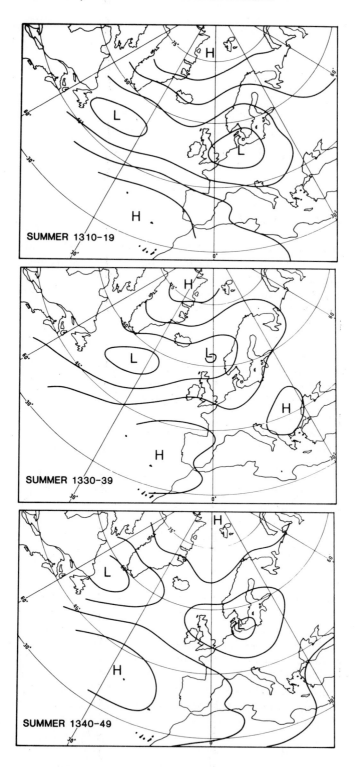

SUMMER 1310-19

SUMMER 1330-39

SUMMER 1340-49

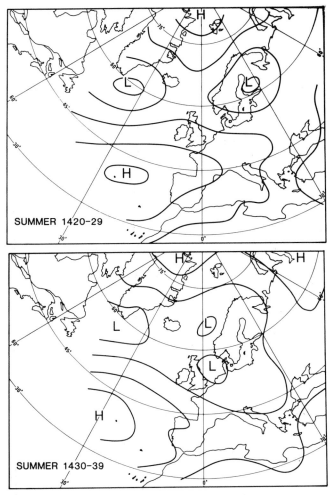

Figure 4.9 Maps of apparent decade average barometer pressure patterns for the summers 1310–19, 1330–9, 1340–9, 1420–9 and 1430–9, derived from similar material to figure 6.7. Note the prevalent cyclonic conditions over Europe.

and followed by great floods in the autumn and spring in England and the Netherlands. In order to bring out the differences from one winter to the next in the 1430s, of which five to eight had a preponderance of severe cold weather in different parts of Europe, the resulting maps are all printed in figure 4.8 *(designed to be read by magnifying glass)*. In only one case, in 1436–7, is there enough discordance among some of the reports to introduce any significant uncertainty into the map analysis. It is clearly stated to have been a milder winter than the preceding one in southern England (Sussex), but there are reports from London, Paris and Antwerp of a long frost. Only

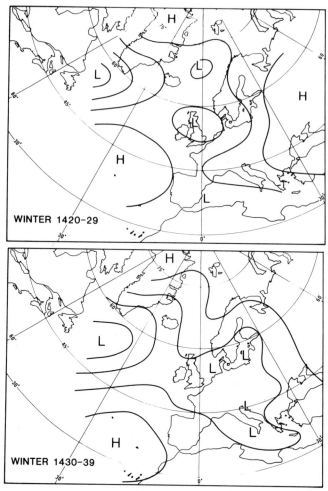

Figure 4.10 Maps of the apparent decade average barometric pressure patterns for the winters of 1420–9 and 1430–9, derived from similar material to figure 6.7. Note the absence of prevailing westerly winds over Europe.

that winter and the next one, which certainly was generally mild and windy, were under-reported for our purpose. The numbers of different places or regions of Europe for which weather descriptions were to hand for these ten winters were 11, 5, 14, 9, 12, 18, 8, 6, 3 and 8 – average 9.4. The low numbers were all from winters which were milder than the rest. The wind patterns prevailing in the mild winters 1430–1 and 1437–8 were quite clear despite the few reports. In addition to the 94 reports used in construction of the maps of these winters, there were six other reports which had to be dismissed as either obviously wrong (not fitting the map) or muddled or otherwise suspect. Sketches of the prevailing average patterns deduced for the summers of the decades 1310–19, 1330–9, 1340–9, 1420–9 and 1430–9 and for the winters of the two decades 1420–9 and 1430–9 are shown here in figures 4.9 and 4.10. These were the decades for which it was judged that enough information was now to hand. The patterns of these maps support the following diagnosis.

The easiest way to explain the characteristics of the period meteorologically seems to be to suppose that the prevailing warmth and anticyclonic regime of the mediaeval period in Europe before AD 1300 gave way at first to situations with cyclone centres (depressions) coming from the Atlantic in latitudes much closer than before to central Europe, and travelling across the continent or through the Baltic in the years around 1315 and in 1340–7; and that this led to, and was succeeded by, a greatly increased frequency of northerly winds drawing cold air and moisture south towards the Alps, in association with repeated lingering cyclonic activity over the Baltic or northern Russia.

With such patterns, it is natural to assume that anticyclones were often present over the Greenland–Iceland region and the European sector of the Arctic. By the fifteenth century, the cyclonic activity may have been far enough south to produce an extraordinary frequency of easterly and northeasterly winds over the north European plain and the Alpine region. Hints that this was so may be seen in the evidence of a heightened frequency of winds from these directions indicated by the reconstructed weather maps

Table 4.2   Prevailing European winds indicated by the reconstructed weather maps

Seasons	East Anglia	German Baltic coast	South Sweden	Swiss Alps
Summers				
Decade averages				
1310–19	C^{ic} N'ly	C^{ic} SE'ly	E'ly	WNW
1330–9	WNW	SW'ly	SSW	Zero (a 'col' area)
1340–9	NW'ly	C^{ic}	NE'ly	WNW to NW
1420–9	WNW	WNW to NW	WNW to NW	Light NW to NE
1430–9	C^{ic} NW'ly	Light S'ly	SE'ly	WNW to NW
Winters				
Decade averages				
1420–9	C^{ic} W'ly	S'ly	SE'ly	Anticyclonic
1430–9	C^{ic} N'ly	C^{ic}	NE'ly	C^{ic} NE'ly
Number of winters, easterly or northeasterly				
1430–9	6–7	5–7	8	5–6

Note: For the decades not listed, there is not yet enough data to hand for mapping. C^{ic}m cyclonic; N'ly, northerly, etc.

(figures 4.8–4.10). The indications of these maps for different parts of Europe are set out in table 4.2. In the regimes depicted, the Italian slopes of the Alps in the Aosta district may have become even drier, in the rain-shelter of the lee-side, than they were in the more anticyclonic period of earlier mediaeval warmth.

If this analysis of the prevailing winds and pressure patterns is approximately right, further development in the same direction could have brought the northern anticyclones systems more and more over Europe between about 1470 and 1540 to 1550. This would, if true, account for the evidence in the Alps of a rather short-lived recession of the glaciers and a tendency for the quick-growing birch and pine trees, which are rapidly spread by their light seed, to spread back again to higher levels on the mountains at that time. Pfister (1985) found that for a few decades before 1564, using data which started in 1525, average summer temperatures in Switzerland were 0.4 degC *above* the twentieth-century level,* probably implying very sunny conditions, and the winters were comparable to today's, although the averages fell sharply after that: between 1565 and 1600 the summers seem to have averaged 0.4 degC below present and the winters 1.2 degC below this century. This sequence of developments suggests a continued need for the irrigation channels in Aosta until AD 1550–60 or thereabouts, with perhaps the most extreme dryness there around the period 1400–50. The dryness at that stage in the Val d'Aosta region, which contrasts so markedly with the wetness north of the Alps in the fifteenth century, tends to confirm the implication of our weather analysis maps that the winds emphasized the rain-shadow effect in that region.

Another part of Europe may also supply some confirmation of the weather analysis between 1310 and the 1430s presented here. There is a historical mystery about why Norfolk, which is in the northern part of East Anglia, in eastern England, exposed to the North Sea to the north and east, suffered a higher proportion of lost and depopulated mediaeval villages and smaller settlements than most of England. Nearly one-fifth of all the settlements in Norfolk listed in the Domesday Book in 1086 disappeared: a quarter of the abandonments had occurred by 1316, and the remainder seem to have taken place mainly in the fifteenth century. Allison (1955), in the most comprehensive survey of the process, shows that the places which disappeared were generally smaller than the surrounding villages, and they

---

* A recent reconstruction (Burckhardt and Hense 1985) of mean April to September temperatures at Basle, using the late-wood density measurements from trees at Lauenen but also wine-harvest data from north-east France (which may, however, be too far from the region we have been considering), also indicates mean temperatures rather higher than today's from as early as about 1450–1560. But these results must be less reliable than Pfister's, because the late-wood measurements should not be used for April to September (only July to September) temperatures.

According to Pfister, 1540 was an outstandingly warm and sunny year. Between March and September there were only six rainy days at Basle. And at Christmas, youths were able to bathe in the Rhine thereabouts.

are shown by the tax records of 1334 to have been notably poor places then. Furthermore, from the year 1432 onwards, reductions of the 1334 tax assessments began to be allowed 'for the relief of poor towns . . . desolate, wasted or destroyed, or over-greatly impoverished places'. Allison lists the reductions of tax allowed in 1449 for about 50 of the villages in Norfolk which were later abandoned. The reductions were commonly between 20 and 40 per cent, in one case 57 per cent. This is a history for which Holmsen (1978) found many parallels in upland Norway. The relevance of this to our maps is that with the westerly and southwesterly winds that are regarded as normal and usual in twentieth-century experience, Norfolk partakes of the generally dry climate of East Anglia, the driest in England; but with winds from between NNW and about E, Norfolk's climate becomes wet, particularly in cold, showery, cyclonic conditions. If, indeed, the severe impoverishment and ultimate loss of these places on the most difficult soils and sites in Norfolk was caused by climatic deterioration, the frequency of northerly cyclonic weather indicated in the maps in this chapter would provide an explanation.

There is a little more specific evidence – so far only fragmentary – which points to a peculiarly wet period in Norfolk, particularly between 1420 and 1450. At three places in the county for which mediaeval records of local legal proceedings survive – Hockham, Kempstone and Rougham, in the area 52.5 to 52.8°N and 0.7 to 0.8°E – proceedings over drainage complaints became prominent at that time. Table 4.3 shows the incidence of such cases from 1380 to 1612, from a list which Mr A. Davison of the Norfolk Research Committee has compiled.

Table 4.3 Legal hearings in the manorial courts at Hockham, Kempstone and Rougham, Norfolk, over complaints about drainage, 1380–1612

1380–99	No cases to list
1400–9	2 cases
1410–19	1
1420–9	4
1430–49	Many cases
1450–9	1
1460–9	1
1470–1515	No cases to list
1516–29	4
1530–51	No cases to list
1552–64	3
1565–1612	1 (in 1612)

Note: Cases in which individual persons were arraigned before the courts for actions which caused a blockage of the drainage have been omitted from the figures in the above table.

Less specifically, the analyses of the prevailing weather situations derived in this chapter, if true, would make it easy to understand certain exceptional tribulations reported elsewhere in northern Europe. In Scotland, the organization of the great estate which had been ruled by the Earls of Mar in the central Highlands broke down in 1433 after a number of difficult years. Poverty rapidly worsened in the region, and in that decade birch bark had to be used to make bread in the Scottish Highlands, as also in Sweden. The origins went back to 1390, when a disaffected son of the king, who had appointed him Earl of Buchan, took advantage of general unrest and the weakness of the royal power in the north to burn and pillage through the Highlands and beyond, destroying the towns of Forres and Elgin and becoming known as 'the Wolf of Badenoch'. Turmoil increased in Scotland, until in 1436 the king was murdered on a hunting expedition near Perth, on the edge of the Highland region. This led to the decision to move the capital from its old seats near Perth and in Dunfermline to Edinburgh, where the castle offered the best security. It was also in this decade that the farming settlement at Hoset, east of Trondheim in central Norway, was abandoned for the second time in its history. In Germany, in 1430, there were damaging frosts as late as 23 May (New Style). And famines were reported in at least three years in the 1430s across Europe from England to Russia. Fern roots were used for bread in England, and there was trouble with hungry wolves in this country (for the last time) and in eastern Europe.

The aspects of the climate development in central Europe which we have examined in this chapter indicate that traffic over several, and possibly over most, of the Alpine passes that had connected the Italian side with the north was sooner or later brought to a halt in the Little Ice Age period in recent centuries. The circumstances which led to abandonment of the routes varied and were haphazard in their incidence. Flood disasters caused by bursting of the ice-dammed lakes in the Swiss valleys were perhaps more important than the increases of the glaciers and the snow higher up. Some of the routes over the Alps used in Roman times seem to have been similarly abandoned during the time of colder climate and glacier advances in the centuries that followed. C. E. P. Brooks suggested long ago that traffic over the Alps had first flourished in the warmth of the Bronze Age and may have virtually ceased in the colder climate which followed, bringing advances of the glaciers from their postglacial minimum stands.

By the eleventh or twelfth century AD, the German-speaking Walser people, taking advantage of the mediaeval glacier recession, had spread over the main Alpine ridge into the high valleys on the Italian side, the movement having actually begun as early as the sixth century. The evidence we have reviewed has also indicated that the drier conditions prevailing during the mediaeval warmth, and consequent water-supply problems in central Europe, were prolonged through the fourteenth and fifteenth centuries on the Italian side of the mountains, presumably through the rain-shadow effect associated with times when more northerly winds and greater wetness seem to have been prevalent in Europe north of the main Alpine ridge. The reporting of the weather in Europe through those times has led to a wind-pattern analysis which is consistent with these human experiences and which brings to light glimpses of the nature of the atmospheric circulation development during an interesting period of the approach to the Little Ice Age.

It should not be overlooked that the highly anomalous atmospheric circulation characteristics of certain decades in the fourteenth and fifteenth centuries indicated by this analysis occurred during the development of the Spörer minimum of solar activity. It seems that sunspot activity was generally falling, apart from the interposition of two great sunspot maxima around 1360 and the 1370s, very much like those around 1560 and the 1570s; and there is auroral evidence of very low sunspot numbers between about AD 1400 and 1500 or 1510. Thus, this time of peculiar wind circulation characteristics in the late Middle Ages (compared with the twentieth century) bears about the same relationship to the Spörer minimum as the seventeenth-century cold climate development does to the Maunder minimum of sunspot activity from about 1600, or more decisively from 1645, to 1715.

References

Abel, W. (1955) *Die Wüstungen des ausgehenden Mittelalters*, Stuttgart, Gustav Fischer Verlag.

Alexandre, P. (1986) *Le climat en Europe au moyen age: contribution à l'étude des variations climatiques de 1000 à 1425*, Paris, Publications de l'école des hautes études sociales.

Allison, K. J. (1955) 'The lost villages of Norfolk', *Norfolk Archaeology*, 31(1), 116–162, Norwich.

Arago, F. (1858) 'Sur l'état thermométrique du globe terrestre', in *Oeuvres complètes, Vol. 8*, Paris, Gide for Acad. des Sciences, and Leipzig, Weigel.

Bell, W. T. and Ogilvie, A. (1978) 'Weather compilations as a source of data for the reconstruction of European climate during the medieval period', *Climatic Change*, 1(4), 331–348, Dordrecht, Reidel.

Beresford, M. W. (1954) *The Lost Villages of England*, London, Lutterworth Press.

Bezinge, A. and Vivian, R. (1976) 'Troncs fossiles morainiques et climat de la période Holocène en Europe', *Étude présentée, Société Hydrotechnique de France*.

Britton, C. E. (1937) 'A meteorological chronology to AD 1450, *Geophysical Memoir, No. 70*, London, HMSO for Meteorological Office.

Bryson, R. A., Baerreis, D. A., and Wendland, W. M. (1970) 'The character of late-glacial and postglacial climatic changes', in *Pleistocene and Recent Environments of the Central Great Plains*, University of Kansas Department of Geology, Special Publication 3, Lawrence, University of Kansas Press.

Bryson, R. A., Irving, W., and Larsen, J. (1965) 'Radiocarbon and soil evidence of a former forest in the southern Canadian tundra', *Science*, 147, 46–48.

Bryson, R. A. and Murray, T. J. (1977) *Climates of Hunger: Mankind and the World's Changing Weather*, Madison, University of Wisconsin Press.

Buchinsky, I. E. (1957) *The Past Climate of the Russian Plain* (2nd edn), Leningrad,Gidrometeoizdat (in Russian).

Burkhardt, Th. and Hense, A. (1985) 'On the reconstruction of temperature records from proxy data in mid-Europe', *Archiv für Met., Geophys., Biokl.*, Serie B, 35, 341–359.

Christensen, A. E. (1938) Danmarks befolkning og bebyggelse i Middelalderen, *Nordisk Kultur*, 2, 1–57, Copenhagen, Oslo, Stockholm.

Comfort, A. (1966) *Nature and Human Nature*, London, Weidenfeld and Nicholson.

Delibrias, G., Ladurie, M., and Ladurie, E. Le Roy (1975) 'Le fossil forêt de Grindelwald: nouvelles datations', *Annales: economies, societies, civilisations*, No. 1, 137–167, Paris, Armand Colin.

Easton, C. (1928) *Les hivers dans l'Europe occidentale*, Leyden, Brill.

von Fels, H. R. (1962) 'Les Rus de la vallée d'Aoste', *Annales Valaisannes*, No. 2–4, 1–16.

Firbas, F. and Losert, H. (1949) 'Untersuchungen über die Entstehung der heutigen Waldstufen in den Sudeten', *Planta*, 36, 478–506, Berlin.

Flohn, H. (1967) 'Klimaschwankungen in historischer Zeit', in *Die Schwankungen und Pendelungen des Klimas in Europa seit dem Beginn der regelmässigen Instrumenten-Beobachtungen (1670)*, H. von Rudloff, Braunschweig, Vieweg-Die Wissenschaft, Band 122.

Galloway, P. R. (1985) 'Annual variations in deaths by age, deaths by cause, prices and weather in London 1670 to 1830', *Population Studies*, 39, 487–505.

Gesnet, J. (1982) *A History of Chinese Civilization*, Cambridge, Cambridge University Press. (Originally published in French as *Le monde chinois*, Paris, Armand Colin, 1972; trans. by J. R. Foster).

Graunt, J. (1662) *Natural and Political Observations made upon the Bills of Mortality*, London, Roycroft.

Griffin, J. B. (1961) 'Some correlations of climatic and cultural change in eastern American prehistory', *Annals of the New York Academy of Sciences*, 95(1), 710–717.

Grove, J. M. (1972) 'The incidence of landslides, avalanches and floods in western Norway during the Little Ice Age', *Arctic and Alpine Research*, 4(2), 131–138, Boulder, Colorado.

Grove, J. M. (1979) 'The glacial history of the Holocene', *Progress in Physical Geography*, 3(1), 1–54, London, Arnold.

Grove, J. M. (1985) 'The timing of the Little Ice Age in Scandinavia', in *The Climatic Scene*, M. J. Tooley and G. M. Sheail (eds), London, Allen & Unwin.

Grove, J. M. and Battagel, A. (1983) 'Tax records from western Norway as an index of Little Ice Age environmental and economic deterioration', *Climatic Change*, 5, 265–282, Dordrecht, Reidel.

Hennig, R. (1904) 'Katalog bemerkenswerter Witterungsereignisse von den ältesten Zeiten bis zum Jahre 1800', *Abhandlungen des Preussischen Meteorologischen Instituts*, Band II, Nr. 4, Berlin.

Holmsen, A. (1961) *Norges historie*, Oslo and Bergen, Universitetsforlaget.

Holmsen, A. (1975) 'Nyrydning og ødegårder i Norge før svartedauen', (New forest-clearances and abandoned farms in Norway before the Black Death), *Heimen*, 16, 481–490.

Holmsen, A. (1978) *Agrarkatastrofen i Norge i middelalderen*, Oslo, Bergen and Tromsø, Universitetsforlaget.
Huitfeldt, H. J. (publisher) (1879) *Biskop Eysteins Jordebog*, Christiania, Oslo.
Janin, B. (1968) *Le val d'Aoste* (esp. ch. 2, 'Le climat et ses consequences'), Grenoble.
Johnsson, F. (1930) *Den gamle Grønlands beskrivelse*, by Ivar Barðarson (or Baardsøn), Copenhagen (trans. by F. Johnsson).
Koch, L. (1945) 'The East Greenland ice', *Meddelelser om Grønland*, Bd. 130, Nr. 3, Copenhagen.
Ladurie, E. Le Roy (1971) *Times of Feast, Times of Famine*, New York, Doubleday. (Originally published in French as *Histoire du climat depuis l'an mil*, Paris, 1967, Flammarion.
Lamb, H. H. (1965) 'The early medieval warm epoch and its sequel', *Palaeogeography, Palaeoclimatology, Palaeoecology*, 1, 13–37, Amsterdam, Elsevier.
Lamb, H. H. (1966) *The Changing Climate*, London, Methuen.
Lamb, H. H. (1977) *Climate: Present, Past and Future. Vol. 2, Climatic history, and the Future*, London, Methuen.
Lamb, H. H. (1982) *Climate, History and the Modern World*, London, Methuen.
Lamb, H. H. (1987) 'What can historical records tell us about the breakdown of the mediaeval warm climate in the fourteenth and fifteenth centuries – an experiment', *Contributions to Atmospheric Physics/Beiträge zur Physik der Atmosphäre*, 60 (2), 131–43.
Müller, K. (1953) *Geschichte des badischen Weinbaus*, Laar in Baden, von Moritz-Schauenburg.
Munch, P. A. (publisher) (1852) *Aslak Bolts Jordebog*, Christiania (Oslo).
Nichols, H. (1970) 'Late Quaternary pollen diagrams from the Canadian Arctic Barren Grounds at Pelly Lake, Keewatin, NWT, *Arctic and Alpine Research*, 2(1), 43–61, Boulder, Colorado.
Patzelt, G. (1974) 'Holocene variations of glaciers in the Alps', in *Colloques internationaux du Centre National de la Recherche Scientifique, No. 219: Les méthodes quantitatives d'étude des variations du climat au cours du Pleistocene*, Paris, CNRS.
Pettersson, O. (1914) 'Climatic variations in historic and prehistoric time', *Svenska Hydrografisk-Biologiska Kommissionens Skrifter*, Häft 5, Gothenburg.
Pfister, Chr. (1984) *Klimageschichte der Schweiz 1525–1860* Band I: *Das Klima der Schweiz von 1525–1860 und Seine Bedeutung in der Geschichte von Bevölkerung und Landwirtschaft*, Band II: *Bevölkerung, Klima und Agrarmodernisierung*, Bern & Stuttgart, Paul Haupt, Academica Helvetica 6.
Pfister, Chr. (1985) 'Veränderungen der Sommerwitterung im šudlichen Mitteleuropa von 1270–1400 als Auftakt zum Gletscherhochstand der Neuzeit', *Geographica Helvetica*, 4, 186–195.
Rey, R. (1984) 'Germanophones en vallée d'Aoste: les Walser', *Jeunesse d'Aujourd'hui*, December.
Röthlsberger, F. (1976) '8000 Jahre Walliser Gletschergeschichte, II Teil: Gletscher und Klimaschwankungen im Raum Zermatt, Ferpècle und Arolla, *Die Alpen*, 52(3–4), 59–152.
Schneebeli, W. (1976) '8000 Jahre Walliser Gletschergeschichte, I Teil: Untersuchungen von Gletscherschwankungen im Val de Bagnes', *Die Alpen*, 52 (3–4), 6–57, Luzern, Swiss Alpine Club.

Schweingruber, F. H., Schaer, E., and Braeker, O. U. (1978) 'X-ray densitometric results for subalpine conifers and their relationship to climate', *Dendrochronology in Europe*, J. Fletcher (ed.), *British Archaeological Reports*, 51, 89–100, Oxford.

Short, T. (1749) *A General Chronological History of the Air, Weather, Seasons, Meteors, etc.* (2 vols), London, T. Longman and A. Millar.

Speerschneider, C. I. H. (1915) 'Om Isforholdene i danske Farvande i ældre og nyere Tid: Aarene 690–1860', *Meddelelser*, Nr. 2, Copenhagen, Danish Meteorological Institute.

Taylor, E. G. R. (1963) 'Shipwrecked in the Arctic in 1432', *Geographical Magazine*, 36(7), 377–383, London.

Thorarinsson, S. (1956) 'The thousand years' struggle against ice and fire', *Miscellaneous Papers No. 14*, Reykjavik Museum of Natural History.

Thoroddsen, Th. (1916–17) *Árferdi á Islandi i thusund ár*, Copenhagen (in Icelandic).

Trevelyan, G. M. (1928) *History of England*, London, Longmans, Green.

Woodbury, R. B. (1961) 'Climatic changes and prehistoric agriculture in the southwestern United States', *Annals of the New York Academy of Sciences*, 95(1), 705–709.

5

The climate theory of race in sixteenth- and seventeenth-century literature and its modern implications

This short chapter provides an interlude in which we are as much concerned with people's impressions of the climate and its effects as with the reality. It is based upon my review of the publication of an Austrian degree thesis (W. Zacharasiewcz 1977).

The theory alluded to is that climate, and the environment it controls, determine the character of peoples – or, in the language of the centuries mentioned, that it determines the prevailing 'humours', the nature and character tendencies of nations. This idea goes back at least to Hippocrates and other medical men of ancient Greece around 400 BC. Aristotle (384–322 BC) already used the notion of prevailing characteristics of people in the cold, warm and middle zones of the Earth to expound the superior qualities of the Hellenes as a master race which, if only they could be united, should be able to rule the world. We should, perhaps, see this theory as having been a proposition by which to learn about the world and human nature; but we must note the dangers to which too popular, glib applications of the theory directly led. Of such applications there have been many.

The climate theory, which is often associated with its presentation by Montesquieu in the eighteenth century, was actually introduced to modern Europe by the Dutch doctor, Levinus Lemnius, in 1561. Its popularity and influence grew from that time until about 1735, when it appeared first in medical and psychological writings, and then soon in political tracts of the times, in travel literature, in prose and poetry, and even in theories on the development of language and pronunciation, in drama and later as a theme for critical examination in scientific writing.

Much of the story is a sad one, showing how quickly this powerful idea served to build up chauvinism, and indeed many varieties of national and religious prejudice, not only in England but in many parts of Europe. Some wry amusement may be found in the sixteenth-century Dutch doctor's enthusiasm for the cleanness of English houses and the peacefulness and patience of the nature of the Irish, not to mention the preaching and advice to young people against travelling to the immoral countries of the Catholic south. One of the objections to the theory, which worried its serious proponents, was of course the changes of character of certain nations since the earliest times: the 'barbaric' state of seventeenth-century Greece, the lapsed state of Italy also compared with classical Rome, and the disappearance of the free democratic practices of the early Germanic peoples among their descendants in the Denmark (possibly even the England) of the seventeenth century. The theory, after all, put everything on environment and nothing on heredity, and no thought seems to have been given at that stage to the possibility of changes occurring in either of these.

It is curious to note the static conception of the world and its climates that ruled the day; but perhaps that was a necessary stage in the sorting out of the new knowledge, *inter alia,* of distant places, that was growing so fast. At most, only among a few exceptionally well-informed people can some awareness of the unusual climatic stresses of the time they lived in have stimulated the interest of seventeenth-century scientists and other observers in climate and in this theory. That century witnessed advances of the glaciers and of the Arctic sea-ice, culminating in probably the severest climate that has occurred since the last major glaciation, but the people living near the advancing glaciers, as at Chamonix and Grindelwald in the Alps, and in parts of Norway, were somewhat alone in their alarm and complaints to authority.

There are other interesting aspects surrounding the climate theory. The greater dramatists of the time, such as Shakespeare, used it as a topic of current interest, often to make fun, without committing themselves to any view of its worth. Their often largely illiterate audiences loved scenes from foreign parts and overdrawn stereotypes of the character of the peoples that inhabited them. This was a taste obviously stimulated by the Age of Discoveries. Other playwrights pandered to the growth of the crassest chauvinism. Most professional writers seemed more concerned with the tastes of their readership than with sound observation or proof. The climate theory was a powerful idea in the seventeenth century, but it gained acceptance in each country through the adaptations of authors whose formulations of it served to bolster the national self-esteem and feed its prejudices.

Another manifestation of the crude forms taken by the widespread and absorbing interest in newly discovered places and strange peoples was the efforts made from time to time by ships' crews to capture a native of this or that country and bring him back to the royal court, almost as a zoological exhibit – a fate believed to have overtaken at least an occasional lonely and terrified eskimo captured in his kayak.

Some concern was apparent among many of the theory's English adherents about the position of their country in a northern latitude, where the peoples were often supposedly dull and lumpish. They might be ready to agree that such descriptions could apply to the Dutch, the Germans, and others, but were at pains to explain their own more favoured regional position. One English *Account of Denmark* in 1694 noted that; 'Island [i.e. Iceland] and Feroe are miserable islands in the North Ocean; corn willl not grow in either of them, but . . . the Inhabitants are great Players at Chess. It were worth some curious mans enquiry how such a studious and difficult Game should get thus far Northward, and become so generally used.' Among the English a pride in their own sense of humour was already firmly established and attributed not only to the benign climate, but to their gross diet. Many were, however, concerned with the nation's alleged tendency to melancholy and spleen. The English climate was cited by William Congreve as a basis for these tendencies and to justify the still widely unacceptable character of Restoration drama as a necessity to cheer up people. This was in answer to those who in 1698 were campaigning to 'clean up' the theatre. A similar point was more moderately put by the *Tatler* in the next century, adding 'Do you not observe that you meet with more affronts on rainy days?'

While all these effusions were being propounded there was, however, a gradual progress towards understanding among the scientists; calmer judgements began to emerge in place of prejudice and preoccupation with breeding the martial virtues. Thus, Sir Henry Blount wrote in 1637 that: 'seeing the customs of men are much swayed by their naturall dispositions, which are originally inspired and composed by the Climate . . . it seems naturall, that to our North-West parts of the World, no people should be more averse, and strange of behaviour, than those of the South-East.' And Joseph Glanvill, one of the first Fellows of the Royal Society, wrote in 1661: 'Thus opinions have their Climes and National diversities: And as some Regions have their proper Vices, not so generally found in others . . . And I take this for one of the most considerable causes of the diversity of Laws, Customes, Religions, natural and moral doctrines.'

We can all supply examples from the twentieth century of the national self-justification and racial prejudices which this theory helped to promote. No wonder that extreme forms of climatic determinism are now discredited. This should not blind us to the way in which climate and environment, and their changes, as well as our heredity affect our lives and outlook.

Reference

Zacharasiewicz, W. (1978) *Die Klimatheorie in der Englischen Literatur und Literaturkritik von der Mitte des 16. bis zum frühen 18. Jahrhundert,* Wiener Beiträge zur Englischen Philologie, 77, Vienna and Stuttgart, W. Braumüller.

6

Climate in historical times and transgressions of the sea, storm floods and other coastal changes

This account was written for a conference in 1978 at the University of Ghent, Belgium, to which representatives from all the countries with low-lying coasts around the southern North Sea were invited. It was published in the book of the conference, as pp. 251–90, in Transgressies en occupatiegeschiedenis in de kustgebieden van Nederland en Belgie *(Transgressions of the sea and the history of settlement in the coastal areas of Holland and Belgium, A. Verhulst and M. K. E. Gottschalk (eds), Ghent, 1980, Belgisch Centrum voor Landelijke Geschiedenis, Nr. 66. (The last part of the original paper, giving weather-map examples of storms which produced floods on different coasts around the North Sea, has been omitted here, since examples will be found in Chapters 7 and 8.) The various papers in the volume are printed in Dutch, German, English or French and cover the history of the coasts and coastal settlements of England and Germany, as well as the Low Countries, over the last several thousand years, with particular attention to the centuries since Roman times.*

Changes around the North Sea

The coasts of the North Sea have been the scene of a succession of storm-flood disasters down the ages among the worst recorded in human history anywhere in the world. The record has been catalogued by various compilers, from F. Arends in 1833 to the three-volume work of Gottschalk (1971, 1975, 1977), which is distinguished for its thorough critical examination of the original manuscripts. It appears that between AD 1099 and 1570, at least 286 towns, villages and parishes, many islands (some of them inhabited), and over 1.5 million human lives were lost in 30 floods for which estimates of the losses are known. In the Christmas Day disaster of 1717, some 12,000 lives were lost, mainly in Emden on the German coast, but also many in Denmark and about 2000 in Holland. But in the earlier mediaeval incidents, 306,000 are reported to have drowned in one flood in north Holland in AD 1212; and about 400,000 altogether are estimated to have died at points along the continental coast from north France to Germany in the flood of 31 October to 2 November 1570, when the sea entered the great cities. Great death tolls are also likely to have occurred when the sea destroyed former great ports on the coasts of England and Holland. And we must suppose that the prehistoric inroads of the sea in early post-glacial times over the inhabited plains which were to become the bed of the North Sea were no less horrifying to the folk of those times.

Maps of the Belgian and Dutch coasts at several epochs during the last two millennia show that a strip of the Belgian coast generally 1–3 km (and occasionally up to 10 km) wide was submerged between about 400 BC and the time of Christ. A somewhat greater transgression, possibly up to 20 km in places near the French border, is indicated around AD 300 to 700. It was in those times that access to the port of Bruges (Brugge) was developed. The times between AD 1000 and 1200 produced renewed submergence of about the same districts (M. Ryckaert, in Verhulst and Gottschalk 1980). The coast of Holland and Friesland between the Hook and the present island of Ameland was probably displaced seaward by 5–25 km in Roman times. But by about AD 900 an inlet was already developing in the Wadden Sea area to become part of the later Zuyder Zee/Ijssel Meer, and by AD 1300 a big sea area had developed, producing the islands Texel, Vlieland, Terschelling, Ameland and others from the line of the old coastal dunes (H. Schoorl 1980).

The enormous reduction in the losses of life from sea-storm floods on the coasts around the southern North Sea since the seventeenth century must be accounted a triumph of the engineering of sea-dike defences, pioneered by the Dutch and extended by them to the English fenland in that century. Nevertheless, breaches of the dikes continue to occur from time to time in great storms, and sea-level in the region has risen since that time. Erosion of the coasts by cliff falls also continues. The rate of sea-level rise varies, but was particularly great in the first 60 years of the twentieth century, when a global rise was taking place with the melting of glaciers in many parts of the world, as the climate became warmer. This seems to have levelled off since. To protect London, the Thames Barrier has been built in recent years. Nevertheless, any continued rise of sea-level must ultimately set a limit to what can be achieved in the way of coast protection. The North Sea storm of 31 January–1 February 1953 took 1600 lives in the Netherlands and a further 350 on the English coast. The storm of 17 February 1962 drowned some hundreds near Hamburg. Later storms have produced higher water-levels, but much smaller numbers have drowned. This is a triumph for the meteorological warning services. The hurricane and tornado warning services have achieved similar success in the USA and in east Asia.

The older North Sea disaster figures quoted may be compared with the drownings in Bangladesh when the Bay of Bengal cyclone struck in November 1970, killing 300,000–700,000 people. The only reportedly greater disasters due to weather may be some of the floods of the great rivers over the plains of central China. The Yellow River (Hoang Ho) floods, which are held to explain the chronic instability of that region during Europe's Middle Ages (Gernet 1972), grew worse from 1327, causing serious famines nearly every year. In 1344 the dikes broke, allowing the river to flood huge areas for five years until the dikes were repaired. During that period it is alleged that 7 million people died. It has been suggested that the deaths among animals and wild life, together with the disturbance of rodents' habitats, may have triggered the great epidemic of bubonic plague which swept the world in the ensuing years as the 'Black Death'.

If we are to understand the varying incidence of sea-floods and coastal erosion, it is necessary to go further into the physical processes at work than we have done in this book so far.

The elements which contribute to sea-floods in the North Sea region include.

(i) *Rises of world sea-level* – due to the melting of glaciers and ice-sheeets. Melting of the ice-sheets of the last ice age is estimated to have raised the general sea-level by about 100–130 m, and a further rise of 50 m might be expected if all the present ice on land were to melt, about 90 per cent of this due to the ice covering Antarctica (Fairbridge 1961; Lamb 1972; CLIMAP Project 1976). It has been noticed that, in the climatic changes of the last two centuries, temperature trends in the Antarctic are sometimes opposite to those prevailing at the same time over most of the rest of the Earth. There seems to be something nearer parallelism between the Antarctic and elsewhere in the case of the longer-term, larger changes in prevailing temperature between ice-age and interglacial (warm) climates, such as the present.

(ii) *Rises of regional or local sea-level* – due to warping of the Earth's crust in this part of the world.

(iii) *Cyclic changes in the strength of the tidal force* of sun and moon, which cause the height range of the tides to vary.

(iv) *Variations in storminess* as the tracks and intensities of the travelling storms vary.

These changes, particularly the last two, also play a part in the other types of coastal change: the erosion of cliffs and the silting up of estuaries, as well as the occasional overwhelming of low-lying coastal features and harbours by blown sand.

The first postglacial inhabitants of the lands around the present North Sea seem to have been coast-dwellers – as at Star Carr in Yorkshire (England) and near Hamburg – using boats for fishing as well as hunting down the grazing animals (reindeer, bison, mammoth, etc.) on the open grassy plains which were gradually conquered by the advancing forest. And in Norway it seems that the first inhabitants must have entered from the lowland plain that is now the bed of the North Sea (see Lamb 1977, pp. 413–4 for further details).

We have mentioned in Chapter 2, Bloch's (1970) argument that the importance of salt, and its manufacture mainly at coastal sites by evaporation of sea-water, as a food-preservative, and also the adaptation of the hunting and fishing folk to the ice-age landscape south of the great ice-sheets, probably meant that the ending of the ice age and rise of sea-level brought great loss of life, one of the rare occasions when the total numbers of mankind were greatly reduced.

Others have supposed that the overruning of the coastal plains – of Africa, Asia, Australia and the Americas as well as Europe – by the postglacial rise of sea-level, even at its fastest average rate of about 1 m per century (or 1 cm per year) was so gradual that no lives would be lost. This is, of course, to be deceived by averages. It is hardly necessary to say to the inhabitants of the countries around the North Sea that the loss of land to the sea with rising sea-level is not a gradual process. It is a history of storm-flood disasters: some of the floods prove to be temporary, others have subsequently been artificially drained – in the case of the Zuyder Zee, after a lapse of seven centuries – and others are permanent losses whereby, for instance, the North Sea has been formed and enlarged over the past 10,000 years.

One of the most graphic reports of this process was given to the Council of Basle in the fifteenth century: that 60 parishes accounting for over half the agricultural income of the then Danish diocese of Slesvig (Schleswig) had been 'swallowed by the salt sea' since AD 1200.

There have also been coastal changes involving loss of life at times of regression of the sea (see Lamb 1977, p. 460), when farmland, settlements, and harbours have been overwhelmed or cut off by blowing sand, and even townships buried under sand dunes which grew to heights of 30 m or more.

Processes which effect coastal change

Changes in world sea-level

The Earth's total water inventory is estimated to be about 1.45×10^9 cubic kilometres, of which about 2.3 per cent is at present 'locked up' (frozen) as ice on land, while only about 0.6 per cent is in lakes and rivers and in the subsoil and porous strata of rocks; only about one ten-thousandth part is in the atmosphere at any given time. At the climax of the last glaciation about 18,000 years ago, approximately 5 per cent of the total was in the form of ice on land. (The sources of these estimates and further details are given in Lamb 1972b.) Consequently, world sea-level is estimated to have been 100–130 m lower 18,000 years ago than now (Fairbridge 1961; CLIMAP Project Members 1976).

One has to be cautious in accepting details of the course of the postglacial changes and fluctuations of world sea-level, because of the difficulty of establishing the world average level to within a metre or so, particularly in remote times. The writer believes that the best outline of these changes is that given in figure 6.1.

The most thorough recent review of the subject which attempts more detail seems to be that due to Mörner (1971, 1976), whose own analysis is based mainly on data from the Kattegat and largely supports the results derived by Tooley (1974, 1978) from north-west England. Both suggest highest world sea-level around 1000 BC, again in Roman times (for which there is some support in Bloch (1965)) and about AD 1000, as well as perhaps around 2000 BC, as in figure 6.1 – in other words, a high stand at the turn of each of the last four millennia. (We shall return to consideration of the possible detail of changes in world sea-level during the last millennium later.)

Changes of world sea-level are not entirely due to the melting or growth of ice-sheets and glaciers on land. Two other factors should be taken into account. First, the volume of water in the sea expands and contracts as its temperature rises and falls. A rise of 1 degC throughout the ocean's depth should raise sea-level by about 60 cm. The estimated cooling of the entire ocean by about 2.5 degC in the last ice age may, if the whole depth of the ocean was in the end equally affected, have contributed 1.5 m to the lowering of world sea-level. But warming or cooling the whole ocean by one or two degrees takes many thousands of years.

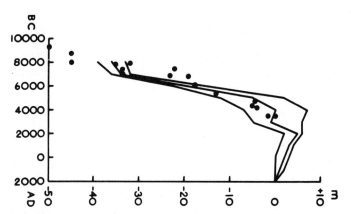

Figure 6.1 The change of world sea-level over the last 12,000 years. The bold printed (middle) curve connects average levels calculated by Schofield and Thompson (1964) at 1000-year intervals from Scandinavian data, allowing for the isostatic variation. All the individual estimates of world sea-level calculated from different localities lay between the limits set by the outer curves. The dots superposed on the diagram are the levels given by Godwin *et al.* (1958), based on data from many other parts of the world.

Additionally, the level of the ocean surface may have been affected by tectonism, i.e. by the Earth's crustal warping and collapses or rises of parts of the sea-floor. Such changes of the ocean-bed are, however, usually localized and produce only very small (nearly unmeasurable) changes of world sea-level.

Regional and more local changes in sea-level

Some changes in sea-level observed over more limited parts of the Earth represent a change in the shape of the planet (or geoid), a redistribution of sea-level by latitude associated with changes in the equatorial bulge of the Earth itself, under changes in the balance between gravitational and centrifugal forces. Such a change, involving a rise of sea-level by 10–15 cm in the higher northern latitudes and perhaps a slight negative change in the tropics between AD 1900 and 1950, has been demonstrated by Maksimov (1971).

Other, more localized, changes are caused by the Earth's crustal warping, under the influence either of the folding associated with continental drift and mountain-building or of isostatic rises and falls, associated with changes of loading of the crust in the area. These changes of loading occur when ice-sheets form or melt, when sediments are deposited or removed, and when the sea overruns or recedes from an area. Progressive downwarping of the North Sea basin has been going on since the Carboniferous. Its rate probably varies. According to one estimate, it may have contributed up to 3 m to the rise of sea-level on the Dutch and East Anglian coasts of the southern North Sea over the last 6500 years, a rate of about 5 cm per century (see comment in Lamb 1977, p. 118, on Churchill 1965). Even greater rates, up to 13 cm per century, affecting London and the Thames Estuary, have been suggested by some authorities (Akeroyd 1972; D'Olier 1972).

Variations of tidal force

The moon's orbit lies in the plane of the Earth's orbit once in about three years. At other times, the orbital planes are tilted at various angles to each other. The Earth's orbital arrangements also vary gradually over thousands of years, so that the distance from the sun, and the times of the year at which this is greatest and least, and the tilt of the polar axis of the Earth, all change. The combined tidal force of the sun and moon varies in consequence. It is greatest when they and the Earth are most nearly in line and the distances between them are least. The latitude on the Earth which experiences the greatest tidal pull also depends on the sun's elevation at the time (a combined effect of the Earth's axial tilt and the time of year). Because all these variations proceed on different time-scales, they rarely culminate together and the various maxima of tidal force are of different strengths.

Big maxima of tidal force, and hence big maxima in the range of the tides, occur at intervals of about 1600 to 1800 years. The last of these great maxima were around 3500 BC, 1900 BC, 250 BC and AD 1433. The next one in the series should be about the year 3300. Correspondingly notable minima of tidal range are indicated around 2800 BC, 1200 BC, AD 550 and AD 2400. Secondary maxima occur at intervals of 84–93 years and a lesser grade of maxima about every nine years.

These variations are not caused by climate but are a matter of the range of the tides, including the range of tidal variations in the depth of any internal boundary surfaces between water masses of different origins and density within the seas. They result from the changes in the tidal force which are of astronomical origin. These variations affect the amounts of the various water masses exchanged over the sills (bottom ridges) separating the Atlantic Ocean from the Arctic and the Atlantic Ocean from the Baltic Sea. Since there are big differences of density between the Atlantic water and the much less saline polar and Baltic waters, these variations can alter the climatic character of the Arctic and Baltic seas to an important extent and, hence, they can affect the climate of much wider regions. (See Pettersson 1914, 1930).

Variations of storminess

The variations of weather from day to day and from spell to spell, and changes of climatic regime over longer periods of time involve changes in the prevailing winds. The places where cyclonic depressions are most frequently generated, the routes along which they travel, and the intensities which they develop also change.

Long histories of the wind circulation patterns each day over the British Isles and over a wide region of the eastern Atlantic and Europe have been published (Lamb 1972a; Hess and Brezowsky 1977).

A study was made by Weiss and Lamb (1970) of the increasing roughness of the North Sea since 1953, measured as a 50–100 per cent increase in the frequencies of wave-heights over 3 and 5 m shown by the lightship records investigated in all parts of the North Sea. The increase has mostly been sustained (and in some areas increased) in the 1970s and since. Analysis showed that the change had been associated mainly with changes in the pattern of wind-direction frequencies. Westerly winds have decreased in frequency over the area; and northwesterly and northerly winds, which blow over long fetches of water into the open north end of the North Sea, have increased in frequency since 1950. This tendency has broadly been maintained in the 1970s, though with more increase of the northwesterly winds and a slight decrease in the frequency of northerly winds.

An updated study (Lamb and Weiss 1979) shows that there has also been an increase in the frequency of cyclone centres passing over the North Sea and in the occurrence of gale and severe gale situations (measured by the daily barometric pressure distributions over the North Sea). This increase since 1950 in the occurrence of gale and severe gale situations is found also to affect the British Isles and neighbouring portions of the Atlantic Ocean between latitudes 50° and 57°N. It is associated with increased frequency of low-pressure systems passing in or near these latitudes.

Measurement of the latitude of the Iceland low and its extension over the eastern North Atlantic (towards northern Europe) on mean barometric pressure maps for January and July shows that since the end of the eighteenth century its 40-year average position has varied by about three degrees of latitude. Its 10-year average position has varied by between seven and eight degrees of latitude. The breadth of the band of latitudes commonly visited by the passing storm centres has also varied.

Bigger displacements than these seem to have taken place between the high Middle Ages and the sixteenth century. Reconstructions based on analysis of the numerous reports of the character of the seasons year by year indicate prevailing storm tracks in the summers of the eleventh century near 67°N in the eastern Atlantic–European sector, compared with 55° to 60°N in the late sixteenth century. The winter positions also suggest a southward shift of about ten degrees of latitude between the respective long-term averages. Mean climatic maps, reconstructed for these periods and for other periods in earlier postglacial times, from which the prevailing winds and storm tracks can be deduced, are given in Lamb (1977).

Thus, the frequencies of the greater wave-heights, and presumably the heights attained by the greatest storm waves, vary with time, owing to changes in the frequency and intensity of low-pressure systems passing over or near the North Sea and changes in the frequency of northwesterly and northerly winds blowing over very long fetches of open water into the North Sea. Both apparently increase at times of cooling or cold climate.

In addition, the combined effects of lowered atmospheric pressure and NW–N storm winds blowing water into the North Sea produce surges of water which have in some cases raised the level as much as 3 m or more

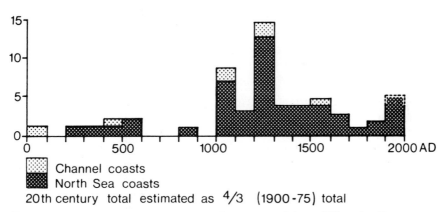

Figure 6.2 Numbers of severe sea-floods (causing much loss of life or land) per century on the North Sea and Channel coasts. (From data tabulated by Lamb 1977.)

above the predicted tide (by over 5 m in the River Elbe at Hamburg in 1634, 1962 and 1976). And in the contrary case, SW–SE storm winds may lower the tide level by 1–2 m. The combination of storm winds from these directions with extremely low tides has been suggested as the explanation of the rare occasions recorded in history, and usually reported as accompanied by strong to hurricane winds, when the River Thames 'ran dry' in London (Aranuva-Chapun and Brimblecombe 1978).

Combined circumstances of coastal changes

Proceeding from the foregoing considerations, we will consider various types of change to the coasts, all linked with fluctuations of climate and weather, and the combinations of circumstances which lead to them. They all constitute threats to human settlements and use of land near the coast, and all are known to have overwhelmed towns and townships around the North Sea.

They are:

(i) *Sea-floods and long-lasting invasions of low-lying coastlands* – these are favoured by high sea-level and climatic regimes which produce and maintain high sea-level by melting glaciers and ice-sheets, by extreme high tides and by storm winds from the directions (NW to N) that blow extra water into the North Sea.

(ii) *Erosion of cliffs and coastal defences both natural and artificial* – these are clearly advanced by any storm winds that bring high waves against the coast, i.e. particularly winds blowing against the shore at times of high tide and, most of all, if the general sea-level is high.

(iii) *Coastal changes by wind-borne and water-borne sand and other material* – these occur in circumstances which are less obvious, but they are clearly favoured by strong winds and strong currents and times of very low water which lay bare unusual expanses of sand. It may be that water movements, such as the long waves of a swell, coming over very long fetches of water from distant storms, must first accumulate sand on the beaches and in river-mouths, from where it is finally caught and transported by violent storms crossing the area itself.

Clearly, the circumstances favouring sea-floods and erosion are much alike, but (i) is the more extreme type of event, demanding combinations of more extreme contributory conditions, including storm winds blowing over the longest paths over the open sea. However, it seems not to be necessary for all possible contributory factors to be in an extreme condition in every case. Gottschalk (e.g. 1975, p. 501) draws attention to several serious storm-floods in the Netherlands which occurred with neap tides, e.g. in 1014, 1421 and 1532. Such cases presumably imply even greater violence of the winds and/or a particularly long and sustained sea-fetch over which the surge and swell build up. Coastal changes by blown sand are also an extreme type of event, and there is evidence that the major events of this kind are accompanied by circumstances that produce extremely low tides (extreme ebbs).

Outline history of climatic changes and coastal events

Early stages in the development of the North Sea, from the opening of a 20-km-wide channel to the widening of the sea between England and Flanders between about 9600 and 8000 years ago, have been mapped, first by Jelgersma (1961) and later by D'Olier (1972). Since that time, the sea has widened perhaps 20 per cent further, as the climate warmed up and world sea-level rose with the general melting of ice-sheets and glaciers. But it has not been a continuous process.

Figure 6.3 shows the course of the temperature changes over the last 12,000 years derived for England from botanical and, in the later periods, from historical and meteorological instrument data (see Lamb 1977). The curves for England are believed to be reasonably representative of the global average over this time-span. The shape broadly parallels that of the world sea-level curves in figure 6.1, except that the sea-level changes show a lag such that the highest level is reached, as it logically should be, about the end of the warmest postglacial times. One must suppose that in all parts of the world, including the Antarctic, a net melting of glaciers and ice-sheets had been going on up to that time. In the last millennium, the estimated (or calculated after 1700) mean temperatures for each century are shown. These show significant variations. Somewhat similar variations from century to century probably occurred in the earlier millennia, though their range may have been less in the warmest millennia and was certainly greater in the Late Glacial cold episode in the ninth millennium BC.

There were doubtless also variations of storminess from century to century and, of course, from decade to decade and year to year.

Jelgersma et. al. (1970) found evidence that the coastline of the Netherlands at times between 5000 and 4000 years ago was about 50 km farther east than now and at one point, 75 km farther east. These 'Calais transgressions' correspond well with the world sea-level maximum indicated in figure 6.1. Between these episodes, during the so-called 'Sub-Boreal' climate period, and thereafter up to Roman times, sand barriers apparently about 10-m high were formed – the 'Older Dunes' – only a few kilometres east of the present coast. There were various periods of dune formation separated by quieter periods, the latter apparently associated with wetter conditions with vegetation (including trees) and soil and peat formation. The intervals with higher water-table may have been related to minor maxima of the sea-level. Whenever dune formation was resumed, the forest growth was interrupted by blowing sand. One of the longest intervals of quiet conditions, with forest becoming established, seems (according to Jelgersma et. al. (1970) to have lasted more or less 11 centuries until about AD 1200).

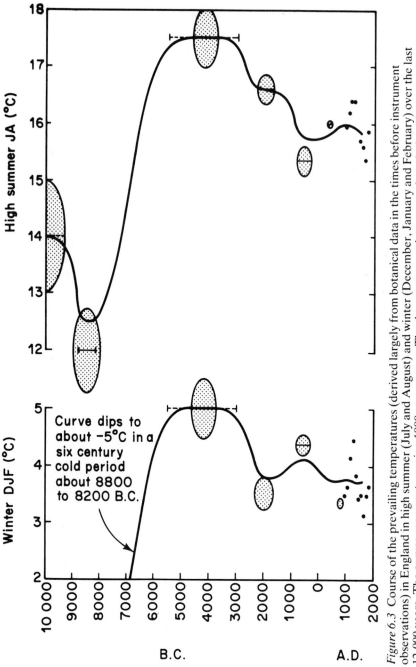

Figure 6.3 Course of the prevailing temperatures (derived largely from botanical data in the times before instrument observations) in England in high summer (July and August) and winter (December, January and February) over the last 12,000 years. The curves represent running 1000-year means. The dots represent the mean values derived for each of the last 12 centuries. Ovals indicate the ranges of uncertainty of the derived temperature values and dates. Horizontal bars indicate the apparent duration of conditions similar to the point plotted.

In those times, particularly after AD 800, the development is complicated by human activities, such as felling the forest and breaching the sand barriers in the course of military operations, though the effects seem to have been fairly localized.

It was during the time between the middle of the first millenium AD and some time after 1000 that Hallam (1959–60, 1965) reports an active (peat-burning) salt-making industry in the English Fenland near the shores of the Wash. This does not necessarily imply a lowered sea-level, and at least around AD 400 and 1000–1100 such was clearly not the case; but the sea-level may have been slightly lowered between about AD 450 and 800 and it probably does imply a long period with rather stable coastline. The method of salt-making practised in the area required the tide to flow over sand which, when saturated with salt, was taken to the drying pans.

A new phase of dune formation – the 'Younger Dunes', growing up to 40 m high – appears to have begun in the twelfth or thirteen century AD, and the coastline was for a time farther west than at present and apparently of more irregular shape. This phase affected other coasts around the North Sea and elsewhere in north-west Europe. Gottschalk (1975, pp. 189, 302) refers to numerous reports of blowout in the Dutch dunes in the fifteenth century and to areas of land and settlements being buried in sand and lost. Once again, it may be necessary, and reasonable in view of the evidence of climatic cooling and extended glaciers, to assume a falling sea-level.

Between about AD 1400 and 1800, blowing sand overwhelmed coastal sites with long-established settlements on the coasts of Wales and the Hebrides, on the east coast of Scotland and in Denmark and buried them under high dunes. At one place in South Wales and another in north-east Scotland, the dune ridges proceeded several kilometres inland. In some cases the precise dates are known and it is recorded that one single storm (see p. 98) covered an area of 20 km^2 with sand (Lamb 1977, pp. 270, 460). But whereas the formation of the Younger Dunes of the Netherlands is suggested to have been completed before the end of the sixteenth century, this epoch of disasters by blowing sand certainly continued until later elsewhere. Two of the greatest incidents of this kind in Scotland took place in the 1690s, and the sanding up of the 'buried church' (*den tilsandede kirke*) and village at Skagen in north Jutland (Denmark) took place in the seventeenth and eighteenth centuries (see figure 6.8).

Dust and sand layers radiocarbon-dated in a peat bog, Fuglmose in eastern Jutland (Bahnson 1972), indicate as the main epochs of sand blowing approximately 600 to 100 BC (perhaps continuing as late as AD 200) and renewed activity, starting perhaps as early as about AD 1000, becoming most serious in the 1600s and continuing until about 1900.

There is ground for asserting that some of the storms in this part of the world in the Little Ice Age period exceeded in violence the probable maximum of the period 1900 to 1940. A meteorological analysis of the

Spanish Armada storms between July and October 1588 (Douglas *et. al.* 1978), using the weather reports in the ships' logs, shows that six times the 24-hour movements of the depression centres indicate jetstream velocities at or somewhat beyond the probable limit for those months in modern times. The energy released by these storms may have been derived from the strong temperature gradient between 50°N and the polar sea-ice which was near Iceland in the 1580s.

For the last 3000 years, within the period that has generally been considered the Nachwärmezeit (or Neoglacial), marked by various signs of cooling since the warmest postglacial times and by various advances of the glaciers in many parts of the world from the postglacial minimum, we come to times for which a preliminary survey, providing a sequence of climatic maps, has recently been performed (Wigley 1978). The compilation includes the widest range of critically examined historical and proxy data. It emerges that each of the last three millennia (and on some evidence also the two millennia before that) has had some sort of fluctuation towards colder climate (probably always complex and consisting of more than one main phase) affecting middle northern latitudes, particularly around the middle of the millennium. There may therefore be an approximately 1000-year cyclical tendency affecting the temperatures (and glacier growth) in these latitudes, but the geographical patterns of these episodes differ: it appears that high northern latitudes were only affected during alternate millennia, and this could be the basis of an apparent periodicity of the order of 2000 years in length reported there (e.g. from the Greenland ice-sheet). Indeed, there may have been a climax of warmth there between AD 600 and 1000, possibly coinciding with a cold phase in Europe. It is not known whether or how far these variations are represented in the region of the world's greatest ice-sheet in the Antarctic, although some slight indications have been reported of a climax of warmth there also around AD 600. The implications for world sea-level cannot be judged with certainty.

The Little Ice Age during the last millennium stands out as the one case in which the departure of the prevailing temperatures from twentieth-century values was in the same direction (colder) in every part of the northern hemisphere for which we have data. The colder climate also affected the southern hemisphere, but whereas forest history and oxygen isotope data from New Zealand indicate a temperature history broadly parallel to that to Europe, in the Antarctic the climax phases were apparently out of step with Europe: coldest conditions seem to have occurred between about 1450 and 1670 and again after 1830 until around 1900, whereas conditions were warmer (the sea-ice limit farther south) between 1250–1450 and 1670–1830. All these dates, except 1900, are however radiocarbon dates for which standard errors between 45 and 100 years are quoted (see Lamb 1977, pp. 100, 431; Spellerberg 1970). This anti-phase relationship south of latitudes 40°–50°S should presumably smooth the curve of sea-level variation. It may suggest that lowest world sea-level should have occurred around 1670 and 1890–1900.

Estimation of the probable variations of world sea-level during the last few thousand years may be tentatively made on the basis of estimates of the global temperature variations and their duration. Many uncertainties remain, particularly regarding conditions in the Antarctic, which we have touched on briefly, and because of a great lack of knowledge of the variations of precipitation globally and over the glaciers and ice-caps, in particular. The probable limits within which the variations should lie are indicated by figure 6.1.

With caution, it may be said that crude estimates and assumptions about rates of ice growth and decay, and the accompanying global fall and rise of world sea-level, during the last major (Weicheslian/Würm) glaciation indicate that sea-level could have dropped 2–2.5 m between about 2000 and 300 BC, possibly to a level between the present one and 50 cm below that. This low stand apparently coincided with an era of dune building and blowing sand on the coasts of north-west Europe.

Harbour works from about 500 BC at Naples and in the Adriatic indicate a sea-level more than 1 m below present, but the Mediterranean region has too much tectonic instability for confidence about detail. Bloch (1965) has suggested a number of fluctuations. It seems certain that there was a progressive rise of sea-level during the time of the Roman Empire (see Lamb 1977, p. 258), conceivably (on thermal grounds) by 60 cm to 1 m, to culminate up to 40–50 cm above present around AD 1000 and 1300–1400, and perhaps for most of the time between those dates (see Funnell 1979), as well as around AD 400.

The cooling which produced the Little Ice Age in recent centuries, although it was apparently world-wide and accompanied sooner or later by advances of glaciers in all parts of the world, probably did not continue long enough to reduce world sea-level from its mediaeval highest stand to more than 10–30 cm below present. This seems to be supported by the tide-gauge records for Amsterdam (since 1682, when the level was 17 cm below that of 1930) and Sheerness in the Thames Estuary (where levels from 1830 to about 1890 were 15–18 cm below those between 1950 and 1970); but these cannot be taken as absolute values because the southern North Sea region encompassing these places is subject to land-sinking. In another, perhaps more representative region of the world ocean, Snitnikov (1969) has indicated a rise of the northern Pacific Ocean from 1807–16 to 1950–7 amounting to 12 cm. The rise of sea-level at the mouth of the River Elbe from gauge readings between 1825 and 1976 is believed to amount to 37 cm (Duphorn 1976), a figure which probably also contains an element due to the regional land-folding. Rohde (see Petersen and Rohde 1977), who has investigated the sea-level rise in the same area and on the German North Sea coasts from Schleswig Holstein southwards, estimates the secular rise there at 25–30 cm per century ever since about 1700. These big rises in sea-level over two to three centuries seem to be a special feature of this particular area. In the German Bight, as also in the Thames, the rise of sea-level seems, however, to have slowed down or halted since 1950–70.

Variations in the numbers of severe sea-floods, with great losses of life and/or losses of land on the North Sea coasts and in the Channel, century by century over the last 2000 years are indicated in figure 6.2. This diagram was constructed from the compilations (Arends 1833; Britton 1937; Brooks n.d.; Brooks and Glasspoole 1928; Hennig 1904; Lowe 1870; Vanderlinden 1924; Weikinn 1960–3) available before the remarkably thorough examination of the original sources for the Dutch and Belgian coasts by Gottschalk (1971, 1975, 1977). Although these earlier compilers are now known to have made many mistakes through uncritical use of secondary sources, any obvious or suspected repetitive reports of the same incident were cut out in compiling figure 6.2; and it is thought that by concentrating on the numbers of the most disastrous class of floods, a useful approximation to the true pattern of variation with time has been obtained. Allowance must, however, obviously be made for the much smaller chance of reports being recorded in writing in the first millennium AD and surviving to our own age.

It is thought safe to conclude that there was a real maximum occurrence of sea-floods in or about the thirteenth century AD, and that there is some suggestion of more severe floods in late Roman times and in our own century than in other periods. All three maxima, if real, came towards the end of a run of some centuries of warm or warming climate.

Schoorl (1981) reports that the Netherlands seem to have become more liable to flooding by the rivers, as well as to sea-storm floods, in and after the mediaeval warm period in the twelfth and thirteenth centuries for another reason. The warmer climate led gradually to desiccation of the peat behind the coastal dunes and of the whole coastal dune region. This was aggravated by the reclaiming of peat marshes in those centuries, and deforestation of the region to allow for agriculture to support the growing population in those healthier times. All this led to the compaction and sinking of the local land surface, to more liability to blowing soil and sand, and to less regular flow of the rivers.

The pattern differs somewhat from that derived by Gottschalk for the coast of Belgium and the Netherlands. This is not just a matter of the latter's acknowledged superior data quality, but results from the inclusion of other coasts of the North Sea in figure 6.2. Gottschalk notes periods when the incidence of flooding seems to have been more concentrated farther north, on the Danish and German coasts, a circumstance associated with somewhat different wind directions and pattern. Nevertheless, when the data are broken down into half centuries, we may agree with Gottschalk that there were notable maxima of storm surge incidence in the early 1400s and in the late 1600s, and that the sixteenth century witnessed a few storms of outstanding geographical range and severity, particularly the storm of 1–2 November 1570 (the 'All Saints flood' as here given on the old style calendar).

These maxima of sea-flood occurrences in the Little Ice Age period, particularly the floods in the sixteenth and late seventeenth century, cannot be associated with high world sea-level and must therefore suggest that the storms concerned were of altogether exceptional severity. We have already referred to a meteorological reason which may have made this possible, namely, the much intensified thermal gradient in these latitudes at times when the Arctic sea-ice advanced south in the East Greenland–Norwegian Sea to near Iceland. There is observational evidence to support this diagnosis in the case of the Spanish Armada storms in 1588, while in the period 1675–1704 the polar water seems to have reached as far south as 61°N, giving an ocean surface about 5 degC colder than now between the Faeroe Islands and Iceland (see Chapter 9).

A listing (Duphorn 1976) of the storm surges exceeding 4 m above predicted tide-level, registered by the tide-gauge at Cuxhaven since 1845, shows that after seven occurrences in the first 11 years (1845–55), there were only eight more occurrences before the end of the century. Four high surges occurred in two years in 1894 and 1895 and again in 1916 and 1917. Apart from these occurrences, and seven between 1921 and 1930, high surges were rare until the 1960s (when there were eight cases in six years) and the 1970s (15 occurrences between 1973 and 1976). These variations must be largely attributable to variations in storminess, i.e. in the paths and intensities of the travelling storm cyclones. The maxima, except in the 1920s, seem to be associated with the times of relatively colder Arctic.

The sequence of observations of occasional or extreme events at any one place – whether it be the mouth of the Elbe or the Thames or the Dutch coast – cannot be expected to be fully representative of the whole North Sea coastal rim. Statistics of the number of gale situations occurring anywhere over the whole North Sea (or any other region) can, however, be obtained by measurement of the pressure gradients shown by the isobars on daily synoptic weather maps. The variations of gale frequency over the North Sea region and over the British Isles (taken as 50°–60°N, 0°–10°E and 0°–10°W respectively) from decade to decade since 1880, using a method and measurements by Jenkinson (1977), are seen in figure 6.4. They correspond with the local Cuxhaven surge figures in so far as they show an increase in the 1960s and 1970s after many decades of less storminess.

The implied changes in the tracks and intensity of the travelling storms in this part of the world are associated with changes in the large-scale circulation of the atmosphere. In particular, they correspond with changes in the frequency of general westerly wind situations over the British Isles (discussed in Lamb 1972a) and of the northerly and southerly, and slow-moving anticyclonic and cyclonic situations that become more frequent when the westerlies decline.

Figure 6.5(a)–(c) shows the variations from year to year and in the 10-year mean frequencies of the westerly, northwesterly and northerly situations over the British Isles during 1861–1977 and for a few years in the 1780s for which daily maps have been constructed. The variations of the north-westerly situations correspond well with the times of generally high and low

GALE INDEX SURVEY BASED ON
PRESSURE DISTRIBUTION

(a) Over the North Sea

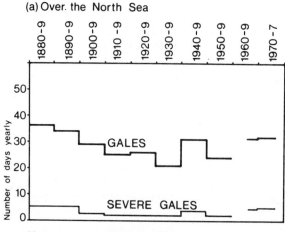

Very severe gales averaged
1 per year in the 1890s
1 in 2 years in the 1880s and 1970s
1 per decade between 1910 and 1939

(b) Over the British Isles / East Atlantic

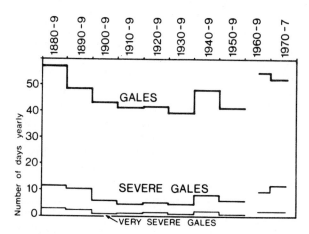

Figure 6.4 Average numbers of gales and severe gales per year over the North Sea region and the British Isles region, by decades, since 1880. (From an index developed by Jenkinson 1977.)

Sea transgressions and storm floods 95

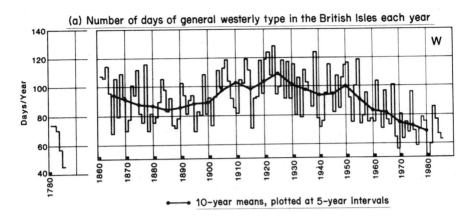

(a) Number of days of general westerly type in the British Isles each year

● 10-year means, plotted at 5-year intervals

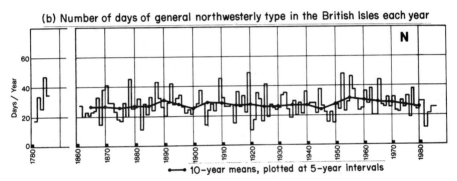

(b) Number of days of general northwesterly type in the British Isles each year

● 10-year means, plotted at 5-year intervals

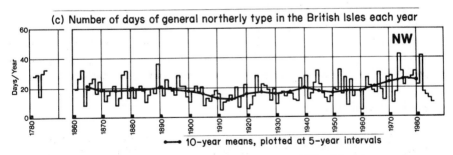

(c) Number of days of general northerly type in the British Isles each year

● 10-year means, plotted at 5-year intervals

Figure 6.5 Number of days each year (1781–5 and 1861–1977) with: (a) general westerly wind situations; (b) general northwesterly wind situations; (c) general northerly wind situations over the British Isles. (Corresponding histories of the frequencies of situations with other wind-flow patterns over the British Isles are shown in Table 11.1, pp. 184–5.)

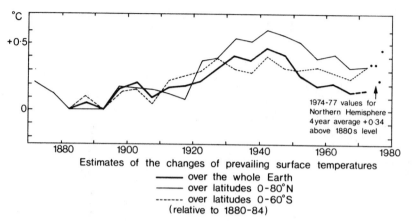

Estimates of the changes of prevailing surface temperatures
———— over the whole Earth
———— over latitudes 0-80°N
------- over latitudes 0-60°S
(relative to 1880-84)

Figure 6.6 Best available estimates of the changes of world temperature over the last 100 years. (For sources see Lamb, 1977): the dots representing complete northern hemisphere values for the four years, 1974–7, are thought to indicate not only margins of error in the assessments for the individual years, but are partly affected by real variations from year to year; the estimates for even the 5-year changes must also have some margin of error, particularly in the case of the southern hemisphere, which is over 80 per cent ocean; the general rise of average temperature up to the 1940s and the fall since 1950 are not in doubt.

frequency of storm surges of Cuxhaven, although the more isolated surges in the 1890s and 1916–30 occurred rather in years when northerly situations were prominent.

There is an unmistakable parallelism between the curve of the frequency of westerly situations in the region and the best available estimate of the changes of average temperature over the Earth and over the northern hemisphere, seen in figure 6.6. This may be linked with the apparent association between increased frequency of storm situations in the North Sea and periods of cold or cooling climate with fewer westerlies.

Concluding summary

We have found reason to associate variations in the incidence of storm surges and North Sea floods with variations of sea-level and storminess, both of which may be linked to overall changes of global temperature. The sea-flood danger may be greatest at times when world sea-level is high after some centuries of a warm epoch, particularly when the regime begins to break down with cooling of the Arctic and intensified storminess in the latitude of the North Sea. This was presumably the case in the thirteenth century AD and it may be the case again since 1950 – as instanced by the great floods in 1953, 1962, 1976 (see figure 6.7) and other examples.

There may have been a secondary maximum of sea-floods and erosion in the sixteenth and seventeenth centuries due to the violence of some of the storms in this region at that time, despite a slightly lowered general sea-level.

The brief notes on a few examples of particular coastal events (not all floods) which follow illustrate our analysis. They include at least one case in which the tidal force situation may also have been important.

(i) *The Zuyder Zee and Jadebusen* (on the coast of Oldenburg in north-west Germany) are examples of extensions of the North Sea formed by flooding produced by great storms in and around the thirteenth to fifteenth centuries. Their formation thus coincides with the time when world sea-level was almost at one of its maxima in the high Middle Ages, and possibly rising until about AD 1350, after some centuries of melting of glaciers and the ice-sheets in high latitudes; storminess had also apparently increased as cooling set in in the Arctic and (later) in middle latitudes of the northern hemisphere.

(ii) *The storm which overwhelmed the mediaeval township of Forvie on the east coast of Aberdeenshire, Scotland with sand in 1413* during the same century in which Gottschalk (1975, pp. 189, 302) reports many similar incidents and blow-outs of the existing dunes along the whole seaboard of the Netherlands from Walcheren to north Holland, presents

Figure 6.7 North Sea flood in the River Elbe breaking through the dikes on to the Haseldorfer marsh near Hamburg on 3 January 1976. (Photograph kindly supplied by M. Petersen and H. Rohde of Hamburg.)

interesting circumstances which call for further information and analysis. Gottschalk reports that in one case in Holland it is obvious that the sand in question was not dune sand, but glacial sand. Others have suggested that the sand involved in the accretions which built the dunes on the Dutch and Scottish coasts was derived from the sea-bed. The storm that ended the township of Forvie is reported to have taken place on 19 August 1413. (The date has been converted to the New Style or modern Gregorian calendar.) This is a strange time of year for winds of the greatest violence to occur; although it is clear that an enormous quantity of sand was transported, and the dunes that bury the place are now over 30-m high. The church, Forvie Kirk, part of which is visible today, had been built in AD 704 at a relatively high point on what then evidently seemed to be a stable coast.

There are further noteworthy aspects, in that the wind concerned was a southerly storm, and winds from that direction lower the level of the North Sea. Moreover, the date is within 20 years of the long-term maximum of the tidal force, and calculation of the tides in the area indicates extreme tides in that month, particularly low ebb tides within 10 cm of the astronomical extreme, on or about the reported date of the storm. Unfortunately, the date of the extreme ebb is calculated as 23 August 1413, i.e. four days *after* the reported date of the disaster. The coincidence is so near that one suspects an error of this amount is due to the calculation or the reporting. The tidal calculations indicate other notable low and high tides in other months of that year, but none so extreme as this one. The circumstances suggest that sand may have been moved towards the area before the event by waves as well as wind, but the final event itself must be attributed to wind transport of sand that had been exposed over exceptionally wide areas by a lowered sea-level.

(iii) *The devastation of a large area of rich farmland on the north-facing coast of Scotland by the Culbin Sands* (near the mouth of the River Findhorn) presents similar circumstances to the case of the Sands of Forvie. The precise date is not known, but can be narrowed to the autumn (the barley harvest, which was probably late) in 1694: the season is given as autumn by all the local historians (notably Mr S. M. Ross of Forres and Mrs A. D. Mackintosh of Kincorth), who have examined available documents (accounts and town archives). It is alleged that there were no large dunes on that coast until the seventeenth century, but storms in 1663 and the 1670s had moved some sand. Local tradition insists that the final event which covered an area of 20 km² including 16 farms, and buried even the farmhouses in sand, was a single storm. It began as a westerly storm (i.e. blowing from the land at the head of Moray Firth near Inverness) which would lower, not raise, the sea-level in the area. And in August and September 1694 the calculated tides again approached within about 30 cm of the astronomical highest and lowest extremes. Only rather less extreme tides again appear in the calculations for mid-November and mid-December 1694.

Extreme wind stengths seem to have occurred repeatedly between about 1660 and 1710, as can be deduced from the reported damage from many storms in England and Scotland. This was undoubtedly the coldest part of the Little Ice Age period in this part of the world. Examples of the severity of the storms of the period include a great northeasterly storm in September 1695 in which at least 70 coal-boats bound for Newcastle from Yarmouth were dashed to pieces on the coast of Norfolk and 13 more ships were stranded on the continental coast between Dunkirk and Boulogne, the wind presumably being westerly in that area (Lindgren and Neumann 1981). There was much damage in London and a ship was overturned in the Thames. Secondly, Daniel Defoe who alluded to the 1695 storm, when writing his *Tour through the Whole Island of Great Britain*, 30 years later, devoted a small book, *The Storm*, to an account of an even greater storm which crossed southern England in early December (New Style) 1703. Another example is the storm in 1697 (Crawford and Swithur 1977; Morrison 1967–8) which overwhelmed with sand a site on the coast of the Hebrides where there seems to have been a continuous settlement

Figure 6.8 The village and church of Skagen at the northern tip of Jutland (Denmark), 'awash' in blown sand, as painted by W. Melbye in 1848.

for four thousand years up to that time. The Danish example at Skagen referred to on p. 89, is illustrated in figure 6.8.

(iv) Severe erosion of the coasts of eastern England in Yorkshire and Suffolk led the North Sea to swallow up the *great mediaeval ports of Ravenspur or Ravensburgh (east of Hull) and Dunwich (south of Southwold)*. The sea began to make inroads in both places some time about 1340, both having been at the height of their prosperity in the decades immediately before that, although the harbour of Dunwich had been choked by blown sand in a storm in 1328. By 1530 or 1540, Ravenspur had disappeared and more than three-quarters of the city of Dunwich had been swallowed up. The remainder was destroyed in a series of storms, for which the dates 1608, 1677, 1680, 1702, 1729 and finally a northeasterly storm in 1739 are mentioned. It seems clear that at both places large-scale destruction began in the time when storminess at the end of the mediaeval warm epoch came with a sea-level probably rather higher than now, and that the situation was apparently sometimes aggravated by extreme tides (of which some direct reporting is believed to have survived). The final stages of the erosion which obliterated Ravenspur and Dunwich must however be attributed to storms, possibly of special severity, occurring during the Little Ice Age period when sea level was somewhat lower than now.

Acknowledgement

I am greatly indebted to Dr J. M. Vassie, of the Institute of Oceanographic Sciences, Bidston Observatory, Birkenhead, for the calculations of the tides at Aberdeen in AD 1413 and 1694.

References

Akeroyd, A. V. (1972) 'Archaeological and historical evidence for subsidence in southern Britain, *Philosophical Transactions of the Royal Society of London, A*, 272, 151–69.

Aranuvachapun, S. and Brimblecombe, P. (1978) 'Extreme ebbs in the Thames, *Weather*, 33(4), 126–131.

Arends, F. (1833) *Physische Geschichte der Nordsee-Küste und deren Veränderungen durch Sturmfluthen seit den Cymbrischen Fluth bis Jetzt* (2 vols), Emben Woortman.

Bahnson, H. (1972) 'Spor of muldflugt i keltish jernalder påvist i høimosprofiler', in *Geological Survey of Denmark, Yearbook*, Copenhagen.

Bloch, M. R. (1965) 'A hypothesis for change of ocean levels depending on the albedo of the polar ice caps', *Palaeogeogr., Palaeoclim., Palaeoecol.*, 1, 127–142.

Bloch, M. R. (1970) 'Zur Entwicklung der von Salz abhängigen Technologien: Auswirkung von postglazialen Veränderungen der Ozeanküsten', *Saeculum*, 21(1), 1–33, Munich.

Britton, C. E. (1937) 'A meteorological chronology to AD 1450', *Geophysical Memoir No. 70*, London, HMSO for Meteorological Office.

Brooks, C. E. P. (n.d.) 'Tabulation from an unknown source, probably work done for Sir Napier Shaw c. 1920–25; found by Meteorological Office Library in 1964.

Brooks, C. E. P. and Glasspoole, J. (1928) *British Floods and Droughts*, London, Benn.

Churchill, D. M. (1965) 'The displacement of deposits formed at sea level 6500 years ago in southern Britain', *Quaternaria*, 7, 239–249, Rome.

CLIMAP Project Members (1976) 'The surface of the ice-age Earth', *Science*, 191, 1131–1144.

Crawford, I. and Switsur, R. (1977) 'Sandscaping and C 14: the Udal, N. Uist', *Antiquity*, 51, 124-136.

D'Olier, B. (1972) 'Subsidence and sea-level rise in the Thames Estuary', *Philosophical Transactions of the Royal Society of London, A*, 272, 121–130.

Duphorn, K. (1976) 'Gibt es Zusammenhänge zwischen extremen Nordsee-Stormfluten und globalen Klimaänderungen?' *Wasser und Boden*, 10, 273–275.

Fairbridge, R. W. (1961) 'Eustatic changes in sea level', ch. 3 of *Physics and Chemistry of the Earth*, New York, Pergamon.

Funnell, B. M. (1979) 'History and prognosis of subsidence and sea-level change in the Lower Yare Valley, Norfolk', *Bulletin of the Geological Society of Norfolk*, 31, 35–44, Norwich.

Gernet, J. (1972) *Le monde chinois*, Paris, Armand Colin. (Trans. by J. R. Foster, *A history of Chinese civilisation*, Cambridge University Press, 1982.)

Godwin, H., Suggate, R. P., and Willis, E. H. (1958) 'Radiocarbon dating of the eustatic rise in ocean level', *Nature*, 181, 1518–1519, London.

Gottschalk, M. K. E. (1971) *Stormvloeden en rivieroverstromingen in Nederland: Deel I – De periode voor 1400*, Assen, Van Gorcum.

Gottschalk, M. K. E. (1975) *Stormvloeden en rivieroverstromingen in Nederland: Deel II – De periode 1400–1600*, Assen, Van Gorcum.

Gottschalk, M. K. E. (1977) *Stormvloeden en rivieroverstromingen in Nederland: Deel III – De periode 1600–1700*, Assen/Amsterdam, Van Gorcum.

Hallam, H. E. (1959–1960) 'Salt-making in the Lincolnshire Fenland during the Middle Ages', *Lincolnshire Architectural and Archaeological Society Reports and Papers*, n.s. 8, 85–112, Spalding, Lincs.

Hallam, H. E. (1965) *Settlement and Society: A Study of the Agrarian History of South Lincolnshire*, Cambridge, Cambridge University Press.

Hennig, R. (1904) 'Katalog bemerkenswerter Witterungsereignisse von den ältesten Zeiten bis zum Jahre 1800', *Abhandlungen des Preussischen Meteorologischen Instituts*, 2, Nr. 4, Berlin.

Hess, P. and Brezowsky, H. (1977) 'Katalog der Grosswetterlagen Europas 1881–1976 (3. verbesserte und ergänzte Auflage), *Berichte des Deutschen Wetterdienstes, Nr. 113*, Band 15, Offenbach am Main.

Jelgersma, S. (1961) 'Holocene sealevel changes in the Netherlands', *Mededelingen Geol. Sticht.*, Serie C-VI, No. 7.

Jelgersma, S., De Jong, J., Zagwijn, W. H., and van Regteren Altena, J. F. (1970) 'The coastal dunes of the western Netherlands: geology, vegetational history and archaeology', *Mededelingen Geol. Sticht.*, n.s., No. 21.

Jenkinson, A. F. (1977) 'An initial climatology of gales over the North Sea', *Meteorological Office Met. 0.13* (unpublished), *Branch Memorandum No. 62*, Bracknell, England.

Lamb, H. H. (1963) 'On the nature of certain climatic epochs which differed from the modern (1900–39) normal', in *Changes of Climate: Proceedings of the UNESCO/WMO Rome 1961 Symposium*, Paris, UNESCO – Arid Zone Research Series XX.

Lamb, H. H. (1972a) 'British Isles weather types and register of the daily sequence of circulation patterns 1861–1971', *Geophysical Memoir*, 116, London, HMSO for Meteorological Office.

Lamb, H. H. (1972b) *Climate: Present and Future. Vol. 1, Fundamentals and Climate Now*, London, Methuen.

Lamb, H. H. (1977) *Climate: Present, Past and Future. Vol. 2, Climatic History and the Future*, London, Methuen.

Lamb, H. H. and Weiss, I. (1979) 'On recent changes of the wind and wave regime of the North Sea and the outlook', *Fachliche Mitteilungen*, Nr. 194, Traben-Trarbach, Amt für Wehrgeophysik.

Lindgren, S. and Neumann, J. (1981) 'The cold and wet year 1695 – a contemporary German account', *Climatic Change*, 3(2), 173–187, Dordrecht, Reidel.

Lowe, E. J. (1870) *Natural Phenomena and Chronology of the Seasons*, London, Bell & Dalby.

Maksimov, I. V. (1971) 'Causes of the rise of sea level in the present century', *Okeanologica*, 11, 530–541, Moscow, Akad. Nauk, Okean. Kom. in Russian.

Morrison, A. (1967–1968) 'Notes on Harris Estate papers', *Transactions of the Gaelic Society of Inverness*, 45, 47–48.

Mörner, N.-A. (1971) 'Eustatic changes during the last 20,000 years and a method of separating the isostatic and eustatic factors in an uplifted area', *Palaeogeogr., Palaeoclim., Palaeoecol.*, 9, 153–81.

Mörner, N.-A. (1976) 'Eustatic changes during the last 8000 years in view of radiocarbon calibration and new information from the Kattegatt region and other northwestern European coastal areas', *Palaeogeogr., Palaeoclim., Palaeoecol.*, 19, 63–85.

Petersen, M. and Rhode, H. (1977) *Sturmflut: die grossen Fluten an den Küsten Schleswig-Holsteins und in der Elbe*, Neumünster, Karl Wachholtz Verlag.

Pettersson, O. (1914) 'Climatic variations in historic and prehistoric time', *Hydrografisk-Biologiska Kommissionens Skrifter*, Häft 5, Gothenburg.

Pettersson, O. (1930) 'The tidal force', *Geografiska Annaler*, 12, 261–322, Stockholm.

Schofield, J. C. and Thompson, H. R. (1964) 'Postglacial sea levels and isostatic uplift', *New Zealand J. Geol. and Geophys.*, 7(2), 359–70, Wellington, NZ.

Schoorl, H. (1980) 'The significance of the Pleistocene landscape of the Texel-Wieringen region for the historical development of the Netherland Coast between Alkmaar and East Terschelling', in *Transgressies en occupatiegeschiedenis in de kustgebieden van Nederland en Belgie*, A. Verhulst and M. K. E. Gottschalk (eds), Ghent, Belgisch Centrum voor Landelijke Geschiedenis, publication Nr. 66.

Snitnikov, A. V. (1969) 'Some material on intrasecular fluctuations of the climate of northwestern Europe and the North Atlantic in the 18th–20th centuries', *Geogrl. Obs. SSSR*, 5–29, Moscow, Akad. Nauk. (in Russian).

Spellerberg, I. F. (1970) 'Abandoned penguin rookeries near Cape Royds, Ross Island, Antarctica and C 14 dating of penguin remains', *New Zealand Journal of Science*, 13(3), 380–385, Wellington, NZ.

Tooley, M. J. (1974) 'Sea level changes during the last 9000 years in northwest England', *Geog. J.*, 140, 18–42, London.

Tooley, M. J. (1978) 'Interpretation of Holocene sea-level changes', *Geol. Fören. Stockholms Förhandlingar*, 100, 203–212.

Vanderlinden, E. (1924) 'Chronique des événements météorologiques en Belgique jusqu'en 1834', *Mémoires de l'Academie Royale Belge*, 2nd ser., No. 5, Brussels.

Verhulst, A. and Gottschalk, M. K. E. (eds) (1980) *Transgressies en occupatiegeschiedenis in de kustgebieden van Nederland en Belgie*, Ghent, Belgisch Centrum voor Landelijke Geschiedenis, Nr. 66.

Weikinn, C. (1960–1963)– *Quellentexte zur Witterungsgeschichte Europas von der Zeitwende bis zum Jahre 1850*, Band 1.1.–1.4., Berlin, Akad. Verlag.

Weiss, I. and Lamb, H. H. (1970) 'Die Zunahme der Wellenhöhen in jüngster Zeit in den Operationsgebieten der Bundesmarine, ihre vermutliche Ursachen und ihre voraussichtliche weitere Entwicklung', *Fachliche Mitteilungen Nr. 160*, Porz Wahn (Luftwaffenamt Inspektion Geophysicalischer Beratungsdienst der Bundeswehr). Also *Fachliche Mitteilungen*, Nr. 194.

Wigley, T. M. L. (1978) 'Geographical patterns of climatic change: 1000 BC–1700 AD', *Interim final report* (unpublished) *to the National Oceanic and Atmospheric Administration*, Washington, DC (Contract No. 7-35207).

7

Some aspects of the Little Ice Age and other periods of cold, disturbed climate

This chapter is mainly concerned with climatic developments with a characteristic duration of a few centuries. They seem to be introduced by abrupt occurrences of short runs of remarkably similar 'bad years', or years which appear anomalous and out of character with the experience of people living at the time. Such occurrences are, of course, not always followed by any noteworthy, longer-lasting change of climate, but they may be a danger signal. This presentation on the subject was written for a volume dedicated to the memory of the great Swedish meteorologist, Professor Tor Bergeron, one of the original members of Vilhelm Bjerknes' Bergen (Norway) School, which in 1917 introduced the concept of fronts and frontal analysis to modern meteorology. Bergeron himself was distinguished for his pioneering work on three-dimensional weather analysis (1928) and for his own most elegant analysis of synoptic weather maps. He was, naturally, intolerant of frontal patterns drawn on weather maps by meteorologists in some countries involving stiff and angular shapes which could not possibly develop in a free-flowing fluid such as the atmosphere. The results of such analysis include lack of logic in the tracing of the fronts' progress, fronts in wrong positions, and, as result, wrong forecasts of their arrival time.

The paper reproduced here was originally published in Pure and Applied Geophysics (PAGEOPH), 119, 628–39, Bergeron Memorial volume, Basle, Birkhäuser Verlag. It has been updated and adapted slightly to avoid duplication of other chapters.

Tor Bergeron, by writing a short paper on the legend of the *Fimbulvinter* (1956), revealed that among his many-sided interests was the past behaviour of the climate, and its possible impacts on human history. The critical aspects of the legend were quoted from Snorre Sturluson in Gylfaginning in the Edda:

då kommer den vinter som kallas fimbulvintern. Då driver snön från alla väderstreck, det er sträng kyla och bitande vind. Solen förmår intet. Tre sådana vintrar följer på varandra och det är ingen sommar emellan.

(then comes the winter that is called the Fimbul winter. Then the snow drives from every quarter. The cold is severe and the winds biting. The sun is powerless against it. Three winters like this follow one after the other and there is no summer in between.)

The quotation goes on to describe the wars and individual strife that ensued.

Bergeron and his co-authors suggest that this legend might well be a long-surviving folk memory in the northern countries about the onset of the colder climate, of which there is much 'fossil' evidence (from glacial advances that left old moraines, from pollen analysis, etc.), during the last millennium before Christ. There can be little doubt that it reports – with some, doubtless rough, approach to accuracy – a real occurrence at some time in the past of an unbroken run of about three bad years with severe winters and poor summers, which had very dire effects on an early population in Scandinavia and severely shocked the imagination of the people, because their previous experience had not suggested that such a sequence was possible.

This seems to be an early example of a phenomenon which has begun to attract some attention in the investigation of the climatic record of more recent times. The phenomenon has been called 'clustering'. It refers to the occurrence, in rather close but not necessarily unbroken succession of groups of years with some specific similar, but otherwise quite unusual, character. Table 7.1, a short list of examples, will show what is meant. The list begins with a case that is still well remembered by many people who are still alive.

One aspect of such clustering can be seen in figure 3.4 (p. 34), which refers to this phenomenon. In that diagram, we see different runs of years at Copenhagen between about 1770 and 1820 when, in some cases easterly winds, in other cases northwesterly winds, and in other cases calms, were at least twice as frequent as in most of the rest of the record. Other examples may be found in figure 6.5 (p. 95): in the 1780s days with general westerly winds over the British Isles were little more than half as frequent as in the first three to four decades of this century or around 1950. In recent years, the frequency has again fallen.

When we switch our attention to evidence of clustering on longer time-scales, at some point we find ourselves considering what is ordinarily thought of as climatic change and examining the stability and homogeneity of each regime while it lasts. This question is raised by the long-recorded history of disastrous North Sea storm-floods over the coastal lowlands bordering that sea. The overall frequency seems to have been highest in and

Table 7.1   Some examples of the clustering of similar climatic events and related atmospheric circulation patterns

---

(1) The three 'war winters', 1939–40, 1940–1 and 1941–2, with mean temperatures (December, January and February) in central Europe (mean of De Bilt, Potsdam, Basle and Vienna) respectively 4.2, 2.2 and 3.4 degC below the 200-year average (Baur 1950), causing freezing of the rivers all over central and western Europe and accompanied by great snowstorms. Only two comparable winters, 1916–17 (−1.0°) and 1928–9 (−4.2°), had occurred since a group 50 years earlier in the 1890s.

(2) A very similar cluster of three successive severe winters affected particularly western Europe in 1878–9, 1879–80 and 1880–1, giving temperatures in central England of 3.0, 1.2 and 1.4 degC respectively, below the 250-year average of 1701–1950 (Manley 1974).

(3) On a somewhat different time-scale, the sequence of three 'skating Christmases' in England with very severe frosts beginning on or just before Christmas Day, with a north European anticyclone and easterly winds over the European plain in 1961, 1962 and 1963. In a moderated degree, with snowy weather and/or a dry frost sufficient for skating between 25 and 28 December, the sequence was continued in 1964 and 1965. This sequence must be viewed against a background of only seven to ten Christmases in southern England during the first 50 years of the century which could in any way (i.e. on the grounds of either white frost or snow cover) be classed as a 'white Christmas'.

(4) A quite similar cluster of years, lasting from 1965 to 1971, in which northerly winds predominated over the British Isles (30 per cent of the days, whereas 10 per cent were westerly) in the first pentad (5 days) of January was a very marked feature at the time. No other run of years for which weather maps are available has shown this feature or any approach to it. Over the 119 years from 1861, when the British Isles daily weather map classification begins (Lamb 1972a), to 1979, the predominant character of the wind-pattern over the same pentad was westerly (32 per cent); northerly situations amounted to only 7 per cent. In a very different cluster of years, 1921–32 inclusive, westerly situations accounted for 73 per cent of the days during the first pentad of January.

(5) The pairs of hot summers in western and northern Europe in 1975 and 1976 and 1983 and 1984.

(6) The run of cold winters in the central and eastern USA from 1977 onwards and the cluster of spells of severe winter weather in Europe in 1979, 1981–2, 1985, 1986 and 1987.

---

around the thirteenth century, when the sea-level may have been a few decimetres higher after some hundreds of years of warmer climate in many parts of the world, and melting glaciers. The reports also strongly suggest that storminess was increasing in the North Sea as the warm regime broke down farther north. But on the coasts of the southernmost North Sea, in Flanders and the Netherlands, the peak frequency of disastrous floods was in and around the sixteenth and seventeenth centuries (Gottschalk 1971, 1975, 1977): that is to say, in the time of almost world-wide cold climate, the so-called Little Ice Age period, when the general sea-level cannot have been high, and the phenomenon must indicate more frequent and more severe storms in the southern part of the North Sea than before, and probably more northerly winds. By contrast, in the twelfth, thirteenth and fourteenth centuries – and in the fifteenth century, apart from a break around 1400 to 1420 and the period after 1470 – the severe flooding incidents were largely concentrated farther north, on the coasts of Friesland, north-west Germany and Denmark, though the coasts of Holland and Belgium were by no means immune.

The North Sea floods on the coasts of England seem, on the whole, to have been less severe than on the continental side, though in England, too, land has been permanently lost and sometimes many people perished. The freqency on the English North Sea coast seems to show two peaks, in and around the thirteenth century and between about 1530 and 1740. In the fourteenth and fifteenth centuries, nearly all the serious floods and losses of land reported relate to the coasts around the Humber in northern England. (It is not clear whether this resulted from some particularly vulnerable formation of the spit of land around Spurn Head and of the coastal dunes about that time, or perhaps represented some strongly repetitive concentration of the paths of the storms. It was in 1364 that the formerly great port of

Table 7.2 Mean temperatures over December, January and February of the seven coldest and seven mildest winters in central England between 1659 and 1979. (Average winter 1850–1950: 4.0 degC)

Winter	1683–4	1739–40	1962–3	1813–14	1794–5	1694–5	1878–9
°C	−1.2	−0.4	−0.3	+0.4	+0.5	+0.7	+0.7

Winter	1868–9	1833–4	1974–5	1685–8	1795–6	1733–4	1934–5
°C	6.8	6.5	6.3	6.3	6.2	6.1	6.1

Note: The severest individual months are closely comparable: mean temperature (January in each case) 1684 and 1795 −3.0 or −3.1°, 1814 −2.9°, 1740 −2.8°, 1963 −2.1°, 1716 −2.0°. Differences in the effects of the winters concerned seem therefore to have had more to do with the length of the frost period than with the temperature of the coldest month. Differences in the amounts of snow covering the ground must also have been important.

Table 7.3 Mean temperatures over June, July and August of the fourteen hottest and fifteen coldest summers in central England between 1659 and 1979. (Average summer 1850–1950: 15.2 degC)

Summer	1826	1976	1846	1781	1911	1933	1947	
°C	17.6	17.5	17.1	17.0	17.0	17.0	17.0	
Summer	1868	1899	1676	1975	1666	1719	1762	
°C	16.9	16.9	16.8	16.8	16.7	16.7	16.7	

Summer	1725	1695	1816	1860	1823	1674	1675	
°C	13.1	13.2	13.4	13.5	13.6	13.7	13.7	
Summer	1694	1888	1922	1812	1862	1698	1890	1920
°C	13.7	13.7	13.7	13.8	13.8	14.0	14.0	14.0

Ravensburgh (or Ravenspur), near Spurn Head, and many smaller places in that neighbourhood are reported to have been taken by the sea.

Clustering of years with exceptional 'blocking', or replacement of the normally prevailing westerlies by meridional or easterly wind flow in slightly different positions in different years over Europe, can be detected in the occurrence of the ten to fifteen mildest and severest winters and warmest and coldest summers shown by the 300-year record of temperatures observed in central England (Manley 1974). The details are given in Tables 7.2 and 7.3. Notice the cases of opposite extremes occurring within a few years of each other, particularly remarkable in the successive winters 1794–5 and 1795–6 and the successive summers of 1674, 1675 and 1676. This is connected with a slight shift from one year to the next in the position of the so-called 'blocking anticyclone', such that the warm southerly winds on the western flank may be over the Atlantic Ocean off the coast of Europe

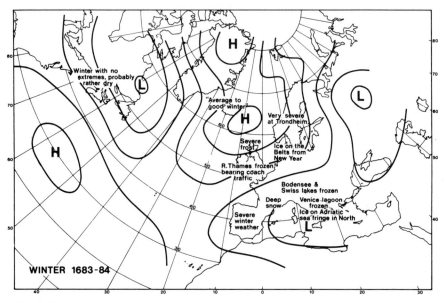

Figure 7.1 Tentative reconstruction of the circulation map (suggested sea-level isobars) for the winter 1683–4.

Note: additional data used: monthly mean temperature for January in Central England −3°C (7 deg C below the modern average) and for the three winter months (December, January and February) −1.2°C (5.4 deg C below the modern average); wind directions observed in London during January 1684; E: 39%, NE: 26%, N: 3%, SE: 3%, W: 29%; frost 29 days during January in London; belts of ice some kilometres wide fringed the Channel coasts of England and France and the coast of Holland; traffic crossed the Zuyder Zee on the ice and also the Sound between Copenhagen and Malmö in Sweden; Dublin, Ireland reported 'a most severe frost'. Note also that dates on figures 7.1, 7.2 and 7.3 have been corrected to the new style (modern)– calendar.

altogether (west of Ireland) in one year and over at least part of the British Isles or western Europe in the next – despite a general similarity of the wind circulation pattern (see figures 7.1 and 7.2).

A related aspect is seen by comparing the long records of snow cover in the area of Zurich and Bern in Switzerland (Pfister 1978) with the temperatures in England. For instance, though the severest winter in the 320-year temperature record for central England was 1683–4 (DJF Mean −1.2 degC, or 4.9 degC below the 1701–1950 average), when ice belts several kilometres wide formed along both sides of the North Sea and English Channel, in Switzerland the next winter, 1684–5, was the outstanding one of the pair. That winter the snow lay for 112 days in Zurich (compare 1963 with 86 days, the longest duration in the last 100 years). Tentative 'isobaric' maps for the Januarys of these two winters are shown in figures 7.1 and 7.2. (The analyses of the monthly maps reconstructed for each January and July of the 1680s were tested by performing an analysis of the same months in the 1880s, first using only the same amount and types of

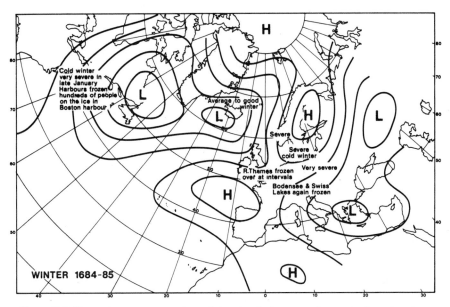

Figure 7.2 Tentative reconstruction of the circulation map (suggested sea-level isobars) for the winter 1684–5.
Note: additional data used: monthly mean temperature for January in central England +0.5°C (3.5 deg C below the modern average) and for the winter +2.7°C (1.5 deg C below the modern average); wind directions observed in London during January 1685: W: 55%, NW: 16%, N: 13% NE; 3%, E: 13%; frost 23 days during January in London. The month of January in England was described as 'at times as cold as any period in the previous winter' with thick ice on the rivers in south and north, but the same month also at times brought rain and storms of wind.

information as in the 1680s, and then comparing maps drawn from actual pressure data.) These winters were probably equalled or exceeded by one other case earlier in the same century: this was in 1607–8 in England and north-west Europe, but 1613-14 in Switzerland, when the snow on the Swiss plateau about Bern lasted for more than 150 days. The third of the snowiest winters identified by Pfister in the Swiss records was 1784–5, with again over 150 days of snow cover near Bern; but in England and north-west Europe this winter, although severe, was surpassed by 1783-4.

Pfister further reports that during the severest phase of the climate in the late seventeenth century, the data for 1683–1700 indicate that March was regularly a full winter month in Switzerland, with temperatures apparently averaging 2.2 to 2.7 degC below the 1900–60 mean. Gisler (1985) has made a detailed study of the meteorological instrument measurements made at a network of places in Switzerland and southern Germany a century later. These indicate another similarly severe phase, with mean temperature for the month of March over the decade 1781–90 some 2.2–2.5 degC below the 1951–80 averages. A striking feature of Gisler's findings is the very rapid spring warm-up that then took place. The Mays of the same decade averaged 0.4–1.1 degC warmer than modern values. Both Gisler and Pfister attribute the cold Marches in the periods named to continual winds from NW and N, bringing cold Arctic air over central Europe. In the eighteenth-century case, this suggestion is substantiated by barometric pressure measurements.

These deviations can be explained by the dominance of blocking patterns in which the persistent cold air supply from the north entered Europe in a slightly different longitude in neighbouring, or nearly neighbouring, years. A slightly greater shift of a similar meridional wind flow pattern might bring predominantly southerly wind instead of northerly or easterly winds in any particular longitude in neighbouring years.

Blocking is presumably the explanation of the occurrences, noticeable in tables 7.2 and 7.3, of opposite extremes within a few years of each other. Other examples of this are found in the 400-year list of dates of opening of the Baltic port of Riga since 1535 (Betin and Preobazensky 1959). The years with least ice and with most ice (1652 and 1653, opening dates 2 and 3 February respectively; 1659, opening date 2 May) all occurred within the same decade. This experience was presumably paralleled in recent years by the contrast between the severe ice-winter in the Baltic in 1965–6 and 1974–5, when there was almost no ice.

Some further points of interest have more recently come to light from the winter which may be regarded in important aspects as the most extreme reported in Europe in historical times, namely, the winter of 1683-4. Most reports say that the frost (probably implying the the period with daily *mean* temperature below the freezing point and hence the freezing of lying water and rivers) lasted 7–9 weeks in southern England, from late December until 17th or 18th February on the modern calendar, and longer than this in Scotland. Snow cover in places lasted some 60 days, or about the same as in

1962–3. There have been other winters with a longer duration of severe weather (sometimes with minor interruptions). An example is the winter of 1657-8, when snow lay for 102 days at one point in southern England, but an Essex diary reports a day of 'unexpected rain' on 30 January. Both the 1430s and the 1690s seem to have been colder decades over all than the 1680s. But the 1684 frost was outstandingly severe. The great frost fair on the River Thames in London began on 26 January, and a dozen or more coaches plied up and down the river.

It is recorded that in early 1684 at one place in eastern Somerset (England) the frost was found to have reached a depth of 3–4 ft (about 1 m) in wet ground, but only 1½-2 ft in dry earth in the same area. The late Professor Gordon Manley also notified the writer (personal communication, 29 June 1972) of reports from 1684 that the frost penetrated 'a yard deep' in Kent and 27 in (about 68 cm) outside Manchester. Comparisons with twentieth-century winters in England seen to indicate that freezing to much more than 1 ft depth (or say 30–40 cm) is quite rare in England. February 1986 provided the first case of this for 30 years at one experimental farm station in Suffolk, and frost penetration barely reached 30–40 cm at the Radcliffe Observatory, Oxford, in the severe winters of 1940 and 1947. But care should be used over such figures as a basis for planning. Water supply pipes were commonly laid at only 3-ft depth in nineteenth-century England, and pipes laid at only half that depth (50 cm) in later years in Guilford, Surrey, froze in the 1962–3 winter. Freezing must have reached 60–70 cm in that case. Great differences may occur over short distances, as were reported in 1684, depending on differences of soil type, wetness and snow cover. (Snow behaves as an insulator protecting the soil beneath. Moreover, on still, frosty nights of clear sky and direct radiation cooling of the ground, great differences of surface air and ground temperature may occur over short distances, depending on the preferred channels of cold air drainage in the absence of any general wind or turbulence. On such a night in December 1981, in almost flat terrain in Norfolk, the writer observed an air temperature of −3 degC in a hedged garden and about the same time −12 degC at a point 150 m away, where cold air with ground mist was draining off a large ploughed field through a gap in the hedge and across a farm lane. The differences of ground level were barely 1 m, but the misty air was draining towards a hollow in another field a little farther down-wind. The ground was thinly snow covered at the time. Such temperature differences if repeated – and they *were* broadly repeated on successive nights and in the following month – must make a considerable difference to the depth to which ground freezes. No one spot can be representative of a whole region or district, when such differences occur over very short distances.

Another report from the 1683–4 winter records a noteworthy extreme for the seas near southern England, France, Belgium and Holland. Ice appeared on the sea and, at its maximum, formed belts perhaps 5-km wide along the coasts on both sides of the Channel and southern North Sea, and, according to one report, 16-km wide off the coast of Holland. The most specific report that has survived seems to be that in a letter written by an observer, Richard Freebody, from Lydd, Kent, on 9 February 1683–4, Old Style, i.e. 19 February 1684 by the modern calendar. This was found in an old Bible and printed in *Notes and Queries*, 2nd series, vol. 11 (January–June 1861), and recently reprinted in a letter contributed by A. Maclean of Stourbridge to Dr G. T. Meaden's *Journal of Meteorology*, 11(105), 18–19, January 1986; it is reprinted here by Dr Meaden's kind permission.

Lyd. ffeby 9th 1683/4
Loving Cossin,
　The frost broke with us last tuesday, which being more noteable than any since the memory of man, take a small account as followeth: the first Instant Mr. Shoesmith told me that the tide for some dayes had not been seen to flow neer folstone towne by 3 leagues by reason of the ice which lay there; that the ice lay some miles off in the sea agst Romny and that there was uppon the topp of the steeples to be seen and (*sic*) Islands of Ice, one to the West of the Light many miles long; but the next day when I was at the light, I took a boat hook in my hand and seeing the ice lying soe thick I went one till I was about 2 rods uppon the sea, soe far that Thomas Smith judged there was 3 faddom of water under me; if I had been there at full sea (which was more than an hour before) I might have gone out a mile the flakes joined so close together and where I put my staff between them I felt ice underneath.
　This was as old Quick judges, about a league in breadth agst the poynt, but at farly poynt it seemed to be at least 3 leagues in length it was as far as I could see from East to West, and 'tis verily believed was the same from dover to the lands end. Old Quick observed some flakes to begin to come about 12 dayes before from the Eastward which increased every day and uppon the fall of the tide went always towards the west, w'ch by reason of the wind never returned again.
　About 2 houres after I was uppon it, I observed that when the wind and tide went together, then all the ice moved as fast as I could ride foot pace along by the side of it, and did drive most part of it from the shore directly towards beachy point. I judg it must come from holland or other eastern pts, w'ch by reason of a continued eastally wind was brought this way. A great deal of it remanes yett to be seen in the sea, but not soe much but that the vessels now pass againe, which was more (as I was told) then the pecquet boats did for some weeks. Rich. Freebody.
　'tis said that (the) ice between dover and callis joyned together within about a league.

The great value of this eye-witness account, as against other reports of the ice in that season, is in making clear the dynamic nature of the situation, with thick ice on the sea moving fast westwards along the English Channel coast, driven by wind and current. It raises the question, put by the writer himself, of where the ice came from. The suggestion that it had formed off the coast of Holland, or perhaps all along the continental shore from Denmark to Belgium, is a reasonable one. The account suggests some similarity with the Baltic ice with broken floes, two or three layers on top of each other, which piled up against the shore of Sweden near Malmö in 1924 – though no actual stacking up of the ice on the English shore is mentioned in the 1684 report. So a source region for the ice similar to the Baltic seems possible at first sight. But the tides and currents in the North Sea are so much stronger than in the Baltic that opportunities for such thick ice to form must be rarer and more localized, apart from the effects of the more oceanic climate. Moreover, the quantity of ice observed in the Channel in 1684 was perhaps too great for any source in the North Sea to account for it. Not only were the ice belts along both coasts of the Channel some miles wide, but they were moving fast to the west and the supply was maintained for several days. The first floes had been seen near the Strait of Dover 12 days before the three- to four-day climax in February and the break-up that followed. The ice belt that passed along the Kent and Sussex shore seems, therefore, to have been 200 miles or more in length. It is not clear that shipping was at any stage stopped in mid-Channel, though the Lydd writer seems to imply that it was. The English ports were certainly closed for some time (according to the reporter, for some weeks).

It seems possible, and in the circumstances which we shall describe in Chapter 9, it may be regarded as likely, that a considerable area of ice detached from the Arctic polar ice may have drifted into the northern North Sea some time before the observations recorded off the Dutch, English and French coasts, and in the Channel, in February 1684. This is entirely consistent with the prevailing winds implied by the map for the winter of 1683–4 in figure 7.1, which was published five years before I knew about the observations of the Lydd writer. As we shall see in Chapter 9, the sea between Iceland and the Faeroe Islands was probaby on average about 5 degC colder than now for 30 years around 1675–1704. This was presumably due to abnormal prominence of a branch of the East Greenland (ice-bearing) current continually supplying cold, polar water and ice towards the region. The same current makes its appearance more briefly from time to time even today. And in 1888, it brought enough of the polar pack-ice south in two tongues, advancing on either side of the Faeroe Islands, to cause the Faeroese fishing fleet to return to port. The situation mapped here in figure 6.1 would probably send more of the ice belt from north-east Greenland towards the coast of Norway and the North Sea than south along the southern half of the east coast of Greenland.

It seems most likely that at some stage in that winter a northerly or northwesterly gale broke away a lot of the ice from the main pack and sent it south into the northern North Sea. There it would probably have broken up further and dispersed a good deal, but would be liable to have converged and compacted again towards the southern end of the North Sea and the Strait of

Dover. With the ice would also have come a great influx of the cold polar water, which has lower salinity and lesser density than Atlantic water, and this would have facilitated freezing, together with any river ice emerging during periods of quiet weather near the coasts of the southern North Sea and the Channel. There would surely be more difficulty for any great quantities of Baltic ice to emerge into the North Sea through the narrower and more tortuous channels between Denmark and the northern countries. And when this does happen, as it did in 1838, the Baltic ice is more inclined to move northwards up the west coast of Norway, although a change of wind might later bring it south.

The apparent increase, since about the 1960s, in the variability of climate from year to year and from one group of a few years to the next, and in the occurrence of many kinds of climatic extremes, has attracted notice in most parts of the world. This is clearly associated with the decline of the westerly winds in the latitude of the British Isles, seen in figure 6.5 (p. 95), and a corresponding increase in the frequency of 'blocked' or 'meridional' wind-patterns as explained on p. 108. It is further illustrated by figure 7.3, which shows the distribution of monthly rainfall over the last 100 years. The frequency of very wet and very dry months has been highest in the later nineteenth century and since 1960. It was probably Wallén (1953) who first pointed out that the standard deviations of the seasonal means shown by the long records of temperature in Europe had varied. In particular, the variance of summer temperatures in the 200-year-long record for Stockholm had been greatest in periods with frequent occurrence of meridional circulation patterns: the figures were 1.25 degC for 1770–89 and 1.23 degC for 1928–47 (the latest years covered) against 0.94 degC for 1906–25. Corresponding changes are now known to have affected the winter temperatures. For example, the standard deviation of winter temperature in central England from year to year was 2.26 degC from 1795 to 1824 (a peak

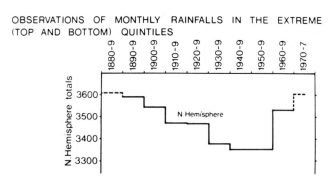

Figure 7.3 Distribution by decades, 1880–1977, of the numbers of reports of 'extreme' monthly precipitation totals, i.e. months falling within either the driest or the wettest 20 per cent of all occasions. The network of 73 stations used was well scattered over the northern hemisphere.

period) and again from 1879 to 1898, and has returned to a similar level from 1938, whereas from 1906 to 1935 it was only 1.45 degC. There is also a noticeable similarity with the history of changes in the frequency of gales and gale situations over the North Sea, shown in figure 6.4.

It is arguable, in connection with different impacts of climate on the biosphere and on the human economy, how far changes of climate, such as that which marked out the so-called Little Ice Age between about AD 1550 and 1850 as different from earlier and later times, are best considered in terms of changes of the mean values of temperature, rainfall, etc., or changes in the frequency of extremes and changes in other measures of variability from year to year. Some meteorologists have even suggested that the latter was the true nature of the Little Ice Age regime, since a skewness of the temperature distribution in the colder months of the year is associated with greater deviations from the long-term mean in the case of the cold extremes.

The severest shocks to the human economy come of course, when two, three or more years tending to the same extreme occur in succession. A prime example is provided by the history of the 1690s in Scotland, when the grain harvest failed in the upland parishes in all parts of the country in seven years out of eight from 1693–1700. The crops, we are told, were blighted by easterly 'haars', or sea-mists; by cold, sunless, drenching summers; by storms; and by early frosts and deep snow in autumn (Graham 1899); and the deaths from famine exceeded those in the great plague (the Black Death) in the Middle Ages. Such sequences of years must also be important in connection with glacier advances. It has been noticed that available accounts from the period of the Little Ice Age seem to imply that most of the advances of glacier snouts took place in just two, three or four disastrous periods of 10–15 years. And Bray (1971) similarly finds from fossil evidence in north-western North America that over half the glacier advances between 1580 and 1900 were concentrated in the years 1711–24 and 1835–49.

Dr O. Liestøl, of the Norwegian Polar Institute, has suggested to me that such sequences of just a few years of cold summers and severe winters may have been sufficient to establish a long-lasting ice cover on some small, high-level lakes (or tarns) in the Scottish Highlands where such was observed, but the water temperature nowadays is as high as 8–10 degC, even in a colder than average summer (as in 1978). The process depends on abnormally great amounts of snow being blown into the water in a cold spring and failing to melt during an ensuing cool, cloudy summer. If the following winter were severe, a thickness of ice might be formed which could not be entirely melted until one or more warmer than average summers returned. (This process has been observed in waters on the Hardangervidda (plateau) in Norway.)

We know from the actual thermometer records in England that from spring 1690 to autumn 1695 inclusive there was an unbroken sequence in which every season of every year was colder than the 1701–1950 average.

Indeed, the sequence lasted with only few breaks from about 1680-1700. Descriptive accounts of the seasons suggest that there were some partly similar experiences in the 1650s and in the early years of the century before 1620; the period probably extended back into the previous century. Moreover, there are strong indications from the reports of sea ice about Iceland (Koch 1945) and of the cod fisheries in the northern seas that polar water from the East Greenland Current had been spreading south and east over the ocean surface for a long time, until from 1675-1704 this watermass dominated the ocean as far as the Faeroe Islands (Lamb 1977, 1979). In the extreme year, 1695, the warm saline North Atlantic Drift water (and its fish stocks) seems to have disappeared from the sea surface as far east as the entire coast of Norway and almost as far south as Shetland. With an ocean surface between Iceland and 60°–62°N, therefore, as much as 5 degC colder than in the twentieth century, prevailing air temperatures in northern Scotland and much of Norway must have been lowered significantly more than England, where the average for the 1690s was 0.8–1.0 degC lower than in the twentieth-century mean.

There exist many reports of permanent snow on the tops of the highest mountains in Scotland in those times and one or two of small, high-level lakes on which there was always ice, 'even in the hottest summer' (e.g. Mackenzy 1675). These seem to require mean temperatures as much as 2.5 degC below modern experience. This figure seems well possible given the anomaly described in the ocean only 500–800 km farther north. Liestøl's suggestion indicates that in such a climate, or one approaching it, just a few colder years in unbroken succession could establish a 'cake' of ice, perhaps 50 cm or more in thickness, on the surface of some small water bodies above about 700 m above sea level which, protected by its own high albedo, could persist for several years and perhaps for many years.

The clustering of recurrences of similar weather patterns and similar timing of seasonal atmospheric circulation developments in successive, or closely following years, as in the examples mentioned in this chapter, must result from the persistence of some elements of the situation which in most cases have not so far been identified. It seems obvious, however, that the great extension of the polar water on the ocean surface to a latitude near 60°N close to Europe in the late seventeenth century must have played an important part in the sequence of cold years in and around the 1690s. This came to light as a result of attempts to map the sea surface temperatures as far back into the Little Ice Age as possible. Mapping of the winds and weather systems, as well as reconstruction of ocean current conditions, may be expected to throw further light on the phenomena discussed here.

By a fortunate circumstance, it has proved possible to produce a series of synoptic weather maps of 60 individual days between May and October 1588, largely from the weather reports of ships of the Spanish Armada, which sailed against England in that year and was partly driven by storms, and partly sailed to escape them, over courses right around the British Isles and at one stage into the northern Atlantic towards Iceland. A historian, Mr

K. S. Douglas, of Belfast, collected the ships' weather reports and supplemented them by such reports as were available on land from the documented history of the occasion. He proceeded, with the help of meteorologists working in Northern Ireland, to produce a series of weather maps. Luckily, the participants in the work up to this stage did not know that the astronomer Tycho Brahe was making and recording daily weather observations on the then Danish Island of Hven (now Ven, 55°55'N 12°45'E) in 1588. It was, therefore, possible to test the preliminary 'isobaric' analysis of the daily weather maps by their success or lack of success in accounting for the weather observed in the Sound between Denmark and Sweden. (This was beyond the eastern limit of the area which the analysts had intended to cover, but the suggested isobars commonly did extend to about that point. The test was therefore a severe one.) The result of the test was that the wind direction and weather observed over the Sound were unquestionably in agreement with the maps on 56 per cent of the days and, if an adjustment were allowed for a position error of up to 250 km of a major front, trough or axis of high pressure, the maps would be considered to be accurate on 72 per cent of the days. On a further 9 per cent of the days, the analysis did not extend so far, and no indication of the winds or weather over the Sound was implied, so no test was possible in those cases. Since only 19 per cent of the preliminary maps failed the test, it seemed worth while to re-analyse the entire series using Tycho Brahe's observations and with the greatest possible care for logical continuity of the analysis from day to day. I therefore did this, and it seems reasonable to deduce that the resulting map series is 80–90 per cent reliable in the Danish area (and probably better than this in the region, mostly around the British Isles, where the observation reports were most numerous). The individual days' final maps generally had either three or four observation points over an area which always included the British Isles and Denmark, and ranged in some cases from Portugal or Biscay in the south to England or Scotland or farther north, and to the west of Ireland. A sample sequence is illustrated in figure 7.4.

Two noticeable features of this sequence recurred commonly throughout that summer: the – by twentieth-century experience – storminess generated exceptionally far south over the North Atlantic and the prominence of high pressure extending south from the Arctic. The maps illustrated in figure 7.4 include 'Midsummer Day' (Sankte Hans); they also conform to the conclusion reached independently by Pfister (personal communication, 1978) from Swiss data that 1588 was probably the wettest summer of the last 400 years in Switzerland. Repeatedly, in each month of the series, the daily weather maps show the fronts pushing up from the south being halted near a line from Iceland to Denmark, presumably by a persistent anticyclone over the higher latitudes of this sector of the hemisphere in that summer, and later being driven south again by northerly winds.

A possibly more important conclusion came from the fact that the weather map analysis fixed the positions of many of the depression centres on individual days and therefore made it possible to measure their speed of

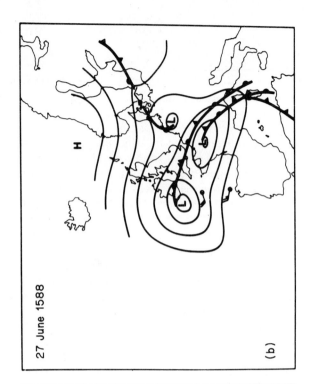

27 June 1588

(b)

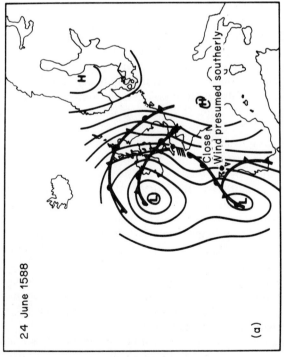

24 June 1588

Close N
Wind presumed southerly

(a)

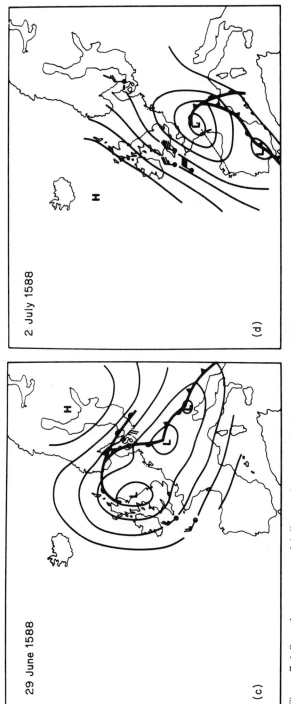

Figure 7.4 Sample sequence of daily weather maps for the summer of 1588, based on observation data reported by the ships of the Spanish Armada and a few other observations (e.g. by Tycho Brahe).

movement. This was a valuable (and hardly to be expected) result owing to the absence of any meteorological instrument measurements. Moreover, the speeds of advance of cyclone centres at the surface bear a statistical relationship to the gradient wind speeds in the warm air aloft, which was already investigated long ago by Palmén (1928), and Chromow (1942, partly based on work by Bergeron and Swoboda), and, through this, to the wind speeds observable at cirrus cloud levels and hence in the jet-stream. The daily weather maps of summer and early autumn 1588 showed that on at least six occasions in this one year the jet-stream strength seems to have reached, or somewhat exceeded, the probable limit of the twentieth-century distribution for the same months. This doubtless confirms the historians' general diagnosis of the reality of the strong storms, which more than any battle, destroyed and dispersed the Spanish Armada. But it also gives valuable insight into the nature of the meteorological developments during this sample of a summer during the onset phase of the Little Ice Age, when the European glaciers, at least in the Alps, were advancing most rapidly and the Arctic sea-ice had extended south (as we know from British ships' reports) to Iceland and to block the Denmark Strait in the three immediately preceding summers of 1585-7. It seems probable that also in 1588 the ice limit was near 65°N across much of the Greenland–Norwegian Sea, and that this played a part by maintaining a strengthened thermal gradient between latitudes about 50° and 65°N in the deduced strength of the jet-streams in this zone.

Could it be that some of the other clusterings, including those in recent years, mentioned in this article, have arisen through similar controls of the jet-stream – and wave developments in it, propagated downstream – associated with ocean temperature anomalies, such as Namias (e.g. 1963, 1969, 1970) has identified, elsewhere in the hemisphere?

References

Baur, F. (1950) 'Langjährige Beobachtungsreihen' in Link-Baur *Meteorologisches Taschenbuch,* Band 1, 2, Auflage, Leipzig, Akademische Verlagsgesellschaft, Geest & Portig K.-G.

Bergeron, T. (1928) 'Über die dreidimensionale verknupfende Wetteranalyse: Teil I – Prinzipielle Einführung in das Problem der Luftmassen und Frontenbildung', *Geofysiske Publikasjoner,* Vol. V, Nr. 6, Oslo Meteorologisk Institutt.

Bergeron, T. Fries, M., Moberg, C.-A., and Ström, F. (1956) 'Fimbulvinter', *Fornvännen* 1, 1–18,

Betin, V. V., and Preobazensky, Ju. V. (1959) 'Variations in the state of the ice on the Baltic Sea and in the Danish Sound', *Trudy* 37, 3–13, Moscow, Gosudarst, Okean. Inst. (in Russian).

Bray, J. R. (1971) 'Solar-climate Relationships in the Post-Pleistocene', *Science* 171, 1242–1243.

Chromow, S. P. (1942) *Einführung in die synoptische Wetteranalyse,* 2nd edn, Vienna, Springer.

Gisler, O. (1985) 'Das Wetter zu Ende des 18. Jahrunderts', *Geographica Helvetica*, 1985, 205–222.

Gottschalk, M. K. E. (1971) *Stormvloeden en rivieroverstromingen in Nederland: Deel 1 – De periode 1400*, Assen, Van Gorcum.

Gottschalk, M. K. E. (1975) *Stormvloeden en rivieroverstromingen in Nederland: Deel II – De periode voor 1400 – 1600*, Assen Van Gorcum.

Gottschalk, M. K. E.(1977) *Stormvloeden en rivieroverstromingen in Nederland: Deel III – De periode 1600–1700*, Assen, Amsterdam, Van Gorcum.

Graham, H. G. (1899) *The Social Life of Scotland in the Eighteenth Century*, London, Black. (A fuller account with citations of original parish records is given in Sir John Sinclair's *Statistical Account of Scotland*, in many volumes, published 1791–9.)

Koch, L. (1945) 'The East Greenland Ice', *Meddelelser om Grønland*, *130*(3), Copenhagen, Danish Meteorological Institute.

Lamb, H. H. (1977) *Climate Present, Past and Future. Volume 2, Climatic History and the Future*, London, Methuen.

Lamb, H. H. (1979) 'Climate variation and changes in the wind and ocean circulation: the Little Ice Age in the Northeast Atlantic', *Quaternary Res. 11*, 1–20.

MacKenzy, G. (1675) 'On a Storm and some Lakes in Scotland', *Phil. Trans. R. Soc.*, *10*(114), 307.

Manley, G. (1974) Central England temperatures: monthly means 1659 to 1973, *Q. J. R. Met. Soc.*, *100*, 389–405.

Namias, J. (1963) 'Large-scale air sea interactions over the Pacific from summer 1962 through the subsequent winter', *J. Geophys. Res. 68*(22), 6171–6186.

Namias, J. (1969) 'Seasonal interactions between the North Pacific Ocean and the atmosphere during the 1960s', *Mon. Weath. Rev. 97*(3), 173–192.

Namias, J. (1970) 'Climate anomaly over the United States during the 1960s', *Science 170*, 741-743.

Palmén, E. (1928) 'Zur Frage der Fortplanzungsgeschwindigkeit der Zyklonen', *Met. Z, 45*, 96–99.

Pfister, C. (1978) 'Fluctuations in the duration of snow cover in Switzerland since the late seventeenth century', *Proc. Nordic Symp. Climatic Changes and Related Problems: Climatological Papers No.4*, 1–6, Copenhagen, Danish Meteorological Institute.

Wallén, C. C. (1953) 'The variability of summer temperature in Sweden and its connection with changes in the general circulation', *Tellus 5*(2), 157–178.

8

Studies of the Little Ace Age: I, The great storms and data available for establishing details of the period

This chapter is based on a paper given in the Second Nordic Symposium on Climatic Changes and Related Problems, held in Stockholm on 16–20 May 1983, and published under the title 'Some studies of the Little Ice Age of recent centuries and its great storms' in the book of the conference, Climatic Changes on a Millennial Basis, N.-A. Mörner and W. Karlén (eds), Dordrecht, Reidel, 1984.

In previous chapters, we have already noticed evidence of increased storminess in the North Sea and around the other coasts of north-west Europe during the Little Ice Age of recent centuries, as well as in earlier times of colder or cooling climate. Examples include the retreat of forests and woodland from the Atlantic coasts of the British Isles, particularly in Scotland and the Hebrides and Northern Isles about 2000 BC or soon after (see Lamb 1977, pp. 416–7); the frequency of sea-floods and of coastal changes brought about by blown sand during some centuries around 500 to 200 BC and between about AD 1300 and 1800; and the storms which demolished the former island in the area of the Goodwin Sands (off east Kent, near 51¼°N 1½°E) in the fifth century AD (and according to legend parts of the coasts of Wales and the Scilly Isles in that or the next few centuries).

It is true that other factors than climate probably played a part in the devastation of these areas. In particular, destruction of the forest by felling to clear land for agriculture or pasture and grazing of the young trees by domesticated flocks and herds may have produced a windier situation near the coasts in which no young trees could survive. In Iceland, too, it has been reported that the early settlers in the ninth and tenth centuries AD found rather bigger birches than have ever grown there since, and that their removal for use in building left a windier situation in which full recovery of the vegetation was not possible. But the rapid retreat of the Canadian forest from its northern limit by 200 to 400 km around 1500 BC, and the advance of the tundra, known from pollen analysis studies (Nichols 1970), cannot be attributed to human activity and was certainly accompanied by a colder and windier climate.

In order to establish whether there was a real increase of storminess and storm severity in the Little Ice Age period – and, by analogy, the probability of a similar increase in earlier periods of cold climate – it is necessary to examine what can be deduced from scientific measurements.

The record of monthly mean air temperatures prevailing at typical sites in the low-lying districts of central England, worked out by the late Professor Gordon Manley (1974) during 30 years of painstaking study of long runs of daily thermometer measurements obtained by careful observers at various places in overlapping periods back to 1659, provides a starting-point. The characteristics of the individual instruments, and their behaviour over the years, as well as the ways in which they were exposed to the atmosphere and shielded from the sun and artificial sources of heat, were carefully studied. The reported readings were checked against reports of frost and thaw, heat difficulties in summer, and so on. They were then reduced to averages for each month and compared with the results for other places, so that standard adjustments could be made to each record to arrive at a single homogeneous record, representative of an ideal site in the region with continuous recording over the more than 300-year-long period. (The dates were all corrected to the modern calendar.) This is the longest instrumental record of temperature for anywhere in the world. Further checks were made by Manley, comparing these central England temperatures for months and years of notably dramatic character with temperatures in Holland from the early eighteenth century (Labrijn 1945) and the reported behaviour of the Alpine glaciers and French wine harvests.

My work has extended Manley's record of central England temperatures by providing half-century averages back to the early Middle Ages (figure 3.1), derived by statistical associations between the incidence of seasons of noteworthy warmth, cold, wetness, and dryness (Lamb 1965), using the summer and winter index values for successive decades briefly described in Chapter 4 (see Lamb 1977, pp. 32–4, for a fuller account). These temperature records can be used to define the Little Ice Age period and the cooling stages that led into it.

Recent study suggests that the individual monthly values of mean temperature given by Manley (1974) for each year since 1659 have a standard error of about 0.1 degC back to around 1730, increasing to about 0.2 degC in the 1720s and perhaps 0.5 degC in the earliest part of the record before 1680. This means that 10-year averages of the monthly values probably have a standard error of about 0.1 degC in the 1690s and no more than 0.2 degC in the 1660s. The 50-year means should be within about 0.1 degC at that stage of history. Temperature records and records of various phenomena, such as glaciers, snow data and wine harvests, that can serve as proxy temperature records from elsewhere in Europe, provide general confirmation, especially of low temperatures around 1700 and in the first half of the nineteenth century (e.g. Labrijn 1945; Matthews 1977; Schweizerische Met. Anstalt 1957).

The extension by the present writer of prevailing central England temperatures over eight centuries before the earliest thermometer records, using statistics of descriptive accounts of months of various well-defined character is certainly also subject to error margins, which are tentatively indicated in figure 3.1, but become impossible to define precisely in the earliest centuries. As there is little difference in the quantity of manuscript source material which reports the seasons in England between about AD 1100 and 1550, the error margins of the temperature values so derived are nearly constant over this span of time. Some decades before AD 1400 are admittedly rather weakly covered by reports, and this has meant that even the 50-year estimates before that date are somewhat open to controversy (Wigley *et. al.* 1986), even though broadly supported by other environmental studies (cultivation limits, wine harvest studies, tree-line changes, and so on). There is no doubt or controversy, however, about the coldness of the period 1550–1700 in this part of the world, nor about the severity of the cooling between the thirteenth and fifteenth centuries.

Iceland and Greenland seem to have had the peak of their mediaeval warmth a century or more earlier, and to have begun cooling sharply by AD 1200. Iceland may have had some temporary recovery of warmth in the fifteenth century, rather than in the early sixteenth century, as in Europe. How far the magnitude of the changes shown by the central England temperature record should be considered representative is another question.

Changes in the ocean

The long continuous record of sea-surface temperature measurements at the Faeroe Islands, going back to 1867, shows that within that time the warmest and coldest 5-year periods differed from each other by 1 degC, i.e. twice the range of the air temperatures prevailing in England or central Europe. This must be largely due to variations in the ocean, particularly the volume of polar water in the East Greenland Current and the branch of it that heads towards, and sometimes passes, eastern Iceland.

The position of the boundary in the ocean surface between this polar current and the warm water of Gulf Stream origin has varied hugely over long periods of time. The reconstruction of sea-surface temperatures between 1675 and 1705, described in Chapter 9, indicates that the ocean surface between south-east Iceland and the Faeroe Islands was on average probably 5 degC colder than in the present century. This surely implies that the 30-year average air temperatures in northern Scotland and southwest

Norway at that time must have been about 2–2.5 degC below the modern averages. In eastern Iceland, the prevailing air temperatures in those years would probably have been 3 degC below today's. It was probably the severest phase of the Little Ice Age, at least in these parts of Europe.

The enhanced thermal gradient between the northernmost Atlantic, around Iceland and the Faeroe Islands, and the regions near latitude 50°N might provide an energy base for the development of occasional storms in European longitudes of greater intensity than in warmer periods.

Evidence of enhanced storminess in the North Sea and surrounding regions

There is evidence of an era of enhanced storminess around the North Sea and the coasts of north-west Europe which roughly coincided with the deterioration of climate (falling temperature levels and increased wetness of the peat bogs that have been studied) that ended the mediaeval warm period and another coincidence with the coldest phases of the climate between 1550 and parts of the eighteenth and early nineteenth centuries. The evidence is of at least three kinds:

(i) Great sea-floods devastating the low-lying coastlands around the North Sea, often with great loss of life and establishing some new water bodies that survived until the twentieth century (Zuyder Zee in the Netherlands, Jadebusen in north-west Germany, the Norfolk Broads in England). A great amount of land with many villages was also lost from the west coast of Denmark – Jutland and the province of Slesvig/Schleswig (which was Danish at the time). Corresponding erosion of the coast of England destroyed two great ports and took all the land around them (Ravenspur or Ravensburgh, which stood east of Hull on land beyond the present Spurn Point, and Dunwich on the coast of Suffolk, which also extended farther east than it does today).

(ii) Storms of blowing sand. These established a whole new range of coastal dunes in Belgium, the Netherlands and north-west Germany from AD 1150 or 1200 onwards, after some centuries during which the older dunes had become stabilized by vegetation. The phenomenon also extended inland in these countries and deposited layers of drifted sand. Similar troubles with drifting sand were reported in the seventeenth century inland in East Anglia. There were many local disasters through sand drift and advancing dunes that changed the coasts in various places in north-west Europe, sometimes closing former harbours.

(iii) Between 1690 and 1703, at least four storms were reported that impressed the people then living as being of exceptional severity, even for those times – in two cases, the precise date has not yet been identified:

(a) reputedly in the autumn of 1694 the Culbin estate, on the south coast of Moray Firth near 57.6°N, in north-east Scotland, which comprised sixteen good farms covering a total area of 15–30 km², was buried by blown sand, reportedly in a single prolonged storm (G. Bain undated; Edlin 1976; Steers 1937); the land remained a wilderness of shifting sand dunes until it was successfully planted with forest in this century;

(b) on 12 September 1695 (possibly Old Style – Julian calendar) a violent storm at night did great damage in London, overturning a ship in the Thames, dashed 70 ships (colliers setting out empty back to Newcastle) on the coasts of Norfolk and the Wash with the loss of most of the crews, drove a fleet of Dutch ships back to the Downs (on England's south coast) and stranded many ships on both sides of the Channel (Lindgren and Neumann, 1981); this seems to be the storm that Daniel Defoe remembered[*] 30 years afterwards as an unparalleled disaster at sea, with another fleet of coal-ships coming south from Newcastle also being wrecked by the NE gale as they approached the coast of Norfolk; altogether 200 ships and more than 1000 lives were lost off Norfolk that night;

(c) in the autumn of 1697, a great storm overwhelmed with sand a settlement at Udal, on North Uist, in the Outer Hebrides, that seems to have known human occupation continuously for four thousand years (Crawford and Switsur 1977; Morrison, 1967–8);

(d) the great storm that crossed southern England on the night of 7–8 December 1703 (New Style – Gregorian calendar) causing enormous damage to buildings and trees, in the cities and in the country, and an exceptional tide in the Thames, has been generally assessed as the severest storm of wind ever reported in England; it was described in great detail by Defoe (1704).

Any account of climatic change which includes as evidence any two of these three types of coastal change must also recognize that risk to the coasts does not only vary with the incidence of storm surges (see Chapter 6). Coasts may also be changed by (a) long-term changes of sea-level, consisting of both eustatic (global) and isostatic (regional or more localized) components; (b) long-term variations of the tidal force (associated with lunar orbit evolutions) and resulting variations in the range of the tides (Pettersson 1914, summarized briefly in Lamb 1972b, pp. 220–2); and (c) any long-term changes in the tides produced by alterations of the bottom topography of a shallow sea such as the North Sea or the Baltic (Nielsen, 1938; I am indebted to Dr E. Rasmussen for drawing my attention to the effects in some regions of changes in the bottom topography, as evidenced in his own 1958 paper). Mörner (1980, and in person, 1983) has interestingly drawn attention to the

[*] See Defoe's *Tour through the Whole Island of Great Britain*, in three volumes (1724–6). Defoe actually dates the storm 'about thirty years ago in 1692, I think that was the year . . .' but the details tally with other accounts of this storm on 12 September 1695.

fact that the detailed histories of sea-level in the north-west European/ north-eastern North Atlantic region (e.g. north-west England, southern shores of the North Sea, Kattegat and north Norway), worked out by numerous contributors, seem to present a regional sea-level history which may differ somewhat from the global average and which agrees in notable detail with the region's temperature–climate history. Such a regional history may, according to Mörner (1980, 1983), indicate changes in the shape of the Earth (geoid), a subject to which Maksimov (e.g. 1971) drew attention many years ago.

Some notable storms analysed: sample cases from the Little Ice Age period

It is to be hoped that some day it may be possible to submit the great storms we have named above to meteorological analysis. In the meantime, other storms within the cold climate period have been analysed.

The stormy summer of 1588

An analysis has been published (Douglas *et. al.* 1978) of the famous Spanish Armada storms in 1588, using the winds and weather reported albeit without instrument measurements by the ships. The first analysis was tested by its ability to explain the weather observed on the then Danish island Hven, in Øresund, daily by the astronomer Tycho Brahe. As this was satisfactory on 72 per cent of the days, despite the fact that the analysis had not been designed to cover the Danish area, the whole series of maps was then re-analysed with the Danish observations incorporated and with the greatest care for logical consistency. The result is a series of synoptic weather maps (two samples showing storms are illustrated in figure 8.1, respectively the severest storms in August and September 1588), with suggested isobars, for sixty individual days between 12 May and 26 October 1588. The positions of depression centres passing across the central part of the mapped area are clear enough to make it possible to measure their speed of travel. Hence, by statistical relationships deduced by Palmén (1928) and Chromow (1942), indications can be gained of probable jet stream and surface gradient wind speeds.

The observation coverage of these maps for storms affecting the British Isles, North Sea and northern Europe is best between late July and end of September 1588. During those weeks in that single summer the study indicates that at least six times the winds at jet-stream level were close to or somewhat exceeded the probable upper limit of the modern (1949–70) frequency distribution for that time of the year. Around 15 September the southwesterly jet-stream over the eastern North Atlantic is indicated as having a probable speed of 130 knots and on 8–9 August, 150–160 knots is indicated as the most probable speed. The strongest surface pressure gradient winds in the postfrontal cold air in the storms of 18 and 21

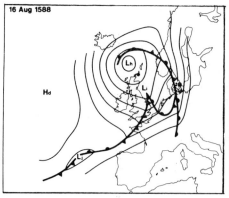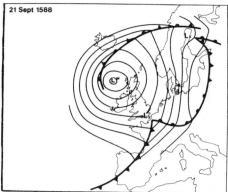

Figure 8.1 Synoptic meteorological analysis for 16 August and 21 September 1588 (both dates adjusted to the modern Gregorian calendar). The latter was the date of the greatest storm which wrecked most of the ships of the Spanish Armada on the west coasts of the British Isles. The positions of the ships which reported the weather, near the Orkney Islands, west of the Hebrides and south-west Ireland are marked, and Tycho Brahe's observations in Denmark are plotted. (The maps for four other days around midsummer 1588 are shown in figure 9.4.)

September on the coasts of the British Isles should probably be estimated at 80–90 knots. Judged by the details reported by the Spanish ships on the west coast of Ireland on 21 September 1588, and the wreckage of most of them, the (westerly) storm on that date (figure 8.1) was one of the severest (Douglas *et. al.* 1978).

The Great North Sea Storms, 21–22 March 1791 and December 1792

I am indebted to my friend Dr K. Frydendahl and the archives of the Danish Meteorological Institute for the most valuable ships' data. A gazeteer of the observing stations, their positions and practices in those times has been built up in the Climatic Research Unit, Norwich, by Mr J. A. Kington, who produced a daily weather map series from 1 January 1781 to 31 December 1786; and this has been added to by the present studies. Data and information about the observations have been graciously given by collaborators too numerous to name individually here but my thanks are due for a particularly big contribution which came from Dr E. Hovmøller working in Swedish archives.

To analyse the wind and weather situations occurring from the 1780s and 1790s onwards, methods very close to those applied to present-day synoptic meteorology can be used. At the date of this storm, observations were available from 50 places in Europe, ranging from northern Iceland to Rome and from Barcelona to St Petersburg and northern Sweden. In addition, occasional reports from ships at sea or in port at useful positions, including the Faeroe Islands, strengthen the analysis. The procedure starts by accepting the barometric pressures reported by a strategically distributed network of stations for which the details known and the quality of the record suggest confidence: these pressures are then corrected to standard gravity and sea-level. On this basis, a map of the mean pressures prevailing over Europe and the eastern Atlantic in that month can be drawn. The average corrections required to make the mean barometric pressures reported at other places fit the map can then be deduced. These corrections are then applied routinely for every day of the month, and the additional stations are entered on the maps. Finally, for every station on the map, including those provisionally accepted as the basis of the monthly average map, the frequency distribution and range of apparent errors of their fit on the daily maps is studied. The standard errors calculated supply an index of reliability for each station, which is a valuable addition to the station gazetteer.

The results of this reliability test may be somewhat flattering to stations at the edge of the mappable area, which tended to be relied on because of their position, so long as the quality of their records inspired confidence; but no such criticism can apply in areas where the observation network was quite dense, even near the edge of the map. From this study it appears that in the 1790s the Swedish stations, Trondheim, Gordon Castle in northern Scotland and the stations used in France were among the best, mostly with standard errors of about 1 mb or less.

A separate test of the reliability of the analysis of the synoptic maps with regard to pressure gradients and the storm winds over the North Sea and surrounding regions was performed by first analysing the maps for the whole month of March 1791 with only four stations (Trondheim, Stockholm, St Petersburg and Copenhagen) in northern Europe north of 55°N, apart from the British Isles stations. These places were as far apart as the places on opposite sides of the North Sea:

		km
e.g.	Stockholm – Trondheim	600
	Copenhagen – Trondheim	830
	Stockholm – St Petersburg	720

compared with

	km
Edinburgh – Hamburg	600
Norfolk – Hamburg	600
Northeast Scotland – Trondheim	1000

Later, the observations at six more places in Sweden and Finland were entered. Unfortunately, the barometer used at Gothenburg was not one of the best, its record being marred by failure to respond fully to the quickest changes of pressure. Eliminating the minority of days, eight out of 31 in that stormy month, when the Gothenburg barometer readings were obviously badly wrong because of this defect, the standard error of the pressures indicated by the first analysis of the maps at that place was 1.24 mb: part of this may still be due to the poor behaviour of the Gothenburg barometer. Overall, the test suggested standard errors of 1–1.5 mb at points in the middle of the biggest gaps in the network over the northern half of Europe. The corresponding errors of the pressure gradients over the North Sea or in gaps of that size within the network would be about 1 mb/200 km, equivalent to about 10 knots (5 m/s) in the gradient winds indicated.

But the maps deteriorate rapidly outside the area of the observations network: bigger errors occurred at no great distance beyond the edge of the network; thus, at Härnösand, 250 km outside the edge of the network used for the test analysis, the standard error was 3.7 mb. A similar test on the May 1795 maps revealed standard errors of the daily maps amounting to 3.8–4.4 mb at points 500–600 km east of the limit of the observation network used, over central-eastern Europe.

In March 1791, the first warning of disturbance to the anticyclonic situation with light winds which had been dominating northern Europe since the 16–17th came with the passage of a cold front southwards over Iceland on the night of the 18–19th, followed by three days of hard northerly wind with bitter frost and blowing snow there. Temperatures as high as 14–16°C in the warm southwesterly air-stream over the continent as far north as Brussels and Hamburg on the afternoon of the 20th, and 8–10 degC in most of Britain, were replaced by northerly winds with wintry showers and afternoon temperatures in England mostly between 3–6 degC on the 21st. A strong northerly gale reached the British Isles on the evening of the 20th, after falls of pressure of 20–30 mb in 24 hours, followed by equally rapid rises. At Hamburg the pressure fell 37 mb in 24 hours between the afternoons of the 20th and 21st; and around 5 am on the 22nd, after a night of storm, the tide in the River Elbe reached an exceptional level (Rohde and Petersen 1977, and contemporary accounts in the old Hamburg newspaper, *Hamburgische Adresse-Comptoir Nachrichten*). This storm maintained gale strength for three days over some parts of the North Sea, probably the longest duration of any of the severe storms of that century, owing to the slow movement of the cyclonic system from the 20th onwards in the increasingly meridional situation.

On this occasion, and during storms in the month of December 1792, geostrophic winds (winds characteristically found in the free air at 300–500 m above the surface in straight wind-streams) reaching 130–150 knots at some points over the North Sea and neighbouring coasts were indicated by the surface pressure map. (Since in all these cases we are dealing with 'straight' – i.e. great circle – wind-streams in the regions where the pressure gradients were strongest, the gradient winds, which take account of the curvature of the wind's path, were presumably also of about this strength). Strongest gusts at the surface over the sea and exposed areas near to the coasts may also have been over 100 knots, perhaps much over 100 knots – the strongest gusts of the surface winds at exposed sites sometimes exceed the gradient wind speed by a substantial margin. This occurred in the storm of 16–17 February 1962, which also produced a disastrous sea-flood in the Hamburg region; 154 knots was measured in Shetland when the gradient wind was only 80–90 knots; and on 9 April 1933, on the mountainous island of Jan Mayen (71°N 8°W), 163 knots was measured by a standard anemometer. No gale appears to have matched this in the twentieth century over the same region, with the exception of the recent storms of 23–25 November 1981 and 18 January 1983, when the strongest gradient winds again reached 130 knots or rather more over the North Sea region.

The meteorological situation that developed on 10–11 December 1792 (figures 8.2 and 8.3) produced the highest sea-flood of the eighteenth century around Hamburg, exceeding the highest tide-level on 22 March 1791 (and earlier flood-levels in 1751 and 1756) by about 10 cm. The strongest gradient winds are estimated as westerly 150 ± 30 knots over the Faeroes–Shetland region on the morning and early afternoon of the 10th and westerly to northwesterly 140 ± 20 knots over the southernmost North Sea and the German coast in the late evening of the 10th till about midnight. (A 9 pm map on the 10th was analysed.)

Tide-levels in the Elbe caused alarm in two earlier storms on the 5th and 7–8th of that month, nearer the time of neap tides. The strongest gradient winds at the critical time were again westerly to northwesterly, 100–140 knots.

It was a remarkable month for the frequency of gales and storm-force winds in many parts of Europe. Although the exact equivalents of the scale of wind force then in use are not known, and the scale may not have been so professionally used as the later Beaufort scale has been, stormy winds were reported by observers in northern Holland on ten days of the month, in Paris on eight days, at Marseilles on seven days, in Moscow on nine and in Hamburg on six. The pressure gradients suggest that Beaufort Force 9 or more occurred on at least 12 days of the month over the sea areas off north Holland, Friesland and northern Germany and also over northernmost parts of the North Sea.

There was a great thermal contrast between the warm and cold air-masses involved in this month. Temperatures up to 11 to 13 degC occurred in the warmest air-masses passing over England and western parts of Europe south of latitude 53°N and up to +6 degC over southern Sweden. Yet, despite the lack of periods of still, clear weather in the cold air-masses an afternoon temperature of −9 degC was registered in a northerly wind at Stockholm on the 12th. And a day or two later the same air-mass produced afternoon temperatures of −13 degC at Trondheim and −20 degC at Härnösand (62°N).

The storm of 8–9 May 1795

This storm is chiefly known for the destruction wrought in the forests of central Sweden. The month opened with a remarkably warm southerly to

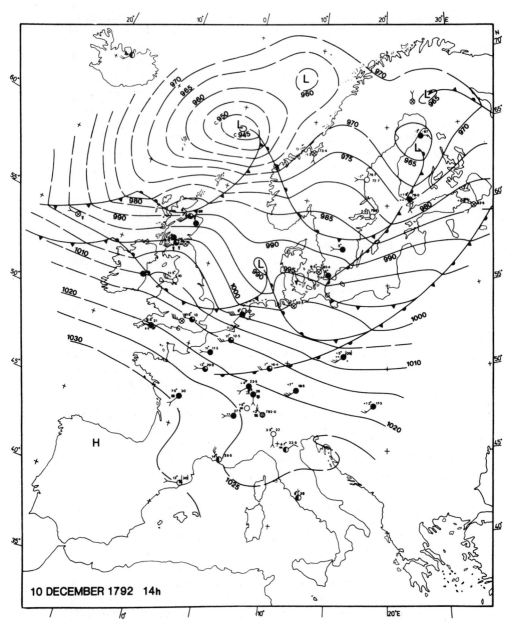

Figure 8.2 Synoptic weather map and analysis for 2 pm on 10 December 1792.

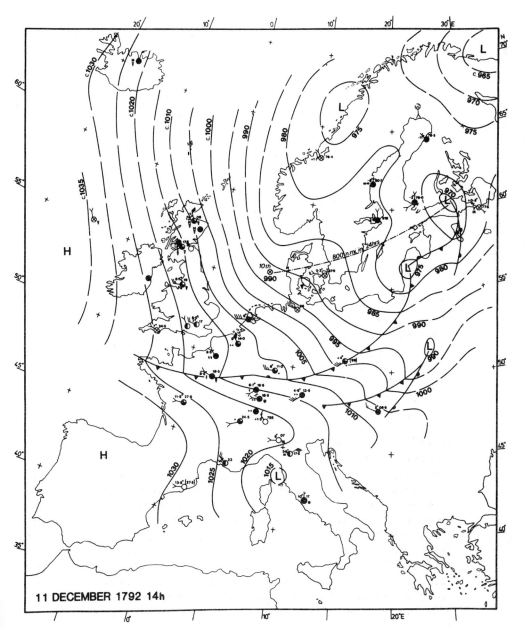

Figure 8.3 Synoptic weather map and analysis for 2 pm on 11 December 1792.

southwesterly air-stream over much of Europe: 26 degC was measured in Stockholm on 1 May. A fresh northerly outbreak from the Arctic appeared over Iceland on the 5th, with prolonged snowfall followed by hard frost. On the 7th–8th a strongly northerly outbreak developed over the whole Norwegian Sea and North Sea, where it rejuvenated a small depression that proceeded to cause strong winds and heavy snowfall over central Sweden for several days. The damage to the forests was undoubtedly partly due to the weight of snow. Geostrophic winds of up to 100 knots are indicated over central, south-western and southern Sweden on the 8th, but the gradient winds (i.e. by implication, the actual winds in the free air 300–500 m above the surface) may have been no more than 60–80 knots. The pressure gradients are particularly well authenticated by the fairly dense network of observations over the area. However, the severest gusts and squalls of wind at the surface near the fronts over Sweden may well have ranged up to about 90 knots.

Conclusion

Owing to the near-impossibility of maintaining constant exposure of anemometer sites – perhaps the only examples of long homogeneous series of wind-strength measurements are from offshore lighthouses – probably the only way of obtaining satisfactory comparisons over long stretches of time of the windiness prevailing is by measurements of the barometric pressure fields (Lamb and Weiss 1979). That is the basis of the studies reported here.

In this research so far, 31 severe windstorms over the North Sea and neighbouring regions of northern, central and western Europe have been analysed. Identification of the severest cases for study in centuries before the present one has been from the greatest reported disasters by coastal floodings, shipwrecks and damage to coasts and inland. This method of selection, particularly because the reports of coastal flooding are generally the most serious and likely to be the most complete, will probably not bring to light all the severest storms but will certainly provide a worthwhile sample list of outstanding storms. Within the present century the list is likely to be more nearly complete, and effort has been made to analyse all the very severe cases, preferably including examples of as many as possible of the different types of meteorological situation that may produce the extremes. This should provide the best possible basis for comparisons with earlier times.

The severe storms analysed so far were distributed as follows:

sixteenth century	2 cases	1900–49	4 cases
eighteenth century	7 cases	1950–83	15 cases
nineteenth century	3 cases		

The results of the analyses so far performed may be tabulated as in tables 8.1 and 8.2.

All the storms which lasted three days or more were either northwesterly to northerly or southeasterly storms.

The month of December 1792 was characterized by remarkably frequent storms, several of them of brief duration because of the rapidity with which even the occluded cyclones were moving. This points to a very strong jet-stream. Using the statistical relationships found by Palmén (1928) and Chromow (1942), already referred to in connection with the 1588 storms, the speeds of movement of ten depressions during 5–19 December 1792 were found to indicate jet-streams moving at over 130 knots in five cases and over 150 knots in three cases, the two strongest cases suggesting figures of 175 ± 25 knots and (on 10–11 December 1792) 260 ± 30 knots. This last figure probably exceeds somewhat the strongest jet-stream observed in the present century over this part of the world, though not equalling the extremes of 300 knots reported over the South Indian Ocean and over Japan. It is to be noted that even the August and September cases in 1588

Table 8.1   Number of severe storms in different categories, 1588–1983

Period	Total numbers of strong storms analysed	Main direction of storm winds at surface					Strongest gradient wind			
		SW	W	NW	N	SE	70–90 kt	90–110 kt	110–130 kt	Over 130 kt
1588–1800	9	1	2	4	2	0	4	1	1	2
1800–99	3*	0	0	1	1	0	2	0	1	0
1900–50	4	0	0	1	1	2	3	1	0	0
1950–83	15	2	3	6	2	1	9	4	1	0
Total	31	3	5	12	6	3				

* In one case the location of the cyclonic centre precluded estimation of the surface wind direction.

Table 8.2   Frequency of different maximum gradient wind strengths as a proportion of the severe storms analysed, 1588–1983 (expressed as percentages)

Period	Strongest gradient wind				Extreme
	70–90 kt	90–110 kt	110–130 kt	Over 130 kt	
1588–1800	44	11	11	22	c. 150 kt (twice)
1800–99	67	0	33	0	c. 120 kt
1900–50	75	25	0	0	c. s100 kt
1950–83	60	27	7	0	100–130 kt

indicated jet-streams of 115-130 knots over the north-east Atlantic–
northern Europe region, also exceeding the probable twentieth-century
maximum for those months.

Gradient winds near the surface of up to 120 knots in a warm sector on
10–11 December 1792 were indicated by the isobars, but as in most other
great North Sea storms, the strongest gradient winds were in the cold air at

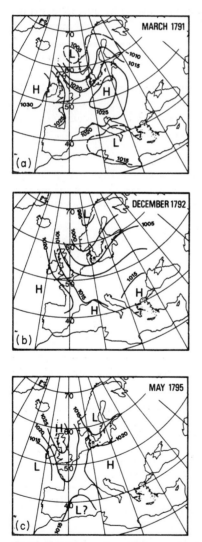

Figure 8.4 Maps of monthly average barometric pressure at mean sea-level for: (a)
March 1791; (b) December 1792; and (c) May 1795.

the rear of the depression. So the probability of speeds of up to 150 knots or thereabouts on that date is confirmed by this method.

The predominance of winds from the quarter between west and north in the severe storms analysed may be a peculiarity of the North Sea region, though the association between outbreaks of air from the Arctic and great storms in these latitudes is certainly a more general experience. Inland and on other coasts in England, the strongest winds generally come from between S and W. This is also a widely-known feature of tornadoes. But in the North Sea and Norwegian Sea, the roughest weather occurs mainly with northwesterly and northerly winds because of their long fetch over the open sea. And these are the directions that tend to produce the worst sea floods about the coasts of the southern North Sea. In various parts of the world, mountainous coasts tend to channel the wind and, through convergence, produce the highest speeds quite near the coasts, in directions parallel to the coast or the flank of the mountains. This is the explanation of the secondary maximum of storm winds in the North Sea coming from SE, in nearly all cases strongest near the coast of Norway. The occurrences of great storms from that quarter seem to be a feature of runs of years when blocking anticyclones over Scandinavia or north Russia are frequent. They also tend to be associated with very cold air from the continental interior channelled between the mountain mass and the warmer air-masses approaching with warm-type fronts from the Atlantic and over the British Isles. These may be called 'frontal funnel' gales.

The association of great storms in this region with outbreaks of Arctic air from the Norwegian Sea is illustrated by the monthly mean sea-level pressure maps for March 1791, December 1792 and May 1795, as shown in figure 8.4. These three months provided seven of the great storms analysed. Indications of meridionality are also seen in the frequencies of the different weather pattern types (Lamb 1972a).

These studies have used observational material never previously submitted to scientific analysis. The meteorological analysis provides a number of indications that some of the storms in the region studied during the Little Ice Age period were indeed more severe than twentieth-century experience, although a tendency towards increasing severity of storms since 1950 has been reported. It is also shown that outbreaks of Arctic air from the north in the Norwegian Sea and an enhanced thermal gradient between latitudes 50°N and 60°-65°N in the east Atlantic – British Isles – central European sector were involved.

Acknowledgements and data sources

The research on the severest North Sea storms of the past was funded by Shell Exploration Ltd from 1975 to 1982, and the author acknowledges his indebtness.

He also expresses his gratitude for generous gifts of basic weather observation data to the late Professor Gordon Manley, who supplied the records for Cambuslang (near Glasgow), Liverpool, Lyndon (Rutland), Stroud (near Gloucester), Modbury (near Plymouth) and London; to Dr Harald Nissen and the University Library, Trondheim, for the Trondheim record; to Dr K. Frydendahl and the Danish Meteorological Institute, for the observations from Copenhagen and from some very usefully placed Danish ships at sea; to Dr E. Hovmøller and the Swedish Meteorological Institute, for observations from ten places in Sweden and Finland; to Dr P. K. Rohan, as Director of the Irish Meteorological Service, for the observations in Dublin and occasionally elsewhere in Ireland; to Mr J. A. Kington, of the Climatic Research Unit, for use of the observations which he had obtained from Paris, Poitiers and Montdidier (in northern France) and from Milan; to the Director of the Schweizerische Meteorologische Anstalt, Zurich, for the observations from four places in Switzerland and from Mülhausen in Alsace; to Professor Manuel Puigcerver, of the University of Barcelona, for the observations from that city; and to Dr G. Zanella, of the University of Parma, northern Italy, for the observations made there.

Most of the other observations mapped came from the collection in the yearbooks for 1791 and 1792 (the last year) of the old Societas Meteorologica Palatina, Mannheim. Gilbert White's weather observation record at Selborne, Hampshire, was kindly made available for copying by the Trustees at Selborne. And a valuable record for Gordon Castle, in north-east Scotland, running from 1781 to 1827, was obtained on microfilm from the Meteorological Office. The author's thanks also go to the descendant relatives and Trustees of persons who kept weather diaries at Kemnay, Aberdeenshire, and at three places in Norfolk (Parson Woodforde at Weston Longville, May Hardy at Letheringsett and the Blofeld family at Hoveton). The agreement between the weather reported in these last three diaries, kept at places 20–30 km apart, lends confidence to the use of such records when it is plain that the diarist was a meticulous person. This agreement was 100 per cent except on days when there was an obvious meteorological explanation of the differences, such as, for example, greater shower activity and windiness near the North Sea.

References

Bain, G. (n.d.) *History of Nairnshire*, ch. 13, Nairn, Scotland.

Chromow, S. P. (1942) *Einfuhrung in die synoptische Meteorologie*, (2nd edn), Vienna, Springer.

Crawford, I. and Switsur, R. (1977) 'Sandscaping and C14: the Udal, North Uist', *Antiquity*, 57, 124–136.

Defoe, D. (1704) *The Storm*, London.

Douglas, K. S., Lamb, H. H., and Loader, C. (1978) 'A meteorological study of July to October 1588: the Spanish Armada storms', *Climatic Research Unit Research Publication No. 6* (CRU RP6), Norwich, University of East Anglia.

Edlin, H. L. (1976) 'The Culbin Sands', in *Environment and Man. Vol. 4, Reclamation*, J. Lenihan and W. W. Fletcher (eds), London, Blackie.

Labrijn, A. (1945) 'Het klimaat van Nederland gedurende de laatste twee en een halve eeuw', *Mededelinger en verhandelingen*, 49(102), De Bilt, Koninklijk Nederlands Meteorologisch Institut.

Lamb, H. H. (1965) 'The early medieval warm epoch and its sequel', *Palaeogeography, Palaeoclimatology, Palaeoecology*, 1, 13–37, Amsterdam, Elsevier.

Lamb, H. H. (1972a) 'British Isles weather types and a register of the daily sequence of weather patterns 1861–1971, *Geophysical Memoirs No. 116*, London, HMSO for Meteorological Office.

Lamb, H. H. (1972b) *Climate: Present, Past and Future. Vol. 1, Fundamentals and Climate Now*, London, Methuen.

Lamb, H. H. (1977) *Climate: Present, Past and Future. Vol. 2, Climate History and the Future*, London, Methuen.

Lamb, H. H. (1979) 'Climatic variation and changes in the wind and ocean circulation: the Little Ice Age in the northeast Atlantic', *Quaternary Research*, 11, 1–20.

Lamb, H. H. and Weiss, I. (1979) 'On recent changes of the wind and wave regime of the North Sea and the outlook', *Fachliche Mitteilungen, Nr. 194*, Traben-Trarbach, Amt für Wehrgeophysik – Geophys. BDBw -FM 194, ISSN 0343–6025).

Lindgren, S. and Neumann, J. (1981) 'The cold and wet year 1695: a contemporary German account', *Climatic Change*, 3(2), 173–187.

Maksimov, I. V. (1971) 'Causes of the rise of sea-level in the present century', *Okeanologiya*, II, 530–541, Moscow, Akad. Nauk. Okean. Kom. (in Russian).

Manley, G. (1974) 'Central England temperatures: monthly means 1659 to 1973', *Quart. J. Roy. Meteorol. Soc.*, 100, 389–405.

Matthews, J. A. (1977) 'Glacier and climatic fluctuations inferred from tree-growth variations over the last 250 years, central southern Norway', *Boreas*, 6, 13–24.

Mörner, N.-A. (1980) 'The northwest European sea-level laboratory and regional Holocene eustasy', *Palaeogeogr., Palaeoclim., Palaeoecol.*, 29, 281–300.

Morrison, A. (1967–8) 'Harris estate papers', *Transactions of the Gaelic Society of Inverness*, 45, 57–58.

Nichols, H. (1970) 'Late Quaternary pollen diagrams from the Canadian Arctic barren grounds at Pelly Lake, Keewatin, NWT', *Arctic and Alpine Research*, 2, 43–61, Boulder, Colorado.

Nielsen, E. Steeman (1938) 'De danske farvandes hydrografi i Litorinatiden', *Meddelelser fra Dansk Geologisk Forening*, vol. 9, Hefte 3, Copenhagen, Reitzees Forlag.

Palmén, E. (1928) 'Zur Frage der Fortpflanzungsgeschwindigkeit der Zyklonen', *Meteorologische Zeitschrift*, 45, 96–99.

Pettersson, O. (1914a) 'Climatic variations in historical and prehistoric time', *Svenska Hydrografisk-Biologiska Kommissionens Skrifter*, Häft V.

Pettersson, O. (1914b) 'Étude sur les mouvements internes dans la mer et dans l'air', *Svenska Hydrografisk-Biologiska Kommissionens Skrifter*, Häft VI.

Rasmussen, E. (1958) 'Past and present distribution of Tapes (Venerupis) Pullastra (Montagu) in Danish waters', *Oikos*, 9(1), 77–93.

Rohde, H. and Petersen, M. (1977) *Sturmflut*, Neumünster, Karl Wachholz Verlag.

Schweizerische Meteorologische Anstalt (1957) 'Der Temperaturverlauf in der

Schweiz seit dem Beginn der Meteorologischen Beobachtungen', *Annalen der Schweiz Met. Anstalt, Nr. 11*, Zurich.

Steers, J. A. (1937) The Culbin Sands and Burghead Bay', *Geographical Journal*, 90, 498–528.

Wigley, T. M. L., Farmer, G., and Ogilvie, A. E. J. (1986) 'Climate reconstruction using historical sources', in *Current Issues in Climate Research*, A. Ghazi and R. Fantechi (eds), Dordrecht, Reidel for the Commission of the European Communities.

9

Studies of the Little Ice Age: II, Changes in the wind and ocean currents in the north-east Atlantic region

This chapter is included at this point to show the evidence we have of fundamental changes in the ocean during the time of colder climate in recent centuries and how the conditions of the seas so close to the coasts of Europe affected the climate. This study uncovers an underlying critical point for the development of human society in the countries close to the north-eastern Atlantic, which has not been recognized. Western and northern Europe today, as is well known, enjoy a climate which is many degrees warmer than the average for the latitudes concerned. And from the Atlantic seaboard to places even as far inland as central Europe the variability of temperature from year to year is also low for the latitude (see Lamb 1972, p. 280, fig. 7.13). This adds up to a sheltered regime, which is widely acknowledged (and taken for granted) and commonly put down to the stabilizing effect of the nearby ocean. What the examination reported in this chapter reveals is that when the ocean currents vary, and cause the boundary between the warm water of Gulf Stream origin and the polar water to shift, the ocean ceases to stabilize the situation and in fact becomes responsible for amplifying the climatic changes in the region. (The substance of this writing was published as a paper entitled, 'Climatic variation and changes in the wind and ocean circulation: the Little Ice Age in the north-east Atlantic', Quaternary Research, 11, 1979, pp. 1–20. The opening section has here been re-written to avoid repetition in other chapters.)

The surface currents in the world's oceans are largely driven by the drag of the winds. Variations must, and do, take place when the prevailing wind-pattern varies. And these variations have effects on the climate through changes in the ocean's heat transport, and hence its thermal and moisture input into the atmosphere. Opportunities to reconstruct conditions in the ocean in the past are more limited than in the case of wind-pattern changes. But in the sea areas with which this chapter is concerned the records of measurements with meteorological and oceanographic instruments are unusually long; and a first search of available fisheries, whaling and naval records suggests that it is possible to derive estimates of prevailing oceanic and atmospheric conditions back to the seventeenth century AD, perhaps even into the sixteenth century in some cases. The results of the survey here presented, which are supported by other kinds of circumstantial evidence, indicate a major anomaly in the ocean in those times in this area, registering ocean current changes, which clearly had important effects on the climate of Europe.

The circulation of the winds provides the most readily observed mechanism of climatic fluctuations, though the ultimate causes presumably lie elsewhere – in the variations of the balance between radiation energy supply and loss, and in the heat stored and exchanged between the ocean and the atmosphere. The ocean circulation is also responsible for an important part of the heat transport that reduces the difference which would otherwise build up between tropics and poles (see Vonder Haar and Oort 1973). As will be seen in a later chapter, the mainstream of the world's wind circulation is a belt of upper westerly winds, often referred to as the 'circumpolar vortex', over either hemisphere. The pattern of the circumpolar vortex varies from day to day, and from week to week and on all longer time-scales as the pattern of heating and cooling of the Earth varies – partly under the influence of the cloudiness, snow cover, swamped or parched ground, and so on, produced by the wind circulation. Anomalies are produced in the ocean surface too, for instance, by the drift of ice or the upwelling of cold water from deeper levels at coasts where the wind blows persistently offshore. And with the changing pattern of the main flow of the upper winds, the development and steering of surface weather and wind systems changes.

So far as the weather at any given time in middle latitudes is concerned, the most important distinctions are: (i) whether the mainstream of the atmospheric circulation is far enough south to leave the place of interest in polar air, and (ii) whether or not the circulation pattern is a mobile one with a general progression of the individual weather systems from west to east. In the alternative ('blocked') type of situation, the systems are more nearly stationary or are steered north or south by great meanders of the upper windstream, large-amplitude waves in the upper westerlies. The persistent northerly and southerly windstreams over different sectors in middle latitudes mark very different, though persistent, spells of weather in those sectors and great differences of heat and moisture transport in the blocked situations. Moreover, the more or less stationary controlling systems of the blocked situation characteristically occupy different positions from one year or one spell to the next. So any given place is liable to experience very different spells of weather – ultimately even opposite extremes of temperature and rainfall – in successive years, when blocking is frequent.

An extreme example of this is provided by the years with most and with least ice on the Baltic as indicated by the 400-year-long record (Lamb 1977, p. 586; from Betin and Preobazensky, 1959) of the duration of ice in the port of Riga: both extremes came in the same decade, the 1650s, in the Little Ice Age period. Also, in the midst of the Little Ice Age period, England experienced the succession of two famous very hot summers that accompanied the (last) Great Plague in 1665 and the Fire of London after the drought and heat in 1666. The frequency of westerly winds is known from numerous daily observation reports made in London and on ships in various south of England harbours (Lamb 1967) to have been rather low in the late seventeenth century, but was perhaps not quite as low as in the 1970s which produced a similar pair of hot summers in 1975 and 1976. The latter was rated as the second hottest summer in the 300-year record (Manley 1974) of monthly temperatures in central England and was immediately followed by a summer that ranked eigtheenth coldest in the record.

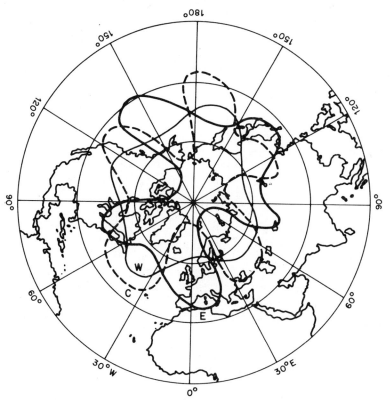

Figure 9.1 Variations of the circumpolar vortex: three different types of flow pattern occurring at different times (according to a type classification defined by Vangeng'ejm and Girs); W = an example of zonal westerly flow over middle latitudes; and C, E = two meridional flow patterns distinguished by different positions of the main troughs and ridges.

Figure 9.2 The prevailing surface currents in the oceans.

Figure 9.1 illustrates schematically, by means of the Vangeng'ejm and Girs (1964) circulation-pattern type classification, the differences between the (mobile) westerly form of the circumpolar vortex, marked W, and two meridional, or 'blocked', situations with large-amplitude waves (meanders) in the upper westerlies. The latter two patterns, the types marked C and E, differ from each other chiefly in the positions of the waves, causing largely opposite wind directions to prevail, for instance, over Europe and the Norwegian Sea in the two cases.

The general world map of ocean currents (figure 9.2) (from Lamb 1972; based on Sverdrup 1945; Dietrich and Kalle 1957; Defant 1961) is very similar to the pattern of the prevailing winds that drive them. And when the wind-flow pattern is distorted by great meanders of the steering current aloft, and corresponding changes in the tracks of the surface weather systems and the pattern of surface wind drag, the ocean current system must tend to be distorted too. The fact that the ocean currents are constrained by the limits of the ocean basins has, however, important effects, which seem to enter into the mechanisms of climatic variation on all time-scales and are an important control of the distribution and magnitude of the climatic shifts observed. It is probably for this reason that the displacements of the mean isotherms of surface temperature north and south during the climatic changes in this century, and between interglacial and ice-age situations, were in general greatest in and about the Atlantic sector.

Variations in the north Atlantic

There are several ways in which the geography makes the Atlantic Ocean liable to greater changes of poleward transport of heat, and therefore of climate, than the other oceans as follows. These are:

(i) The roughly north–south arrangement of the limits of the Atlantic Ocean guides the currents into directions which have prominent northwards and southwards components.
(ii) There are wide open connections with both the Arctic and Antarctic Oceans. Sverdrup (1945), earlier in this century, estimated the inflow and outflow of water exchanged between the Atlantic and the Arctic as approximately ten times as great as the flow through the Bering Strait from the Pacific. There are, however, indications that the amount of water flowing through the Bering Strait from the Pacific Ocean into the Arctic has varied significantly; it is thought to have been greater in the Little Ice Age 300 years ago than since.
(iii) The coast of Brazil near 5°S stands in the course of the South Equatorial Current like a mighty wedge and the current divides there, so that the water between the equator and 5°S which would otherwise be turned south by the Earth's rotation is deflected into the North Atlantic where it ultimately feeds the Gulf Stream. A shift of the ocean current by only 1° or 2° of latitude north

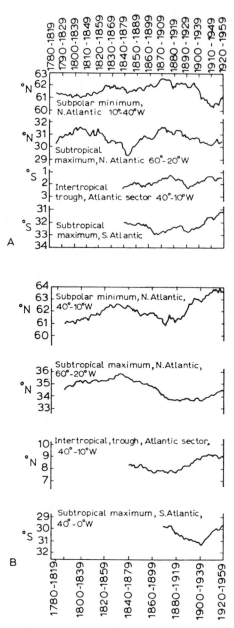

Figure 9.3 Varying latitude positions maintained by main features of the atmospheric pressure distribution (and corresponding surface wind pattern) over the Atlantic Ocean: (A) January; and (B) July (running 40-year averages).

or south in relation to this feature should affect the warmth and volume flow of the Gulf Stream. So much was suggested by Brooks (1949) for its probable climatic significance in the course of continental drift and polar wandering over tens of millions of years. But the zones of the wind circulation shift north and south in this region by up to 10° of latitude with the changing seasons, and we now know that they have indeed shown long-term shifts of 1° to 2° of latitude over periods lasting several decades within the past 100 years (figure 9.3) (see Lamb and Johnson 1959, 1961, 1966, for details).

(iv) Another – probably less important – distinctive feature is that with the present geography of the Earth very much more river water (and ice) drains into the North Atlantic and Arctic than into the other oceans of the world (see, for example, Olausson 1967).

Among the changes of prevailing air temperature in the course of the present century some of the greatest have been over the northernmost Atlantic and those parts of the Arctic reached at times by warm saline North Atlantic Drift water of Gulf Stream origin and the winds that have blown over it. This is clearly seen in figure 9.4 which surveys the recent ten-year periods of greatest warming and cooling. Both in the warming and cooling periods around 1920 and 1960 respectively the changes were particularly great along an axis through the Norwegian Sea and Barents Sea, increasing towards a maximum over the farthest limits reached by the warm saline Atlantic water at the surface in the eastern Barents Sea. This surely implies some secular changes in the strength of the North Atlantic Drift.

Analysis of the numerous ocean-bed sediment cores examined in the CLIMAP programme in the USA (McIntyre *et al.* 1976) makes it clear that the greatest depression of prevailing ocean-surface temperature at the height of the last ice age about 18,000 years ago was in the North Atlantic generally north of 40°N (figures 9.5 and 9.6), notably across the central, eastern and north-eastern parts of the ocean. The greatest departures from present conditions in the British Isles and Biscay region, averaged over the year, amounting to −10 to −12 degC, were greater than in most land areas of the Earth. Again a great shift of the ocean current pattern clearly must have been involved, with the North Atlantic Drift passing eastwards across the ocean south of 40°–42° N. McIntyre and others (1972) have shown that this has happened repeatedly in many of the great cold climate stages of the last 225,000 years (see Lamb 1977, figure 15.6, p. 328).

Ruddiman and Glover (1975), examining the traces in the ocean-bed cores of volcanic debris too large to have been carried by the winds, have deduced the presence 9000 to 10,000 years ago of an ice-bearing ocean current passing southward from the Iceland region to mid-Atlantic (see figure 14.4).

What can we find out about the variations in this region over periods intermediate in duration and in date of occurrence between this extreme and those of recent years?

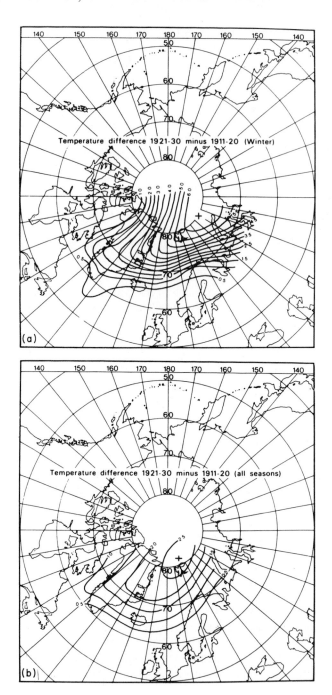

Temperature difference 1921-30 minus 1911-20 (Winter)

(a)

Temperature difference 1921-30 minus 1911-20 (all seasons)

(b)

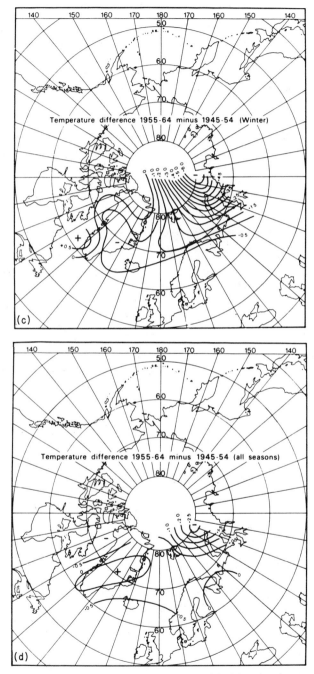

Figure 9.4 Temperature changes from one decade to the next: (a), (b) During the general warming – average air temperatures for 1921–30 *minus* those for 1911–20: (a) winter, (b) year. (c), (d) During the general cooling – average air temperatures for 1955–64 *minus* those for 1945–54: (a) winter, (b) year. (After Kirch 1966.)

The East Greenland and East Iceland Currents

Oceanographic measurements and calculations in recent years (Aagard 1969; Malmberg *et al.* 1972) have shown that the volume of water transported southward by the East Greenland Current in the region between about 75° and 65°N, can vary by a factor of 10. In 1965 the average flow of the current between 75° and 72°N, where it runs strongest, was 35 × 10^6 m^3/s as compared with Sverdrup's figure of 3.35 × 10^6 m^3/s for the average during the earlier part of this century. This flow is not all polar water. Icelandic water temperature and salinity observations conducted in June each year from 1948 onwards indicate that up to 1961 the sea area between 67° and 69°N north-east of Iceland was uniformly dominated by warm, saline North Atlantic Drift water (salinity 34.8–34.9 parts per thousand) and temperatures mainly 0 to +2 degC). By contrast, after 1962

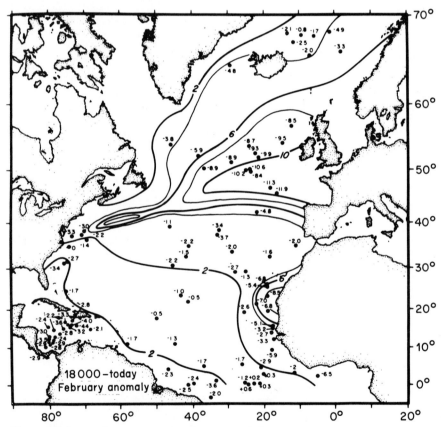

Figure 9.5 Sea-surface temperature depression (°C), below modern values, prevailing at the last glacial climax about 18,000 years ago in February. (From McIntyre *et al.* 1976.)

this area was penetrated to a varying extent by polar water (salinity varying down to 33.5 parts per thousand and water temperature down to −1.8 degC). The East Iceland Current, the branch which heads southward around the east side of Iceland not only varies in strength, but has been observed to undergo a correspondingly great change of water-mass content (Malmberg and Stefansson 1969). In the warmer years before 1960, it consisted largely of Atlantic water that had come around the north of Iceland in winter. In 1965 and the following years, the cold tongue of the East Iceland Current spread farther south and east than before, particularly in 1968 and 1969 and again in 1975, and acquired the characteristics of a polar-water, ice-bearing current. (Ice had last been observed in it in 1918 and 1919 and had occurred oftener before 1890.) Figure 9.7 shows the polar water approaching the Faeroe Islands in April 1968, i.e. penetrating the middle of the broad channel between Scotland and Iceland which had been

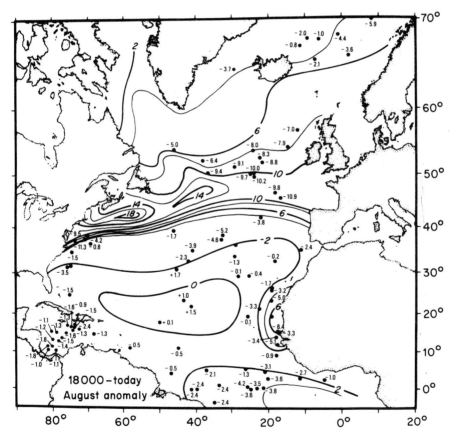

Figure 9.6 Sea-surface temperature depression (°C), below modern values, prevailing at the last glacial climax about 18,000 years ago in August. (From McIntyre *et al.* 1976.)

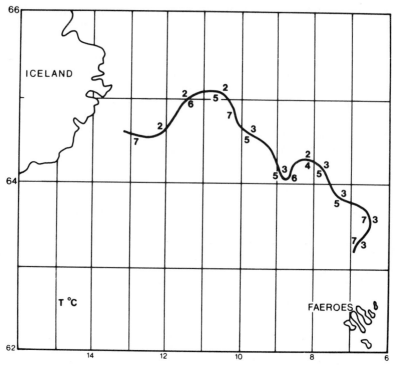

Figure 9.7 The polar front in the ocean between Iceland and the Faeroes and prevailing water surface temperatures (°C) on either side of it, surveyed by an airborne radiation thermometer, on 9 April 1968. (From Pickett and Athey 1968.)

the main path of the warm North Atlantic Drift water for most of this century.

From the Faeroe Islands we have one of the longest series of sea-surface temperature measurements in the world, starting in 1867 (P. M. Kelly, personal communication – see also Smed 1978, p. 210). This record seems to be an indicator of the interplay between the contrasting water masses in this sensitive area, even though the years since 1867 probably showed fewer and smaller penetrations of the polar water than some earlier times. The 100-year average sea-surface temperature at the Faeroe Islands was 7.7 degC. The coldest years, at the beginning (1867–9) and near the end of the series (1965–9), and 1888 and 1891, produced mean temperatures of 6.5–6.9 degC. The warmest years, 1894 and 1951, produced means of 8.5 and 8.9 degC respectively. The means for the extreme years in this recent period thus differed by 2.4 degC. The warmest and coldest five-year means since 1867 differed by 1.0 degC, a difference twice as great as occurred in the prevailing air temperatures over land in central England and in the best available estimates for the average over the whole Earth. That occasional intrusions of the polar water from the north were involved is shown not only by the temperatures surveyed (figure 9.7) in April 1968, but by the fact that an ice barrier, and advancing tongue of the polar ice, was observed both east and west of the Faeroe Islands in 1888.

Figures 9.8 and 9.9 illustrate the presently available picture (largely from Koch 1945; more recent data supplied by courtesy of *Veðurstofan Islands* (Iceland Weather Bureau) of the variations of the incidence of polar sea-ice at the coasts of Iceland over the last 1000 years and more widely over the waters here discussed in the last 200 years. The reporting may be incomplete before about 1780, and is certainly incomplete before 1600. It comes in those earlier times to a considerable extent from secondary sources of data, some of which have recently been shown to be suspect (Bell and Ogilvie 1978; T. M. L. Wigley and D. J. Underhill, personal communication). Figure 9.8 can therefore only be treated as a provisional outline of the history of the ice before 1600 and 1780.

It may be possible to survey a much longer period of sea temperature history in these regions with the aid of the records of the fisheries. In particular, the abundance of cod seems to be limited by the 2 degC water temperature isotherm. This is apparently a firm physical limitation, in that the cod's kidneys do not function in colder water, so although the optimum conditions for cod are in waters that are not very much warmer (at temperatures of 4–7 degC), cod can only occur briefly and in very small numbers in water below 2 degC (Beverton and Lee 1965). It must be a rarity to find even single specimens there. There were many years during the Little Ice Age period when the cod fishery failed at the Faeroe Islands, Svabo (1782) recounted from his visit to the Faeroes in 1781–2 verbal reports that the cod fishery did not fail so often before 1600–1650. The first hints of

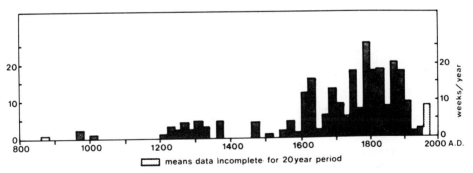

Figure 9.8 Average number of weeks a year with Arctic sea ice at the coasts of Iceland, by 20-year periods since AD 860 up to 1960–75. (After Koch 1945, updated; owing to the general scarcity of reports before AD 1600, the diagram can only be treated as a very rough indication of the variations of the ice before that date; in particular, it is now clear that there was much more ice in the late sixteenth century and it is probable that there was more in the 1680s and 1690s than Koch knew of.)

impoverishment of the Faeroe Islands fisheries were seen in complaints between 1615 and 1624 of unfair encroachment by English and Scottish fishing boats; total failures of the fishery followed in 1625 and 1629, and from 1675 onwards there were no cod for many years. Despite some improvement after about 1700, there was another total failure in 1715 and between 1709 and 1788 there were only four years (around 1750) when there was a significant surplus for export. But after 1828 (up to 1856), there was only one year, 1839, in which there was no significant export. Figure 9.10 is a map of the earliest actual sea temperature observations (data reprinted in West 1970) available in this region (shown here as departures from modern average values). It is an interesting sample, being from a slightly warmer

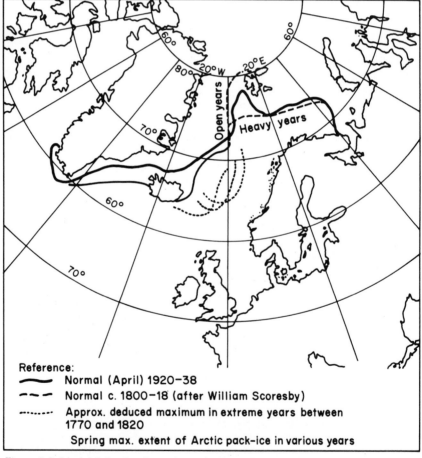

Reference:
━━━ Normal (April) 1920–38
━ ━ ━ Normal c. 1800–18 (after William Scoresby)
······· Approx. deduced maximum in extreme years between 1770 and 1820
    Spring max. extent of Arctic pack-ice in various years

Figure 9.9 Limit of the Arctic sea ice at its late-winter maximum in various periods of years.

than average summer (1789) in the Little Ice Age period. In that year, warm water from the south was reaching south-west Iceland, but conditions were colder than in the present century across the whole width of the Norwegian Sea. However, Kristjansson (1971) reports that just in the severest phase of the Little Ice Age between 1685 and 1704 the cod fishery failed also off south-west Iceland. Presumably the warm Irminger Current from the south was failing to reach so far north as any part of the Iceland coast in those years.

It may be possible to reconstruct further details than this. There are untapped Icelandic and other fishery records still in existence (Gisli Gunnarsson 1978, and personal communication) which seem to the present writer to indicate that even in the late seventeenth century the continuation of the East Iceland Current along the south coast of Iceland took the form of a tongue of cold water, whose length fluctuated greatly and which narrowed

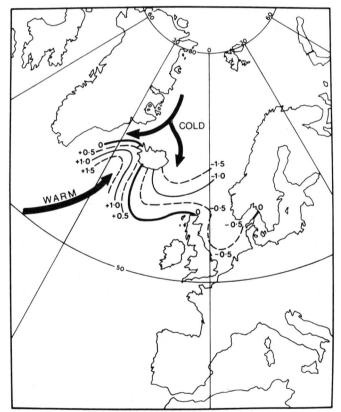

Figure 9.10 Sea-surface temperatures observed in the summer of 1789, as departures (°C) from modern average values. The arrows indicate the implied ocean current anomalies.

westwards generally to no more than 20 km in width near the south-west of the country. Cod were found by foreign vessels farther out from the coast, and we may perhaps deduce that the warm, saline Atlantic water which was so prominent in 1789 (figure 9.10) only narrowly failed to reach the south and west coast of Iceland between 1685 and 1704.

The worst years in the region seem to have been the 1690s, particularly 1695, when the sea-ice surrounded Iceland – with the exception of one point on the west coast – for some months. It is recorded that in most places open water could not be seen from the tops of the highest mountains and ships could not approach the country. From information kindly supplied by the Fisheries Research Institute (Havforskningsinstitutt), Bergen, and details of local history found in the library of the Stavanger Museum, it is now known that in that one extreme year, 1695, the fisheries also totally failed along the whole coast of Norway south to Stavanger, with the exception of a small cod stock apparently surviving cut off in the inner part of Trondheim fjord, perhaps in an isolated pocket of old North Atlantic Drift water. The implication seems to be that, after some years of increase, the polar water had in 1695 spread across the entire surface of the Norwegian Sea and south past the Faeroe Islands. In that same year, the cod fishery also became sparse in the region of the Shetland Islands (G. Manley, personal communication, 1977). The immediately following years were, however, already somewhat better in Shetland and along the Norwegian coast.

If, as seems probable, these reports imply that water temperatures below 2 degC prevailed over the years mentioned, this means that the polar water dominated the region as far south as the Faeroe Islands for periods of up to 20–30 years and was prominent for most of the time between about 1600 and 1830. Between 1675 and 1700, the water temperatures prevailing about the Faeroe Islands presumably were on overall average 4–5 degC below the average of the last 100 years, an anomaly four to five times as great as that shown by the thermometer observations of the time in central England, where the coldest decade (1690s) averaged about 1.5 degC below the warmest decades in the earlier part of this century and the 50-year mean was depressed by about 0.8 degC.

(This is a part of the background to our study in Chapter 7 (pp. 110–14) of the exceptionally cold winter of 1683–4 in Europe which led to the explanation suggested there of the extensive ice that appeared on the sea on both sides of the southern North Sea and the Channel.)

Discussion: independent evidence and controls

The great lowering of sea-surface temperatures which we deduce over a wide area extending south and east from Iceland in the seventeenth century seems to be supported by hitherto unexplained (and almost unnoticed) observations from Scotland at that time. English travellers describing visits to eastern Scotland between about 1600 and 1700 mostly referred to permanent snow on the tops of the Cairngorm Mountains (1200–1300 m above sea-level). Today there are only a few semipermanent snow-beds in this or that hollow or gully on the north-east, north and northwest slopes: Manley (1949) takes the mid-twentieth century (theoretical) snow-line in west-central Scotland (Ben Nevis) to be about 1600 m above sea-level. The apparent lowering of the snow-line by 300-400 m in the Little Ice Age would imply an average temperature level 2–2.5 degC below that of the mid-twentieth century, an anomaly rather more than twice as great as in central England, but one that could be explained by the low temperatures which we derive from the ocean within 500 km to the north. The contemporary depression of temperatures prevailing over the low ground in northern Scotland need not have been more than about 1.5 degC though about twice as great as in England, and in southern Scotland the change may well have been closer in magnitude to that established by Manley (1974) for central England.

Confirmation of the substantial lowering of temperatures in the upland areas of central and northern Scotland is found in a report (by Mackenzy), dated 1675, that describes a 'little lake' on which there was always ice, at least in the middle, even in the hottest summer: the 'little lake' is fairly certainly identified from the description as a tarn about 750 m above sea-level at 57°22'N 5°02'W, screened on its southern, eastern and western sides by high mountain walls. The present annual mean temperature there is presumably a little below the value of 3 degC, believed typical for that level in northern Scotland. So again a 2–2.5 degC lowering in the seventeenth century might explain the observation.

One must suppose that the seventeenth-century lowering of prevailing air temperatures in southern Norway below modern values was similar to that in Scotland. Matthews (1977) has used a tree-ring series from upper Gudbrandsdalen in central-southern Norway, calibrated against glacier variations and thermometer measurements in the area, to estimate the temperature variations there back to 1700, finding that the mean prevailing summer temperatures in the first quarter century (1701–25) were probably 1.6–1.7 degC below the modern (1949–63) average. There can be no doubt that in the 1690s they were still lower, probably 2 degC or rather more below present. Dr O. Liestøl of the Norsk Polarinstitutt reports (personal communication, 1977) that there are a few small glaciers on the Hardanger-vidda plateau in south Norway (e.g. Omnsbreen 60°39'N 7°30'E at 1460 to 1570 m above sea-level) which seem to have formed in the Little Ice Age and exist only as relicts of dead ice today. In Iceland (particularly eastern and northern Iceland) the overall average temperature lowering in the last quarter of the seventeenth century, 1675 to 1699, was probably greater, perhaps 3 degC at the upper levels and 2–2.5 degC near sea-level.

Recently discovered weather diaries, which include reports of days with snow lying, from 1683 to 1738 in Switzerland (Pfister 1978) may indicate that the effects spread even farther into Europe, if, as seems probable, northwesterly and northerly winds were commoner in the late 1600s than now, although they could be taken rather to indicate the general coldness of that time. The average number of days with snow lying in Zurich from 1683 to 1700 was 70 days/year (1691–1700 75 days/year) compared with 42 in the decade 1705–14 and 35 to 42 in the least snowy periods of the present century. The average was about 55 days between 1710 and 1738 and again in the late nineteenth century.

Other implications

The increased gradient of temperature between Europe near 52°N (central England) and latitudes 57° to 63°N in the northeast Atlantic region presumably explains the evidence of some greater cyclonic activity in those latitudes, in particular the occurrence of occasional peculiarly intense windstorms, during the Little Ice Age period (see Chapter 8).

The synoptic weather maps of the North Sea storms already analysed of 21–22 March 1791, 10–12 December 1792 and the summer storms reported by the Spanish Armada in 1588 may be the strongest evidence so far available for believing that some windstorms in the Little Ice Age period exceeded all twentieth-century storms in intensity. The storms in December 1703 and at Christmas 1717 seem likely to provide further evidence. In this century, the nearest equivalents seem to have been the storms and sea-floods on 31 January–1 February 1953 (with 2000 lives lost and great damage to the sea dikes in Holland and eastern England), 16–17 February 1962 (which caused 500 deaths on the German coast) and 23–25 November 1981 (with floods and coastal defence damage in Denmark and Germany and which caused two North Sea Rig platforms to break away). The gradient winds in all these recent cases were, however, 20–50 knots less than those estimated from the measured pressure gradients in the 1791 and 1792 storms. Another storm, which produced similar gradient wind strength to the others in recent years mentioned, passed across southern England on 2 January 1976 on a path similar to the 1703 storm and affected the southernmost North Sea: the damage it did in England was estimated by the insurance industry at £50–£80 million. This aspect of the climate of the Little Ice Age period in north-west Europe seems not to have received much attention hitherto, but is dealt with in Lamb (1977).

The changes here studied in the northernmost Atlantic seem to have been accompanied by an appropriate shift of the Gulf Stream–North Atlantic Drift system according to a recent historical study by Stolle (1975). Assembling observations from 1577 to 1974, Stolle deduces that already in 1577 the Gulf Stream was turning east away from the American coast on a more southerly track than has been normal in this century, although at that date one strongly marked branch of the North Atlantic Drift was reaching southwest Iceland from the south. Whatever variations took place during the seventeenth and early eighteenth centuries, for which Stolle seems to have no data, by the mid-eighteenth century the warm current was still farther south, turning east near Nantucket Island at 41°–42°N and then east–south–east to 45°–50°W where it bifurcated, one branch going north-east towards Iceland and the other branch south-east towards the Azores. (The extreme southern position established by Stolle's data was as late as 1794; further details are given in Lamb 1977, p. 512.)

Earlier development of the situation in the northernmost Atlantic

Regarding the earlier part of the last millennium much less is known, though again it seems possible that more could be established from historical records of the fisheries and ice, etc., perhaps also from studies of deposits on the ocean-bed, in fjords, and in estuaries, particularly at points where sedimentation is rapid but undisturbed by later events.

It is well known from Icelandic records that the East Greenland ice began to be more prominent about AD 1190–1200, after apparently being absent (or rarely seen) south of the Arctic Circle for several centuries before that. There is evidence to suggest (Lamb 1977, pp. 183–184) that the water temperatures in the fjords of south-west Greenland around AD 1000 at least sometimes reached values not less than 4 degC warmer than in the warmest part of the present century. And although light winds and summer sunshine doubtless played an important part in producing the extremes, it is reasonable to suppose that an anomaly of about that magnitude may have characterized the open sea areas occupied by the southernmost extension of the East Greenland Current. By AD 1250, however, the East Greenland ice constituted a formidable difficulty between Iceland and southern Green-land, and by 1342 the old sailing routes had been abandoned in favour of passages farther south. After about 1410, there was no longer any regular communication with any part of Greenland because of the ice until the 1720s, and no vessels are known to have reached the east coast of Greenland between 1476 and 1822.

The course of history seems to have been significantly different in the eastern part of the Norwegian Sea, where the fisheries off the coast of Norway seem not to have been adversely affected before about 1550. It may even be that increased southward flow of the polar water off east Greenland was compensated by increased northward drive of the warm Atlantic water up the coast of Norway. Between AD 1300 and 1500 the fisheries' records are indeed believed to hint at this. However, in the short-term fluctuations in recent years near Iceland, Malmberg reports (personal communication, 1976) that the volume transports in the Atlantic water current (Irminger Current) and the cold East Greenland Current appear to be uncorrelated. There are years in which both are strong or both weak but also years in which either one is strong and the other weak. The reports by Ljungman (1882) and Ottestad (1942) of a cyclic variation of about 50 years in length in the

Norwegian herring catches may well be related, through corresponding fluctuations of strength of the North Atlantic Drift in this region, to a quasi-periodicity of the same length (Lamb 1977, p. 451) in the occurrence of blocking patterns in the atmospheric circulation and of cyclonic situations over the British Isles.

Conclusions

The region of the north-east Atlantic here studied is of great interest because of the sharp variations of prevailing sea-surface temperatures – and the movement of ice and fish populations, etc. – that accompany any intrusion of the polar water, and because of the proximity of these changes of the marine environment to Europe. With a wealth of miscellaneous recorded data going back to the seventeenth century, it has been possible to provide estimates of the variability of temperature in the surface layer of the ocean in this region over the last 300 years. Circumstantial evidence from the nearest surrounding countries tends strongly to support the finding that the ocean surface in the region of the Faeroe Islands, and between there and Iceland, was about 5 degC colder in the 30-year mean for 1675 to 1704 than the twentieth-century average.

The ocean surface temperature variations between the warmest and coldest five-year periods even within the last 100 years have been twice as great as the variations of air temperature over central England.

The evidence alluded to of the changed environmental conditions in the countries nearest to this ocean region around 1700 appears as a testimony to the important effects of oceanographic changes upon the climate.

It should be possible ultimately to construct a far more complete picture than that given by the present study. Besides the untapped Icelandic fishery records referred to earlier, there are extensive Dutch, English and North American maritime records of the time (D. J. Underhill, personal communication). From these records it should be possible to trace changes in sea temperature and to obtain direct observational evidence of the movements of the polar sea-ice.

There is a case for attempting to assemble a collection of daily weather observation data from Iceland, Scandinavia, the British Isles, continental Europe and ships at sea for the most extreme years between about 1680 and 1740 to produce a synoptic meteorological analysis of situations at that time.

The advance of the polar water in the upper layer of the ocean south around Iceland and to the Faeroe Islands, and in the extreme year east to the whole of the west coast of Norway, which our reconstruction indicates, appears to be a minor example of a fluctuation of the kind which carried the same water mass south over the eastern Atlantic to 42° to 46°N in the last major glacial period. Lesser instrusions of the same water mass into the Faeroes/south-east Iceland region, lasting a few weeks, have been demonstrated in recent years.

There may not be many sea areas in the world where similar opportunities exist for reconstructing the conditions that prevailed in the ocean in the last 300 or 500 years, during the Little Ice Age. But this study shows that any such opportunities should be exploited and could be expected to improve our understanding of the climatic regime of those times. No less interesting would be any possibilities of tracing the North Atlantic Drift system, and the flow in and out of the Arctic, in the early mediaeval warm epoch and in the warmest postglacial times.

Acknowledgements

The author acknowledges his gratitude to the North Atlantic Treaty Organization's Scientific Affairs Division, a grant from which enabled him to visit the Iceland Weather Bureau, Reykjavik; the Arkeologisk Museum, Stavanger, Norway; the Havforskningsinstitutt, Bergen; and the Historical Institute of the University, Trondheim, where lectures on the theme of this paper led to useful discussions and the inclusion of supporting data previously unknown to the writer. He also expresses gratitude to Shell Exploration Ltd., which has supported his investigations in the Climatic Research Unit of the severest North Sea storms of the past, and to his colleagues Dr T. M. L. Wigley, D. J. Underhill and others in the Climatic Reseach Unit for helpful discussions.

References

Aagard, K. (1969) 'The wind-driven circulation of the Greenland and Norwegian Seas and its variability', *ICES Dublin 1969 Symposium on Physical Variability in the North Atlantic,* International Council for the Exploration of the Sea, Copenhagen.
Bell, W. T. and Ogilvie, A. E. J. (1978) 'Weather compilations as a source of data for the reconstruction of climate during the medieval period', *Climatic Change* 1, 331–348.
Betin, V. V. and Preobazensky, Ju. V. (1959) 'Variations in the state of the ice on the Baltic Sea and in the Danish Sound', *Trudy Gosudarstvennogo Okeanograficheskogo Instituta,* 37, 3–13 (in Russian).
Beverton, R. J. H. and Lee, A. J. (1965) 'Hydrographic fluctuations in the North Atlantic Ocean and some biological consequences', *The Biological Significance of Climatic Changes in Britain,* C. G. Johnson and L. P. Smith (eds), 79–107, London, Institute of Biology and Academic Press.
Brooks, C. E. P. (1949) *Climate through the Ages* (2nd edn), London, Ernest Benn.
Defant, A. (1961) *Physical Oceanography,* Vol. 1, Oxford, Pergamon.
Dietrich, G., and Kalle, K. (1957) *Allgemeine Meereskunde,* Berlin, Borntraeger.
Douglas, K. S., Lamb, H. H. and Loader, C. (1977) *A Meteorological Study of July to October 1588: the Spanish Armada Storms,* Norwich, CRU University of East Anglia, Climatic Research Unit Reseach Publications.
Gunnarsson, G. (1978) 'The limitations of climatology as an explanatory factor of human institutions', in *Proceedings of the Nordic Symposium on Climatic*

162 Weather, Climate and Human Affairs

Changes and Related Problems, K. Frydendahl (ed.), Danish Meteorological Institute, Climatological Papers No. 4, 27–34, Copenhagen.

Kirch, R. (1966) Temperaturverhältnisse in der Arktis während der letzten 50 Jahre. *Meteorologische Abhandlungen*, 69(3).

Koch, L.(1945) 'The East Greenland ice', *Meddelelser om Grønland*, 130(3).

Kristjansson, L. (1971) 'Ur heimildahandra a seytjándu og atjándu aldar', *Saga*, Reykjavik.

Lamb, H. H. (1967) 'Britain's changing climate', *Geographical Journal*, 133(4), 445–468.

Lamb, H. H. (1972) *Climate: Present, Past and Future. Vol. 1, Fundamentals and Climate Now*, London, Methuen.

Lamb, H. H. (1977) *Climate: Present, Past and Future. Vol. 2*, London, Methuen.

Lamb, H. H. and Johnson, A. I. (1959) 'Climatic variation and observed changes in the general wind circulation', I,II, *Geografiska Annaler*, 42, 94–134.

Lamb, H. H. and Johnson A. I. (1961) 'Climatic variation and observed changes in the general wind circulation, III', *Geografiska Annaler*, 43, 363–400.

Lamb, H. H. and Johnson A. I. (1966) *Secular Variations of the Atmospheric Circulation since 1750*, Geophysical Memoir 110, London, HMSO for Meterological Office.

Ljungmann, A. (1882) 'Contribution towards solving the question of the secular periodicity of the great herring fisheries', *United States Commission on Fish and Fisheries*, 7(7), 497–503.

McIntyre, A., Kipp, N. G., Bé, A. W. H., Crowley, T., Kellogg, T., Gardner, J. V., Prell, W. and Ruddiman, W. F. (1976) 'Glacial North Atlantic 18,000 years ago: a CLIMAP reconstruction', in *Investigation of Late Quaternary Paleoceanography and Paleoclimatology*, R. M. Cline and J. D. Hays (eds), 43–76, Geoglogical Society of America Memoir 145, Boulder, Colorado.

McIntyre, A., Ruddiman, W. F. and Jantzen, R. (1972) 'Southward penetrations of the North Atlantic polar front: Faunal and floral evidence of large-scale surface water mass movements over the last 225,000 years', *Deep-Sea Research*, 19, 61–77.

Mackenzy, Sir G. (1675) Report and notes, *Philosophical Transactions of the Royal Society of London*, 10(14), 307, in *Hutton's Abridgment of the Philosophical Transactions of the Royal Society*, Vol. II, 1672–1685.

Malmberg, Sv. A., Gade, H. G. and Sweers, H. E. (1972) 'Current velocities and volume transports in the East Greenland Current off Cape Nordenskjöld in August–September 1965', in *Sea Ice Conference Proceedings, Reykjavik 1972*, 130–139, Reykjavik.

Malmberg, Sv. A. and Stefansson, U. (1969), 'Recent changes in the watermasses of the East Icelandic Current', *ICES Dublin 1969 Symposium on Physical Variability in the North Atlantic*, International Council for the Exploration of the Sea, Copenhagen.

Manley, G. (1949) 'The snowline in Britain', *Geografiska Annaler*, 36, 179–93.

Manley, G. (1974) 'Central England temperatures: monthly means 1659 to 1973', *Quart. J. Royal Meteorological Society*, 100, 389–405.

Matthews, J. A. (1977) 'Glacier and climatic fluctuations inferred from tree-growth variations over the last 250 years, central southern Norway', *Boreas*, 6, 13–24.

Olausson, E. (1967) 'Marine sediments', in *Encyclopaedia of Oceanography*, Marine Geological Laboratory, Gothenburg memorandum, New York, Reinhold.

Ottestad, P. (1942) 'On periodical variations in the yield of the great sea fisheries and possibility of establishing yield prognoses', *Fiskeridirektoratets skrifter Serie Havundersøkelser*, 7(5).

Pfister, C. (1978) 'Fluctuations in the duration of snow-cover in Switzerland since the late seventeenth century', in *Proceedings of the Nordic Symposium on Climatic Changes and Related Problems*, K. Frydendahl (ed.), Danish Meteorological Institute, Climatological Papers No. 4, Copenhagen, 1–8.

Pickett, R. L. and Athey, G. L. (1968) 'Iceland Sea-Surface Temperature Survey, April 1968', Naval Oceanographic Office: IR 68–46, Washington DC.

Ruddiman, W. F. and Glover, L. K. (1973) 'Counterclockwise circulation in the North Atlantic subpolar gyre during the Quaternary', in *Mapping the Atmospheric and Oceanic Circulations and Other Climatic Parameters at the Time of the Last Glacial Maximum about 17,000 Years Ago*, Norwich, Climatic Research Unit, CRU RP 2.

Ruddiman, W. F. and Glover, L. K. (1975) 'Subpolar North Atlantic circulation at 9300 years BP: Faunal evidence', *Quaternary Research*, 5, 361–89.

Smed, J. (1978) 'Fluctuations of the temperature of the surface water in areas of the northern North Atlantic, 1876–1975', in *Proceedings of the Nordic Symposium on Climatic Changes and Related Problems*, Danish Meteorological Institute, Climatological Papers No. 4, Copenhagen, 205–211.

Stolle, H. J. (1975) 'Climatic change and the Gulf Stream', memorandum, University of Wisconsin, Madison.

Svabo, J. C. (1782) *Indberetninger fra en Reise i Faerøe 1781 og 1782*, Copenhagen (republished 1959 by Selskabet til Udgivelse af Faerøiske Kildeskrifter og Studier).

Sverdrup, H. U. (1945) *Oceanography for Meteorologists*, London, Allen & Unwin.

Vangeng'ejm, G. Ja. (1964) *Catalogue of Macroscopic Processes according to the Classification of G. Ja. Vangeng'ejm, 1891–1962*, M. Š. Rolotinskaya and L. Ju. Ryzakov (eds) (in Russian), Arctic and Antarctic Institute, Leningrad.

Vonder Haar, T. H. and Oort, A. H. (1973) 'New estimate of annual poleward energy, transport by northern hemisphere oceans', *Journal of Physical Oceanography*, 3(2), 169-172.

West, J. F. (1970) *The Journals of the Stanley Exhibition to the Faeroe Islands and Iceland in 1789*, Introduction and Diary of James Wright, Torshavn, Føroya Frodskaparfelag.

10

Climate and human, animal and crop diseases – an example: The Great Irish Potato Famine of the 1840s and some lessons for today

This contribution springs from a research study carried out by the author and Dr A. Bourke, formerly Director of the Irish Meteorological Service and sometime President of the World Meteorological Organization's Commission for Agricultural Meteorology, of the Irish potato famine in the last century and possible lessons to be learnt from it for the modern world. The research was carried out for the Commission of the European Communities between 1982 and 1984.

We have noticed in Chapter 4 how diseases affecting crops, and others attacking animals and the human population more directly, made parts of the fourteenth century and probably most of the fifteenth century an unhealthy period in European history, as appears very clearly from various works on the Black Death (e.g. Twigg 1984, Ziegler 1970). This also is obvious from the literature of the time even so long before the keeping of medical statistics, but seems to be substantiated by such studies as those of burial remains in England which indicate a reduction of the average length of human life from 48 to 38 years between AD 1276 and 1400 (Comfort 1966). We have discussed briefly in Chapter 4 the breakdown of the earlier mediaeval warm climate. Further details of the meteorology of the breakdown are examined in Lamb (1987). Diseases, and the vector organisms which transmit them, each have their distinctive optimum conditions for existence and for flourishing. A majority of the moulds which affect crops grow at temperatures between 15 and 30 degC, but dryness limits their growth. No mould appears to grow in stored grain that has a water content of 13 per cent or less, hence the great importance for the economy and for human health of the modern drying of cereals before storage (Matossian 1984a).

Probably the most devastatingly toxic of these fungal poisons of crops is the ergot blight (*Claviceps purpurea*), which readily infects rye and some wild grasses. Bread with only 2 per cent ergot is likely to cause an epidemic of ergotism, the disease also known as 'St Anthony's fire', which attacks the central nervous system of human beings and produces symptoms ranging from gastric disorders to paralysis and abortions, poisoning of mothers' milk and either convulsions and hallucinations, or dry gangrene leading to loss of limbs and, ultimately, to death. Matossian (1984b) has recorded how the very severe ergot infection of the rye crop in Brittany and other parts of France in July 1789 was accompanied by widespread hallucinatory panics among the French landowners and the already undernourished peasantry about brigands coming to seize property and crops: this was a complication, coming on top of a severe winter and previous dearth, which undoubtedly affected the course of the French Revolution.

The steady increase of population in Europe, which began in some countries early in the eighteenth century and has continued since, set in with the adoption of potatoes and white (wheaten) bread instead of rye and other cereals, and in Germany with control measures against the ergot danger. The change came early in Ireland with the introduction of the potato in the seventeenth century, later in Britain and later still in France. Ergotism was, however, still a problem in Russia in the early part of the twentieth century (Matossian 1986).

Another rot of the grain crops, particularly winter rye, and grasses is due to the parasite *Fusarium nivale*, which is active under a snow cover, particularly when this lies longer into the spring than usual. This can cause crop losses and damage the seed. It is troublesome in Scandinavia and north Germany (where it is known as *Schneeschimmel*) and in agriculture at the highest levels in the Alpine region. In the Little Ice Age period, it occurred in the main agricultural regions at lower levels in Switzerland and led to harvest failures, notably in 1709 and 1740, and the slaughter of cattle (Pfister 1975, p. 115). *Fusarium* toxins in fodder are directly poisonous to livestock, liable to produce haemorrhages and nervous disorders and tumours in surviving cattle and their progeny even many years after the initial fatalities and the weather which caused them (Schoental 1981). It seems likely that in the occasional wet, cold summers of recent years (e.g. 1974, 1980 and 1985 in western Europe, the last-named being particularly wet in Ireland, Wales, western England and south-west Scotland), there would have been substantial losses of livestock due to cattle 'murrains', as in the late Middle Ages, and sheep liver-fluke but for modern veterinary medicine (P. A. Tallantire, personal communication, 17 October 1985).

A climatic sequence which clearly affected the incidence of diseases

From soon after 1700, the cold climate development of the Little Ice Age seems to have begun to wane. (The dates of the climax phase of the Little Ice

Age show an interesting correspondence to the prolonged minimum of sunspot activity, the 'Maunder minimum', from about 1640 to 1715.) But there were many setbacks, such as the cold year 1740 followed by a couple of famine years in parts of Europe, the wet European summers of the 1750s and 1760s, and the prominence of the polar sea-ice in the Iceland region sustained from 1780 to about 1840, coinciding with another minimum of the long-term average temperature level in Europe. In northern Europe, the glaciers generally were at their maximum extent between the late 1600s and about 1730 to 1780. These dates also seem to apply to the Canadian and Alaskan Rocky Mountain glaciers. But in the Alps, in Chile and perhaps more generally in the southern hemisphere, the greatest maxima of the glaciers seem to have been registered (from dating of their moraines) around 1820 or 1850 – albeit this maximum was more or less matched in the case of the Alps around 1600–50. Some of the setbacks to the recovery of the prevailing temperature level, including also the cold years in Europe and eastern North America between 1836 and 1843–5, seem to have coincided with screening of the sun's radiation by dust and aerosol veils maintained in the atmosphere by notable successions of great volcanic eruptions. Most of

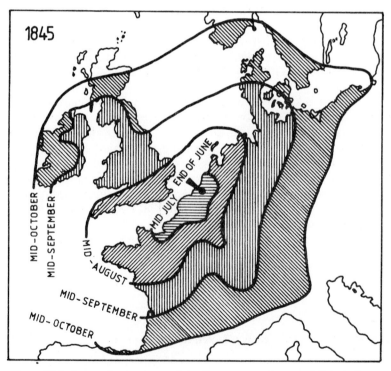

Figure 10.1 Approximate dates of the first reports of potato blight in Europe in 1845.

these setback phases, like the cold climax around the 1690s, seem to have been followed by sudden warmings in the record of temperatures measured in England and western Europe (e.g. central England overall average 1690–99 8.1 degC; 1702–9 9.3 degC; 1808–17 8.7 degC; and 1821–28 9.6 degC), including at least one or two individual years of remarkable warmth and southerly wind or anticyclonic character in western and central Europe. One of the founding members of the Royal Meteorological Society, E. Sabine (1846), recorded that the Gulf Stream water advanced much nearer to the coasts of Europe in 1845–6 than had been usual in those times. Many parts of Europe endured famines due to failures of the grain crops in the 1690s (notably Scotland and Scandinavia – see Lamb 1977, 1982), 1740–2 and 1780–2 (the same regions), and in 1816–18 (very widespread – see Post 1973). Ireland may have been spared the disaster of the 1690s because of the very early adoption of the potato there to meet the problem of a country where the summers were generally rather too wet and cool for wheat to ripen well. The sturdy root became known in the eighteenth century as 'Irelands' bulwark against famine' and was spoken of elsewhere in Europe as 'the Irish potato'. So Ireland's population began to grow earlier and faster than the population of Britain and other neighbouring countries. There was, however, one bitter lesson of inexperience along the way. In the usually mild Irish climate, and especially in the mild period between about 1717 and 1738, the Irish poor acquired the habit of not harvesting the potatoes in the autumn, preferring to dig them out of the ground as needed during the winter. Bourke (1984) records that the exceptionally bitter winter of 1739–40 put a stop to this. There was such mortality among the peasantry – almost rivalling the famine 100 years later – that 1740 became known as 'the year of the slaughter'. From 1741 on the potatoes were harvested and safely stored either in pits or in clamps.

After that, the population increase in Ireland continued and, by 1845, the situation had become precarious to the point of being ripe for disaster: the total number of souls living in the island had reached 8.5 million, an overwhelmingly rural population living on 700,000 farms, nearly half of which (45 per cent) were of less than 3 acres (1.2 hectares). The potato was by that time the only crop capable of producing the bulk of food needed to feed the families living on such tiny plots (see Bourke 1964, 1968).

The European potato disaster in the 1840s

The potato blight (*Phytophthora infestans*) which struck in the 1840s had not been known in Europe before. It was first reported in Belgium in June 1845 and seems to have spread from there in all directions, reaching all southern England and northern France in August and Ireland in September of that year. By October, it had arrived in every part of Ireland, southern Scotland, and south and southeast Norway, and went on to reach Poland, southern France and northern Italy (figure 10.1). Some contemporary accounts may

.nts may now be seen to indicate that the disease first appeared in
pe in potatoes grown in western Flanders in 1844, where new varieties
irted from the Americas were being tried out, in order to counter the
u.c..ining vigour of those previously grown in Belgium. The exact origin of
the trial potatoes imported is uncertain. The varieties named in the first
reported attacks of the blight seem to have been Peruvian, but the blight first
appeared in the north-eastern states of the USA in 1843, and the
consignment of potatoes used for the Belgian trial could have come from
there. By 1845, an area from Pennsylvania to the Canadian maritime
provinces and the Great Lakes was also infected.

The spread of the disease during the summer of 1845 is readily explained
by the winds. An example of a situation in July of that year with easterly
winds all the way from the Gulf of Finland, and across the Netherlands and
Flanders, to the British Isles and the Atlantic is shown in figure 10.2.

There was weather suited to development of the blight that summer in all
the countries where the blight was later reported, including in the case of
Ireland in many weeks before the blight first occurred. Presumably the
infection had not yet arrived at that point. From laboratory study (Crosier
1934), the minimum conditions required for development of the blight are:

  (i) a period of at least 12 hours with relative humidity always 90 per cent or
       above and air temperature never below 10 degC;
 (ii) free water on the potato leaves for at least a further 4 hours.

Condition (i) allows the formation of disease spores outside the leaves of
infected plants and their transport by air currents to new sites, while
condition (ii) permits newly arrived spores to germinate and produce further
infection. These observations are the basis of a potato blight warning system
now used in Ireland and of the reckoning, for comparative studies, of the
total duration of Effective Blight Hours in different areas and different
years.

In the spell of weather sampled by the map (figure 10.2), conditions over
central Europe were warm, with high humidity, night fogs and scattered
thunderstorms by day; this, then, would be an ideal blight development
situation wherever the infection had arrived. The fact that the blight never
reached the northern half of Scotland in 1845, and that farms in Ireland
which were high enough up in the hills to have significantly lower
temperatures than the low ground escaped infection, can certainly be
explained by the frequency of northerly winds and the coldness of that
summer in those regions. This has been paralleled by the decline of potato
blight in Iceland in recent decades, after it had given serious trouble there in
the warmer climate during the 1930s, 1940s and early 1950s. The temperature map for September 1845 (figure 10.3) bears witness to the cold
influence of repeated northwesterly winds over the British Isles–North Sea
region in that month. In July and August 1845, there had been temperature

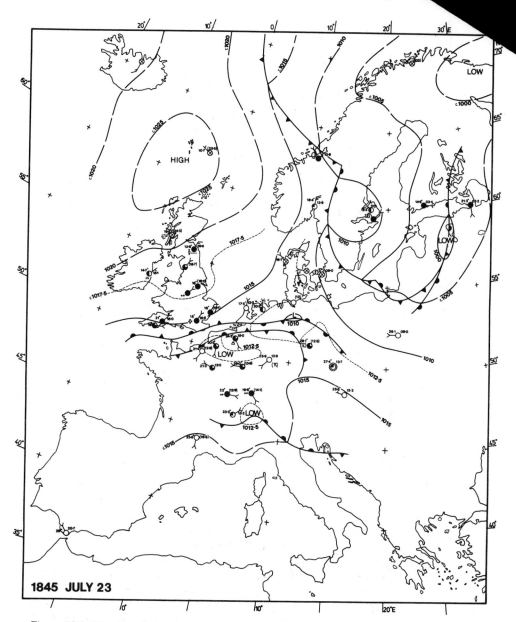

Figure 10.2 Weather situation over Europe at about 2 p.m. on 23 July 1845, with easterly or northeasterly breezes all the way from Finland, across the Low Countries and England to Ireland.

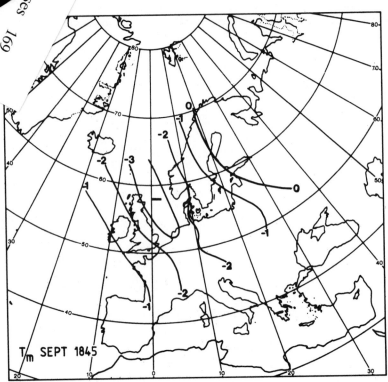

Figure 10.3 Mean temperatures prevailing in September 1845, as departures in degC from the 100-year averages, 1851–1950.

deficiencies of 2 degC and more below the 1851–1950 average centred over the British Isles.

The following year was different [A historic mild winter doubtless helped to keep the infection alive in any diseased potatoes of the previous season left lying in the fields, and in those stored in clamps rotting was found to have continued. Very warm seasons followed in the spring, summer and autumn of 1846: in England, it was among the four warmest years, and produced the second warmest summer, of the nineteenth century. June was as much as 3–4 degC warmer than the 100-year average nearly everywhere in Britain and Ireland. After that, the weather broke down on the western fringe of Europe, giving frequent rains in Ireland, Scotland and Norway, and the potato blight was much more serious in those regions than in 1845. In the Hebrides and west and north-west Highlands of Scotland, the situation became almost as bad as in Ireland; by the end of 1846 it is estimated that three-quarters of the crofting population were without food (Hunter n.d.). In Norway, the situation seems to have been less serious because of more diversity of crops and other food resources. Figure 10.4 shows the

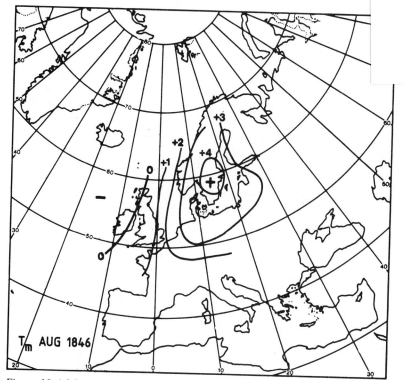

Figure 10.4 Mean temperatures prevailing in August 1846, as departures in degC from the 100-year averages, 1851–1950.

temperature pattern for August 1846. In Dublin, there were 24 days with rain in that month, and 64 between July and September inclusive. In London, the corresponding figures were 16 days with rain in August and 39 between July and September. Most of Europe was freer of potato blight than in 1845, but there were occasions when humid air and thunderstorms penetrated far east into the continent and some districts were quite seriously affected by the blight.

The disaster was by far worst in Ireland because of the population development mentioned earlier. Over 1 million people in Ireland died of starvation or of diseases encouraged by malnutrition* over the next few years, the blight continuing. Many more emigrated. By 1851, the population had fallen by a quarter; before very long it had dropped by a half, and so remained until at least 1950. It has never again approached the 1845 level.

The analysis on behalf of the Commission of the European Communities (Bourke and Lamb 1984, unpublished), showed that although the years

* In this connection, Galloway's (1985, 1987) analyses of the diagnosed causes of deaths in years of malnutrition in London from 1670 to 1830 are interesting.

concerned in the 1840s produced much weather in Ireland and western Europe that was conducive to blight, those years were neither unique nor extreme in that respect. The weather in 1958 and 1981, in particular, was more frequently favourable to blight in various parts of Europe, including Ireland and the Low Countries, producing up to 390 effective blight hours against highest totals in Ireland of about 300 in 1846. (Brussels with 360 EBH had more in 1845 than Ireland in either 1845 or 1846.) The disaster in the 1840s had to do with various factors besides the overpopulation of rural Ireland. The varieties then grown in Ireland, especially the watery Lumper, had been chosen for their high yields and seem to have been especially vulnerable to the disease (Bourke 1971). Moreover, everyone concerned was taken by surprise, faced with a previously unknown or new disease attacking what was in some regions the only crop. In recent summers, with much 'potato blight weather', the situation has been eased because more resistant varieties of potato are now grown and crops are sprayed when blight warnings are issued.

Problems of the present day

This case study shows how a continent-wide epidemic of crop disease was weather-dependent, though its major social impact was controlled by other factors which made Ireland's people particularly vulnerable. The results are important to this day: a greatly reduced population in Ireland and great Irish communities established in the USA and several other countries.

There are also lessons in this for the world today and how future climatic vagaries might affect agricultural yields. The risk of serious damage by plant parasites has increased over recent decades precisely as a result of practices aimed at increasing production. Perhaps the greatest danger lies in devoting huge areas to a single crop, and even to a single variety, instead of the much more diverse patterns of the agriculture of the early twentieth century. This is a return to the one-crop economies ('monoculture') that were smitten with the worst harvest failures and famines of past centuries. What it can do in a modern setting was illustrated in the Netherlands with a newly developed strain of wheat, known as Heines VII. Such was the success in the yields of this strain, and its resistance to the then known forms of yellow rust disease, that three years after its introduction in 1952 80 per cent of the wheat sown in the Netherlands was Heines VII. Then a new variety of the yellow rust appeared and destroyed two-thirds of the Dutch winter wheat sown for the 1956 harvest. Thus the rapid and widespread adoption of new strains of seed prepared the way for a weather-induced disaster. This was, in essence, a new disease in the same sense as the potato blight in the 1840s. Such things could occur anywhere in the world, but seem most liable – depending on the agricultural practices used – to lead to disasters in poor countries of the Third World, or on poor soils. They seem also to be a distinctive risk for extensive regions of similar terrain, such as the great plains of North and

South America, the Ukraine and central Asia, the African savanna belts, and parts of China and Australia.

Acknowledgement

I am indebted to Dr Austin Bourke for much valuable information collected in the course of our collaborative study of the Irish Potato Famine for the Commission of the European Communities. His knowledge of the situation in Ireland had already been built up from a collection of documentary reports and published writings of the time, including a generous appraisal of the relief measures attempted by the authorities then ruling and notably by the Society of Friends (Quakers).

References

Bourke, P. M. A. (1964) 'Emergence of potato blight, 1843–46, *Nature*, 203, 805–808, London.

Bourke, P. M. A. (1968) 'The use of the potato crop in prefamine Ireland', *Journal of the Statistical and Social Inquiry Society of Ireland*, XXI, 72–96, Dublin.

Bourke, P. M. A. (1971) 'Disease resistance of European potato varieties in the 1840s', Science, 171, 955–956, Washington, DC.

Bourke, A. (1984) 'Impact of climatic fluctuations on Europoean agriculture', *The Climate of Europe: Past, Present and Future*, H. Flohn and R. Fantechi (eds), Dordrecht, Reidel for Commission of the European Communities, 269–314.

Bourke, A. and Lamb, H. H. (1984) 'The spread of potato blight in Europe in 1845–6, and the accompanying weather patterns', Commission of the European Communities, Contract No. CLI 065 UK (H) (unpublished).

Comfort, A. (1966) *Nature and Human Nature*, London, Weidenfeld and Nicolson.

Crosier, W. (1934) 'Studies in the biology of *Phytophthora infestans* (Mont.) de Bary', Cornell University Agricultural Experimental Station, Memoir No. 155.

Galloway, P. R. (1985) 'Annual variations in deaths by age, deaths by cause, prices, and weather in London 1670 to 1830, *Population Studies*, 39, 487–505.

Galloway, P. R. (1987) 'Population, prices and weather in pre-industrial Europe,' Berkeley, California, University of California Ph.D. thesis, Feb. 1987. Partly published as 'Long-term fluctuations in climate and population in the pre-industrial era', *Population and Development Review*, 12(1), 1–24. (1986.)

Hunter, J. (n.d.) *The Making of the Crofting Community*, Edinburgh, John Donald.

Lamb, H. H. (1977) *Climate: Present, Past and Future. Vol. 2, Climatic History and the Future*, London, Methuen.

Lamb, H. H. (1982) *Climate, History and the Modern World*, London, Methuen.

Lamb, H. H. (1987) 'What can historical records tell us about the breakdown of the medieval warm climate in Europe in the fourteenth and fifteenth centuries – an experiment?' *Contributions to Atmospheric Physics*, 60(2), 131–43.

Matossian, M. K. (1984a) 'Mold poisoning and population growth in England and France, 1750–1850', *J. Economic History*, 44(3), 669–686.

Matossian, M. K. (1984b) 'The time of the great fear', *Sciences*, 38–41, New York Academy of Sciences.

Matossian, M. K. (1986) 'Climate, crops, and natural increase in rural Russia, 1861–1913', *Slavic Review* 45(3), 457–469.

Pfister, C. (1975) *Agrarkonjunktur und Witterungsablauf im westlichen Schweizer Mittelland 1755–1797*, Bern, Geographisches Institut der Universität.

Post, J. D. (1973) 'Meteorological historiography', *J. Interdisciplinary History*, 3(4), 721–732. Cambridge, Massachusetts.

Sabine, E. (1846) 'On the cause of the remarkably mild winters which occasionally occur in England', *London, Edinburgh and Dublin Philosophical Magazine and Journal of Science* (April).

Schoental, R. (1981) 'Relationships of *Fusarium* toxins to tumours and other disorders in livestock', *J. Vet. Pharmacol. Therap.*, 4, 1–6.

Twigg, G. (1984) *The Black Death – a biological reappraisal*, London, Batsford.

Ziegler, P. (1970) *The Black Death*, Harmondsworth, Penguin.

11

The recent increase of storminess in and around the North Sea region and related changes since the twentieth-century climatic optimum

The now well-known twentieth-century warming of climates seems to have reached its main climax in the first half of the century, around the 1930s and 1940s, a sort of climatic 'optimum', at least in the northern hemisphere. The great prevalence of relatively simple, zonal westerly, wind patterns over middle latitudes, with which its development was associated, culminated earlier, in the first three decades of the century, particularly the 1920s in the northern hemisphere, though the North Atlantic westerlies returned to a later, somewhat lower, peak again around 1950. This is the commonest individual wind pattern that can be recognized in this part of the world, probably at all times since the last ice age, but its frequency undergoes significant variations. An apparently corresponding climax of the prevailing westerlies over the middle latitudes of the southern hemisphere is shown by the indices used by Lamb and Johnson (1959, 1961, 1966) around 1900–25. It is the overall average temperature level at any one time, and the temperature difference between tropics and polar regions, that is fundamental to the condition of the climate. But the strength and patterns of the global wind circulation may provide the readiest index of it.

The warmest run of summers of the century in Europe was registered by the average temperatures for 1933-52 (von Rudloff 1967), and the highest level of the overall yearly average temperature was about that time. The warmest run of winters came in the decades between 1910 and 1930. There was another striking run of mild winters from 1971 to 1976 in England, and which continued even two years longer in central Europe to 1978. Figure 11.1 shows how these variations are registered in the ten-year averages shown by the very long record of prevailing temperatures in cental England from 1659 (homogenized by Manley 1974), updated to 1985. A corresponding succession of mild winters in eastern North America in the 1970s was

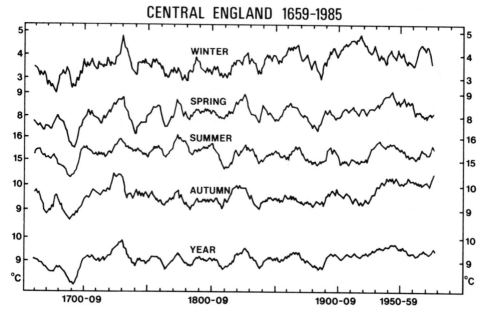

Figure 11.1 Temperatures prevailing in central England: ten-year averages for each season and for the year, from the work of the late Professor Gordon Manley, updated in the Climatic Research Unit, Norwich, in collaboration with the Meteorological Office:

Winter is taken as	December, January and February;
Spring,	March, April and May;
Summer,	June, July and August
Autumn,	September, October and November.

briefer and was followed from 1977 by a sequence which showed at least three outstandingly cold winters and three very mild ones.

The decline of temperature since the 1930s and 1940s peak has been less than the rise over the previous 50 years, and the twentieth century remains warmer than the previous two centuries. This seems – e.g. from the evidence of glaciers and Arctic sea-ice, as well as thermometer measurements – to be true for the world as a whole. It is still difficult, and probably always will be, to get an accurate survey of temperatures for the whole Earth by more direct means because of the great ocean areas (71 per cent of the whole, 81 per cent of the southern hemisphere) and the error margins to which all methods of sensing ocean temperatures from the air are liable. In England, the 1900-84 temperature average is 0.8 degC above that derived for the late seventeenth century (1660-99) and about 0.25–0.3 degC above the averages of the eighteenth and nineteenth centuries respectively. In parts of the southern hemisphere – e.g. New Zealand and especially the Antarctic – the warming was very marked from the 1930s to the 1970s. Indeed the oceans of the southern hemisphere seem to have continued warming into the 1980s, although it is difficult as yet to be sure how far this impression may be due to the exceptional magnitude of the El Niño warming of the equatorial Pacific in 1982–3. The El Niño, as already explained elsewhere in this book, is a cyclic phenomenon of the ocean circulation which affects the wide Pacific, producing temporary warmth while the cycle lasts over the vast area of the equatorial Pacific, particularly just south of the line. The 1982–3 warming was greater than previously experienced and was sufficient to raise the overall average surface temperature of the oceans of the world by an

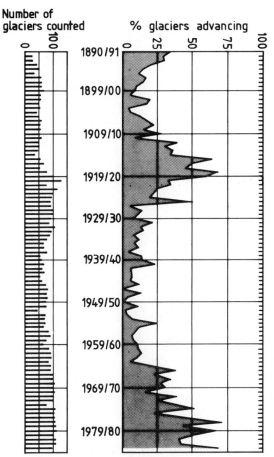

Figure 11.2 Changes of position of the ends of the glaciers in the Swiss Alps, 1890–1984.

    Adapted from a diagram published by M. Schüepp and G. Gensler in *Geographica Helvetica*, 1986, Nr. 1, based on the 105th Report of the Swiss Glaciers Commission of the Naturforschende Gesellschaft.

Figure 11.3 The Upper Grindelwald glacier, Switzerland, seen in June 1986, after more than 20 years of advance which have reconstituted the ice-fall in the picture.

The advance is reported to have begun around 1960 and has often averaged 40 to 50 m a year (in 1972, 120 m).

estimated 0.3–0.4 degC. It was (inevitably?) followed by some cooling of the world average.

In Europe, including England, the Octobers (and, as a result, the whole autumn averages) are generally still at or near their warmest level. How far the net warming of climate in the twentieth century should be attributed to the increase of carbon dioxide in the atmosphere is not clear. It seems to be at least partly due to natural processes, notably a long lull in volcanic activity. And the recent cooling is almost certainly from natural causes. We shall return to these matters in Chapter 18. The recent cooling of the northern hemisphere means, for example, that the latest ten-year average (since the mid-1970s) temperatures in England are 0.25–0.3 degC below the 1930s and 1940s level, mainly because of cold springs. The winters have also been generally colder than the earlier twentieth century peak around 1920. In Iceland and the east Atlantic-European sector of the Arctic, the deterioration is greater and has affected agriculture as well as the ranges of fish and bird species. In Europe, the glaciers have mostly ceased their long recession and gone over to advance since about 1965, as illustrated in the case of Switzerland in figures 11.2 and 11.3.

There are other developments which are very hard to reconcile with the concept of a world warming under the influence of carbon dioxide. Using the

Table 11.1 Months in which most of the Arctic in the North Polar region of 70°N was warmer or colder than the 1931–60 averages

Year	Number of months		Year	Number of months	
	warmer	*colder*		*warmer*	*colder*
1956	4	2	1970	0 or 1	8 to 9
1957	4	3	1971	1 or 2	7 to 9
1958	1	6 or 7	1972	1 or 2	7 to 8
1959	7	3	1973	0 or 1	7
			1974	1 or 2	6
1960	4	5	1975	0 or 1	8 to 10
1961	1	8	1976	1 or 2	6 to 8
1962	0	6	1977	2	6 to 7
1963	0	9	1978	0 or 1	7 to 9
1964	0	11	1979	0 or 1	11
1965	1 or 2	7 or 8			
1966	1	8 to 11	1980	5	6 or 7
1967	1 or 2	7 to 9	1981	5	4 or 5
1968	0	10 to 12	1982	0	8 to 10
1969	2	6	1983	0 to 2	7
			1984	3 to 5	5 to 6
			1985	5	5
			1986	2	6
					(to August)

hemispheric maps published each month in the *Berliner Wetterkarte* and *Grosswetterlagen Europas*, issued by the Deutscher Wetterdienst, Frankfurt am Main, a count year by year of the number of months in which most of the Arctic area north of latitude 70°N showed temperatures above or below the 1931–60 average yields the figures in table 11.1. This reveals a more than four-to-one preponderance of colder months in the Arctic from 1960 to 1986 compared with the previous 30 years. Professor H. Flohn of Bonn published a similar series covering the years from 1949 to 1978 based on more precise analysis of the upper air temperature over all the Arctic regions north of 65°N. He found an almost identical result from analysis of the upper air temperatures everywhere north of latitude 50°N (see diagram in Lamb 1982, p. 259). The temperature prevailing after 1961 never approached the levels maintained in ten of the previous 13 years. There may have been a suggestion in these figures that the Arctic has been coldest in, or just after, the years when the equatorial Pacific has been warmest: a suggestion in keeping with the correlation between the Southern Oscillation and anticyclones in high latitudes which was first noted many years ago. According to a diagram published by Newell and Chiu (1980), the El Niño years of the last 30-year period were 1957–8, 1961. 1963–4, 1965, 1969–70, 1972, 1976–7 and, of course, 1982–3.

The discrepancy between the figures here shown and the findings of Wigley *et al.* (1985) in detecting the climatic effects of increasing carbon dioxide that 1981 and 1983 were the warmest years ever known at the Arctic as a whole (and for the northern hemisphere generally) presumably has to do with the notably high temperatures reported in certain areas indicating intense positive temperature departures there. This may be a feature of the most modern stations in the Arctic, and perhaps elsewhere, associated with the artificial warmth generated locally by the camp surrounding the observation site. See also Wigley (1986). Some doubt over the findings of a recent warming of climates seems, however, to be thrown by comparisons carried out between the temperature trends indicated by the highest grade of meteorological stations in the United States, which are mostly at the airports of large cities, and the more or less opposite recent trends revealed when one considers all available weather service stations in every sort of location, five thousand in number as compared with the three hundred high grade stations. The latter could be thought liable to artificial sources of heat. These comparisons were published by Kukla *et al.* (1986). They suggest there may be a case for re-assessment of the alleged northern hemisphere trend of recent years.

The present writer has drawn attention to the remarkable incidence around the world of seasons with great extremes of one sort or another since 1960 (Lamb 1982, p. 257): up to 1986 the cases marked by extremes of cold outnumbered the extremes of warmth by about two to one. And in the centre of the North American continent, at one sample station where the temperature records have been closely analysed, namely, West Lafayette,

Indiana (Agee 1982; cited also by Watt 1986), a similar trend is seen. From 1901 to 1918, there were about equal numbers of new cold and warm extremes. Between 1919 and 1946, which was the great period of warming of the northern hemisphere (and particularly the Arctic), new warm extremes

Table 11.2   Correlation coefficients indicating associations in world weather (Lamb and Mörth 1978)

Items and period surveyed	Units of time (years)	Number of values in series compared	Correlation coefficient	Significance level (%)
Iceland ice and British Isles westerly days, 1861–74	5	23	−0.54	99
Iceland ice and days with SW surface wind at London, 1780–1976*	10	20	−0.47	95
Iceland ice and central England temperature, 1870–1974	5	21	−0.53	about 99
Iceland ice and northern hemisphere temperature, 1870–1974	5	21	−0.61	99
Iceland ice and world temperature, 1880–1974	5	19	−0.64	99
Days with SW surface wind at London and central England temperature, 1730–1976*	10	25	+0.39	about 95
Central England temperature and Northern hemisphere temperature, 1870–1974	5	21	+0.71	99.9
Central England temperature and world temperature, 1880–1974	5	19	+0.67	99
British Isles westerly days and world temperature, 1880–1974	5	19	+0.31	—
British Isles westerly days and rainfall in the Sahel zone of Africa (10°–20°N), 1900–73†	1	74	+0.56	99
Days with SW surface wind at London and snow at the South Pole, 1760–1957*‡	10	20	+0.75	99.9

* Incomplete final decade taken as if complete.
† From Winstanley (1974).
‡ Data from Giovinetto and Schwerdtfeger (1966).

outnumbered cold extremes by 2½ to 1. But from 1946 to 1980 the position was reversed and cold extremes outnumbered warm cases by more than two to one. See, however, the postscript on p. 192.

We can use the number of days each year on which general westerly winds blow across the whole British Isles as an index of the development of the middle-latitudes westerlies over the northern hemisphere. These islands are in the middle of the zone of prevailing westerly winds, but here in the sector between Greenland and Scandinavia, along with the Alaskan–Canadian sector, is where other patterns – the so-called blocking patterns – are most liable to develop. There seems to be a tendency to parallelism between the changes of prevalence of the westerlies and the general temperature level in England and Europe, in the northen hemisphere and, perhaps, over most of the world. The linkages seen in the list of correlation coefficients in table 11.2 seem to indicate that the state of the westerlies in the north-east Atlantic–British Isles region may be an index of the more general condition of world climate. Links of the westerlies with the amount of Arctic sea-ice reaching Iceland, with prevailing temperatures in England and possibly the world average, and with rainfall in Africa and snowfall at the South Pole, are apparent in the list.

We see from the long record of the yearly frequency of westerly wind days over the British Isles (see figure 6.5, p. 95) that there has been a great decline of the number of westerly days since 1950, which corresponds, at least roughly in its timing, with the cessation of the main twentieth-century warming but is far more marked than the amount by which prevailing temperatures have fallen. There have been increases in a variety of other wind patterns in the neighbourhood of the British Isles, all of them representing less mobility of the travelling weather systems than with the westerly days – in other words, corresponding to various degrees of 'blocking'. The most notable increases have been in different years in the frequencies of northwesterly, easterly and southerly winds and changes in the frequencies of anticyclonic and cyclonic centres over the British Isles (see figures 11.4 and 6.5).

For the details, table 11.3 surveys the frequencies of the main wind-flow patterns recognized over the British Isles, showing how recent decades have differed from the experience of the previous 113 years, 1861-1973 (the figures in brackets show how the average numbers of each wind pattern per year have differed from the previous long-term average). In the 1960s the compensation for the fewer westerly days was made up of modest increases of NW to N, easterly, cyclonic and unclassifiable (U) days. In the 1970s the biggest increase was in northwesterly days. But in the 1980s so far, the most notable increases have been in cyclonic and unclassifiable days, with a possible noteworthy increase also of southerly days. This latter detail could be linked to events just 200 and 400 years ago, to be mentioned later in this chapter.

With the decline of the westerly-type situations and increased frequency of other wind patterns have come changes in the distribution of rainfall over the British Isles, with decreases in the west – a noteworthy incidence of drier summers in the western Highlands of Scotland, where the 30-year average summer rainfall, 1951–80, was more than 10 per cent below that in 1916-50, and also in Wales, south-west Scotland and part of the English Lake District – and some increase in rainfall in the east and north-east (e.g. over 10 per cent increase in Shetland). The summer rainfall 1951–70 showed increases of 10 per cent and much more over much of central, eastern and extreme southern England attributable to increased thunderstorm and convective activity in the lighter wind situations prevailing. But after 1972, the frequency of anticyclones led to mostly dry summers there also. For the year as a whole, a similar pattern of rainfall changes, though somewhat weaker, could be recognized already in the mid 1960s. The rather cold, wet summer of 1985 in the British Isles, apart from Shetland where it was fine, seemed the more exceptional, being a break from recent remembered experience. There have been important rainfall changes elsewhere in the world as well, e.g. in Africa, which we shall deal with in Chapter 12.

Changes in the frequency of different wind circulation patterns, such as discussed here, naturally affect other things besides the temperatures and rain- or snow-fall. Among the other things that change are the incidence of storminess – as we saw in Chapter 6 – and the drift of pollution and, of course, of acid rain. My colleagues, Davies and others (1986), at the University of East Anglia have demonstrated that the acid deposit from precipitation in south-west Scotland is less in years with more frequent westerly winds, although the rainfall is high, and is greater with cyclonic situations which circulate air from the industrial areas in England and on the continent round over the area. And in the cold January of 1987 the east winds brought west such high concentrations of pollution from the lignite (brown coal) fires and factories of eastern Germany that people in the cities of west Germany were requested to shut their windows and stay indoors and polluted snow, grey in colour, fell in eastern England.

Table 11.3 Where is the compensation for the decline of the westerlies in the British Isles/north-west Europe region?

Years	Average numbers of days per year							
	W	NW	N	E	S	AC	C	U
1861–1973	93	17	27	28	31	91	64	14
1960–9	80	19	29	34	31	84	69	19
departures	(−13)	(+2)	(+2)	(+6)	(0)	(−7)	(+5)	(+5)
1970–9	73	26	27	31	34	94	64	16
departures	(−20)	(+9)	(0)	(+3)	(+3)	(+3)	(0)	(+2)
1980–5	67	19	23	25	37	93	79	22
departures	(−26)	(+2)	(−4)	(−3)	(+6)	(+2)	(+15)	(+8)

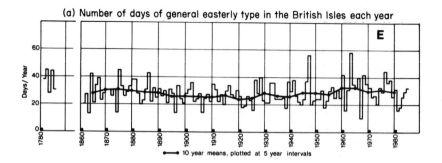

(a) Number of days of general easterly type in the British Isles each year

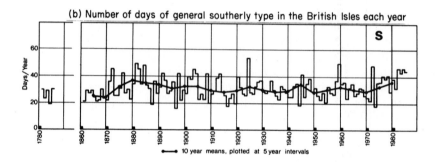

(b) Number of days of general southerly type in the British Isles each year

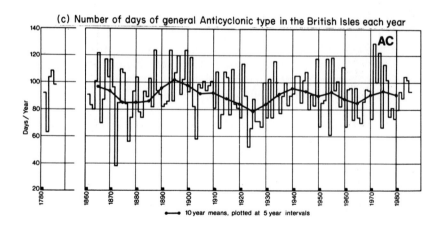

(c) Number of days of general Anticyclonic type in the British Isles each year

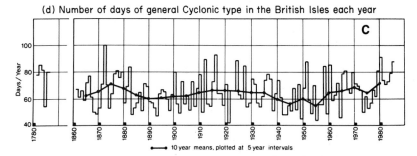

Figure 11.4 Number of days each year with: (a) general easterly wind situations; (b) general southerly wind situations; (c) anticyclonic centres or generally anticyclonic situations; (d) cyclonic centres or generally cyclonic situations, over the British Isles.

For the frequencies of days with other wind-flow patterns over the British Isles see figure 6.5.

There has been an increased incidence of various types of extreme in these recent decades of more frequent 'blocking' – e.g. increases in the occurrence of very dry and very wet, very warm and very cold, months and seasons in the British Isles (as illustrated in Lamb 1982, pp. 255–6). The succession of winters in North America in the 1970s and 1980s, already referred to, provides another example. These occur when the controlling weather systems remain for periods of many days or weeks without much movement. As in previous centuries, despite the warmth of the twentieth century, there has also been a greater incidence of cold winters (with mean temperatures in England below about 3 degC) in the 1940s and later decades than in the earlier part of this century. Tables 11.4 and 11.5 show this tendency, which hints at an underlying 100-year periodicity of which there is other evidence in climatic records. Months with mean temperature in England below 0 degC have been rare in the twentieth century, but occurred in 1940, 1947, 1956, 1963 (twice), 1979 and 1986. There may indeed be some virtue in admittedly simplistic forecasts based on these apparently cyclical recurrences.

Any such forecasts must be made and used with great circumspection, for it is clear that the so far unidentified influence* producing this shadowy cyclical tendency is only one of many influences on the climatic development (and is also not precise to the individual year). Moreover, some things are clearly different in the twentieth century. As an example, most of the previous centuries since the Middle Ages seem to have produced more wet summers in England and Wales in the second half. In the eighteenth and

---

* It is possible that the apparent roughly 100-year periodicity should be seen as a 'harmonic' (sub-multiple) of the 200- year and 400-year cycles (of presumed solar origin), discussed later in this chapter.

Table 11.4 Coldest winters* of each century in western Europe, from the compilation by C. Easton (*Les hivers dans l'Europe occidentale*, Leyden, Brill, 1928)† updated

Century	3rd	4th	5th	6th	7th	8th	9th	10th	11th	12th	13th	14th	15th	16th	17th	18th	19th	20th
	296 (10)	359 (10)	401 (21)	545 (21)	695 (10)	760 (21)	822 (10)	913 (10)	1033 (21)	1125 (10)	1205 (10)	1303 (21)	1408 (4)	1511 (21)	1608 (4)	1709 (4)	1830 (4)	1917 (1.5)
			411 (10)	554 (10)		764 (10)	845 (21)	928 (21)	1044 (21)	1143 (21)	1210 (10)	1306 (10)	1423 (10)	1514 (10)	1621 (10)	1716 (21)	1838 (20)	1929 (1.7)
			432 (21)	556 (21)			856 (21)	940 (21)	1068 (21)	1150 (10)	1217 (21)	1316 (17)	1432 (17)	1544 (21)	1656 (17)	1740 (8)	1845 (20)	1940 (1.5)
			462 (10)	593 (21)			860 (10)	975 (21)	1074 (10)	1179 (21)	1219 (17)	1363 (21)	1435 (5)	1546 (21)	1658 (10)	1784 (11)	1871 (19)	1947 (1.1)
							874 (21)		1077 (10)		1225 (10)	1364 (10)	1443 (17)	1565 (4)	1667 (21)	1789 (10)	1880 (12)	1963 (−0.3)
							881 (21)				1236 (10)	1394 (21)	1458 (17)	1569 (17)	1672 (21)	1795 (11)	1891 (8)	1979 (1.4)
											1270 (10)	1399 (21)	1465 (21)	1571 (10)	1677 (21)	1799 (21)	1895 (16)	
													1481 (17)	1573 (17)	1684 (17)			
														1587 (21)	1695 (21)			
														1591 (21)				
														1595 (10)				

* The winters are numbered according to the year in which the January falls; much of the sparseness of entries in the earlier centuries of the table must of course be attributed to lack of information.
† Easton's ratings of the temperatures implied by the manuscript descriptions are given (in brackets) according to his index, which increases with the temperature level, so that the mildest winters are rated over 80 on the scale; since in the twentieth century the Easton index is lacking, the central England temperatures averaged over December, January and February are given instead.

Source: reprinted from Lamb (1982).

nineteenth centuries, there were respectively seven and ten summers with more than 140 per cent of the 1916-50 average rainfall in London, five or more of them in each case in the second half of the century. But in the present century, there were four such summers between 1912 and 1931, three more between 1946 and 1958, and none since.

Another aspect of the changed tendencies of the general wind patterns has been an increase in the frequency and severity of gale and storm situations, particularly affecting the seas, in the region about the British Isles. This is seen in figure 6.4 (p. 94), covering the years up to 1977. It may be noted that the gale frequencies over the North Sea and British Isles region were generally lowest in those decades when days with general westerly winds were most frequent, and the storminess increased in the decades with more 'blocking', notably in the 1880s, the 1940s and since 1960. The 1970s and 1980s have produced some very great damaging gales, in their turn, notably on 2 April 1973 in the Netherlands, 2–3 January 1976 on both sides of the southern North Sea, 24 November 1981 and 16-17 January 1983 in Denmark. There were very high storm tides and sea-floods in the 1976 case and another in 1978, but the death tolls of such occurrences in earlier years have been avoided by improved sea defences and warning systems.

In 1969–70, and again in 1979, the German navy commissioned studies of the causes of the increasing tendency from about 1953 onwards of the bigger wave heights, reported by their vessels in the North Sea and German Baltic

Table 11.5 Numbers of coldest winters in western Europe falling in the different decades of each century (averaged)

Decade	00–09	10–19	20–29	30–39	40–49	50–59	60–69	70–79	80–89	90–99
Percentage of the winters with Easton index under 22 listed in table 8	7.6	10.9	7.6	7.6	13.0	7.6	14.1	12.0	7.6	12.0
Average number of freezing months (mean temperature below 0°C) in central England since 1659	0.3	0.7	0.7	0.7	1.3	0.7	1.2	1.5	1.5	2.7

Source: reprinted from Lamb (1982).

waters as well as in the southern Norwegian Sea, together with informed opinion about the outlook. Two reports were written (Weiss and Lamb 1970; Lamb and Weiss 1979), of which this chapter is in part a digest.

Up to the mid–1960s the increased frequency of wave-heights greater than 4 or 5 m – at some of the lightship and weather-ship positions a 50–100 per cent increase – was not paralleled by any increase in the frequency of gales and stronger storms. The diagnosis attributed this to the increased frequency of northwesterly winds, blowing over long fetches of open sea into the North Sea, at the expense of the westerly and southwesterly winds. However, in subsequent years the number of observations of gales and storms with Beaufort force 8 or more, though always varying widely from year to year, showed an unmistakable increase at most of the ship positions in the southern North Sea south of 57°N and western part of the German Baltic coast. Apart from the German Bight area, the increase was general and in most cases amounted to a doubling of the number of gales of these strengths. The Swedish meteorological service reports that the number of days a year with storm-force winds (Beaufort Scale 9 or over) at the observation points along the south-west coast in 1966–80 showed a 86 per cent increase over the previous 15 years, 1951–65. See also the postscript on p. 192.

These changes of storminess seem to have to do with changes of the prevailing latitudes in which most of the storm centres move. An analysis of the long record of observations at the Danish island Fanø, off the west coast of Jutland (Peterson 1983), from 1872 to 1980 showed that the number of cyclones with pressure below 990 mb at Fanø was lowest in the 30 years centred on 1933, more than 20 per cent below the late-nineteenth-century level. By 1954 to 1983 it had already risen by about 10 per cent. It seems, therefore, that the frequent westerly winds of the 1900–54 period in the latitude of the British Isles and North Sea were produced by big cyclones centred far away to the north – a conclusion supported by the present writer's own studies of the weather maps.

Since the development of the general westerly winds over the middle latitudes of the hemisphere is so deeply involved in the state of the climate, it is fortunate that we can reconstruct a longer history of this item. Figure 11.5 gives a reconstructed record of the average (roughly 10-year average) frequency of southwesterly surface winds in England since 1340. From 1669 this is a direct count of the number of days with southwesterly wind observed in London. The earlier part of the graph is sketched from indirect indications, including actual weather diaries; the earliest of these is from Lincolnshire about 1337–44*, but others come from observations on the continent in about this latitude. Other parts of the curve are derived from analysis of individual years' and seasons' patterns of reported weather. This graph shows four main peaks of the frequency of the general westerly winds,

---

* A diary kept by the Rev Fr William Merle, a Fellow of Merton College, Oxford, apparently mostly in his parish at Driby, Lincolnshire.

which are best registered by southwesterly winds at the surface over England. The first peak is seen around 1340, when the frequency is confirmed by the weather diary mentioned, and later peaks are well marked around 1530, 1730 and 1930. In this instance, therefore, we sense a cycle or periodicity of close to 200 years in length. Each of the high points of the westerly wind frequency was followed by a sudden decline to little more than half the previous level at some stage within the next 50–100 years and then, after some more decades, a slower recovery towards the next peak. However, some of the decade-to-decade variations superposed on this simple pattern must also be important.

There may be a valuable indication of the origin of this apparent 200-year recurrence tendency, in that the sharp declines of the southwesterly wind indicated in the late 1300s, the 1560s, 1740s–1770s and now, in each case, fell at about the end of a sequence of sunspot cycles which built up to periods of exceptionally great solar disturbance (around 1360–80, the 1570s, the 1770s, the 1950s and more recently). The frequency maxima of the southwesterly wind, and evidence of warm climate periods in Europe sustained over several decades, all bear a similar relationship to these variations of the sun's activity. This roughly 200-year cycle in disturbance of the sun is traceable in solar and auroral data over at least a thousand years past, and is related to an opposite variation in (negative correlation with) the amount of radioactive carbon in the Earth's atmosphere. (This is known from the incidence of errors in radiocarbon dating of objects of known age.) The radiocarbon fluctuation in cycles of either 200 years – or, in earlier millennia BC, 400 years – in length can be recognized back over the last 7000 years (Stuiver 1961). This gives rather more ground for believing that a continually recurring cycle of events in the sun is involved. It is the impact of emissions from the sun – the magnetic fields associated with streams of solar corpuscles

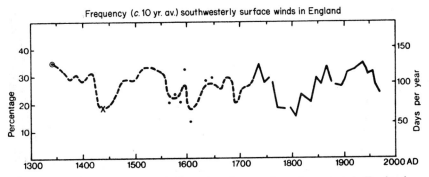

Figure 11.5 History of the frequency of southwesterly surface winds in England, 1340–1978; 10-year averages of daily observations in London from 1669. The earlier part of the curve is a sketched reconstruction from various indirect indications, including some weather diaries, among them one in eastern England (Lincolnshire) in 1340–4 and one in Denmark 1582–97.

during disturbances on the sun – on the cosmic rays approaching the Earth from elswhere in the galaxy that modulates the production of radioactive carbon (C^{14}) from the nitrogen in our atmosphere.

For forecasting it is, of course, awkward that this many times repeated cycle of approximately 200 years was in much earlier times replaced by a 400-year cycle. But the situation in the late twentieth century seems to be at an easily recognized point in the 200-year recurrence cycle, at or near the low point of the development of the westerly winds. Confidence in this diagnosis would be increased, if it turns out that the next short-term (11-year) sunspot cycles show weaker activity, as occurred around AD 1600 and 1800.

With these considerations in mind, it may not be too simplistic to use the history of the southwesterly wind over England as a basis for forecasting in the manner indicated in figure 11.6. Here the history is repeated in the broken curve, moved 200 years on. This shows what we should expect in the evolution of the wind circulation, if the course of the last previous (and therefore best-known) 200-year evolution is broadly repeated. This forecast indication was first issued in the Weiss and Lamb (1970) report and 16 years later still seems a reasonably good guide to the course of events. It implies a continuance until well into the twenty-first century of a low frequency of the westerly winds in this part of the world and a high frequency of blocking patterns, with prominence of northwesterly winds over the North Sea and other wind patterns, including anticyclone and depression centres, also more frequent than in 1900–50. The enhanced storminess may also, therefore, be likely to continue in most decades till about AD 2050.

One detail of the last cycle must be mentioned, although points of detail may be less reliably used as a forecast than the generalized trends. There was

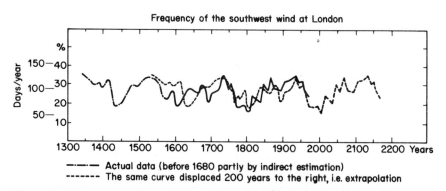

Figure 11.6 The same history of the southwest wind as seen in figure 11.5 with a tentative forecast (broken line) obtained by simply moving the whole curve 200 years to the right, i.e. the forecast implied by accepting the apparent 200-year recurring oscillation shown by the series.

a period of little more than a decade in length – perhaps 15 years – at the extreme low point of the westerlies in the last cycle, around 1786 to 1800, when instead of the prominence of northwesterly, northerly and north-easterly winds over the North Sea, the highest frequency was winds from the southeast and south there, while the northerly winds were transferred farther west (Brooks and Hunt 1933). After 1800, the high frequencies of northwesterly, northerly and northeasterly winds returned.*

The frequency of gradient winds from different directions over London, derived from the daily surface wind reports, decade by decade from the 1720s to the 1870s is shown in table 11.6.

Table 11.6   Apparent decade frequencies of gradient wind directions over London from reported surface wind directions, veered one point (average numbers of days/year)

	W	NW	N	NE	E	SE	S	SW	N + NW + NE
1723–9	103	42	26	25	40	25	16	35	93
1730–9	**123**	34	**62**	15	40	23	**33**	26	111
1740–9	101	32	**56**	15	**59**	17	**29**	12	103
1750–9	**114**	38	38	30	**51**	17	19	30	106
1760–9	104	38	33	**40**	39	16	17	39	111
1770–9	71	122	39	**40**	38	22	17	**10**	201
1780–9	68	**64**	25	23	37	**57**	17	30	112
1790–9	69	**68**	8	4	16	**88**	33	70	80
1800–9	59	**58**	16	**40**	41	40	16	38	114
1810–9	80	**57**	37	34	42	32	19	50	128
1820–9	75	**70**	26	30	36	**50**	17	41	126
1830–9	70	**61**	21	30	36	39	14	42	112
1840–9	79	40	24	37	37	30	14	42	101
1850–9	99	28	21	**39**	42	24	16	40	88
1860–9	**124**	40	20	30	38	30	19	35	90
1870–9	103	42	16	31	37	33	19	38	89

The figures must be treated as provisional; it cannot be foreseen when it will be possible to carry out a thorough critical appraisal, though it should be possible, after sufficient research, to present a fully homogenized series from 1669 to the 1970s; it is immediately clear, however, that there are some variations in the figures for individual decades which are so strong that they can hardly be modified significantly by any further research and must be accepted as broadly true.

---

* The daily weather observations made by the astronomer Tycho Brahe on Hven Island in the Sound between Denmark and Sweden, between 1583 and 1597 also show southeasterlies as the most frequent wind. This may mark a phase of the previous cycle corresponding to the period 1786–1800.

Postscript

Despite the colder Arctic, at least in the Atlantic sector, and the overweight of cold cases among the extremes elsewhere referred to on pp. 179–82, it is noticeable that the coldest and most disturbed summers in central England since 1950, e.g. 1954 (June, July, August average 14.0°), 1977 (14.3°) and 1985 (14.6°), have not matched the coldest in earlier periods, e.g. 1816 (13.4°), 1922 (13.7°). And the warmest recent cases have been in the very top class, e.g. 1976 (17.6°) only equalled in 1826 and July 1983 (19.6°) the warmest month in the central England record back to 1659. These examples would be seen by some as hints of global warming by carbon dioxide and the other 'greenhouse effect' gases. On the other side, the increased storminess in this same part of the world since 1950 – witnessed by some severe cases, e.g. the great storm on 16th October 1987 in southern England and others in 1953 (Scotland and North Sea), 1962 (North Sea and Germany), 1981 (mainly North Sea), and 1983 (North Sea and Denmark) – is in line with the periods of sharply cooling climate just 400, 300 and 200 years ago. The safest conclusion may be that we have entered a time of great climatic instability.

References

Agee, E. M. (1982) 'Diagnosis of twentieth century temperature records at West Lafayette, Indiana', *Climatic Change*, 4, 399–418.

Brooks, C. E. P. and Hunt, T. H. (1933) 'Variations of wind direction in the British Isles since 1341', *Quart. J., Roy. Meteorol. Soc.* 59, 375–388, London.

Davis, T. D., Kelly, P. M., Brimblecombe, P., Farmer, G. and Barthelmie, R. J. (1986) 'Acidity of Scottish rainfall influenced by climatic change', *Nature*, 322, 359–361.

Giovinetto, M. B. and Schwerdtfeger, W. (1966) 'Analysis of a 200-year snow accumulation series from the South Pole', *Archiv für Meteorologie, Geophysik und Bioklimatologie*, A15, 227–250, Vienna.

Kukla, G., Karl, T. R. and Gavin, J. (1986) 'US versus hemispheric temperature trends', *Proceedings of the Eleventh Annual Climate Diagnostics Workshop*, pp. 114–128, Champaign, Illinois (Illinois Water Survey for National Weather Service).

Lamb, H. H. (1982) *Climate, History and the Modern World*, London, Methuen.

Lamb, H. H. and Johnson, A. I. (1959) 'Climatic variation and observed changes in the general wind circulation: parts I and II', *Geografiska Annaler*, 41, 94–134.

Lamb, H. H. and Johnson, A.I. (1961) 'Climatic variation and observed changes in the general wind circulation: part III, *Geografiska Annaler*, 43, 363-400, Stockholm.

Lamb, H. H. and Johnson, A. I. (1966) 'Secular variations of the atmospheric circulation since 1750', *Geophysical Memoir* No. 110, London, HMSO for Meteorological Office.

Lamb, H. H. and Mörth, H. T. (1978) 'Arctic ice, atmospheric circulation and world climate', *Geogr. J.*, 114(I), 1–22, London.

Lamb, H. H. and Weiss, I. (1979) 'On recent changes of the wind and wave regimes of the North Sea and the outlook', *Fachliche Mitteilungen,* Nr. 194, Traben-Trarbach, Geophysikalischer Beratungsdienst der Bundeswehr.

Manley, G. (1974) 'Central England temperatures: monthly means 1659 to 1973', *Quart. J., Roy. Meteorol. Soc.,* 100, 389-405, London.

Newell, R. E. and Chiu, L. S. (1980) 'Climatic changes and variations: a geophysical problem', in *Climatic Variations and Variability: Facts and Theories,* A. Berger (ed.), Dordrecht, Reidel.

Peterson, E. W. (1983) 'A study of the weather record from Fanø (1872–1980) including an analysis of climate variations', *Risø National Laboratory (Risø-R-483),* Roskilde, Denmark.

von Rudloff, H. (1967) *Die Schwankungen und Pendelungen des Klimas in Europa seit dem Beginn der regelmässigen Instrumenten-Beobachtungen* (1670), Braunschweig, Vieweg-Die Wissenschaft, Band 122.

Stuiver, M. (1961) 'Variations in radiocarbon concentration and sunspot activity', *Geophyseal Research,* 66(1), 273–276, Washington, DC.

Watt, K. E. F. (1986) 'Tree mortality, acid rain, carbon dioxide and the greenhouse effect: a case study on the relations between scientific methods and the structure of scientific paradigms', unpublished memorandum by Professor Watt, Department of Zoology, University of California, Davis, Calif.

Weiss, I. and Lamb, H. H. (1970) 'Die Zunahme der Wellenhöhen in jüngster Zeit in den Operationsgebieten der Bundesmarine, ihre vermutliche Ursachen und ihre voraussichtliche weitere Entwicklung, *Fachliche Mitteilungen,* Nr. 160, Porz-Wahn, Geophysikalischer Beratungsdienst der Bundeswehr.

Wigley, T. M. L., Angell, J. K. and Jones, P. D. (1985) 'Analysis of the temperature record', in *Detecting the Climatic Effects of Increasing Carbon Dioxide,* M. MacCracken and F. M. Luther (eds), Livermore, Calif., Lawrence Livermore National Laboratory, US Department of Energy, DOE/ER–0235, 55–90.

Wigley, T. M. L. (1986) 'Testimony to the US Senate on the greenhouse effect', *Climate Monitor,* 15(3), 69–77, Norwich, University of East Anglia Climatic Research Unit.

Winstanley, D. (1973) 'Rainfall patterns and general atmospheric circulation', *Nature,* 245. 190–194, London, 28 September.

Winstanley, D. (1974) 'Rainfall and river discharges in the sub-Sahara zone (10°–20°N) of Africa. *Background paper for United Nations Food and Agricultural Organization mission on inland fisheries in West Africa.*

12

Drought in Africa: the climatic background to a threatening problem of today's world

In September 1983, the international relief agency Oxfam took the exceptional step of issuing a 'Weather Alert', announcing to the nations that 'drastic changes in the world's weather are causing havoc around the world'. A special appeal, which was soon echoed by other relief agencies, announced that 'droughts and floods are currently affecting over forty countries in Africa, Asia and Latin America, causing hardship and suffering to millions of people'. Experienced relief officials, it was reported, could not recall any period when so many people had been faced by disaster over such a large area of the globe. Oxfam already then was confronted by more demands for emergency aid than at any time in its 40-year history, and this was 'in large measure caused by unprecedented weather conditions that had devastated agricultural production'.

In February 1984, the National Geographic Magazine publicized the emergency in a well-informed article by its assistant editor, Thomas Y. Canby, on the exceptional weather of 1982–3 which linked the serious anomalies in many parts of the world to an El Niño ocean current event, a phenomenon which affects the warmth of the wide Pacific in the equatorial zone, and has known repercussions on world weather, as explained in Chapter 1. The 1982–3 El Niño was the strongest occurrence ever measured, with ocean temperatures for some months up to 7 degC warmer than normal in the equatorial Pacific as against the 3 degC anomalies familiar in such cases. The phenomenon is associated also with the so-called Southern Oscillation, mentioned elsewhere in this book. Canby's article expounded the global scale of the effects on the atmosphere and its circulation. It listed droughts and bushfires in Australia, droughts causing crop failures in Indonesia, in the Philippine Islands and in India, and droughts in Africa; in sub-Saharan Africa north of the Equator this was only an intensification of the already long succession of drought years. Near the Equator Ecuador and *northern Peru experienced excessive rain, flooding and landslides, while the direct effect of the warm water off the coast ruined the anchovy fishery which had already been very severely hit in similar manner by the 1972 El Niño occurrence (that was associated with the first severe phase of the Sahel–Ethiopian drought in Africa). Farther from the equator there was drought in Mexico, as in the same latitudes in Africa; while exceptional storms and rains in the lower-middle latitudes caused floods and destroyed crops in California and Colorado, in Cuba, and in the Gulf coast states of the USA. From all these afflicted regions Canby's estimates indicated a world-wide cost of US $8.65 billion.*

The anomalies continued into 1984. At least so far as the countries of Africa along the southern fringe of the Sahara were concerned, worse was to come when the drought of that year proved the most extreme yet. Crop failures set millions of starving refugees flocking to urban centres in search of sustenance, particularly in Ethiopia and the Sudan. More than 10 million are estimated to have been affected by the famine (Oxfam report 1986), and probably more than a million died. Television news reporting shocked the world, which had never witnessed a natural disaster on such a scale at the time of the occurrence, and produced the 'Live-Aid' relief fund collection organized by Mr Geldof.

The extent and distribution of the emergency in Africa was certainly affected by the civil war in Ethiopia, and by political factors in various countries. But underlying it was the drought, which had already then continued over nearly 20 years. This is an example of a climatic shift, which few governments or administrators before the 1980s seemed able to take serious cognizance of. It seems to be still regarded largely as something previously unheard of in human affairs. But in poor countries, with rapidly growing populations expanding on to areas with poor soils and always marginal (arid) climate, such changes are menacing and inevitably affect international affairs.

In recognition of this, the Royal Geographical Society held a one-day discussion meeting in London on 25 November 1985 on 'The geographical background to Africa's crisis'. The rest of this chapter reprints the invited contribution made by the writer.

Introduction

People in the developed world have grown accustomed to taking the climate for granted. This seems to have more to do with the high proportion of our lives which most of us spend in artificial climates indoors, and to such things as the mechanization of agriculture and crop-drying, together with the rest of the technology that reduces the risks of harvest failure in the advanced countries, than with any sound basis in the natural world. In reality, we live in an ever-changing environment. Changes come from natural causes as well as from human activities. In the case of climatic fluctuations and changes, most of these still seem to be of natural origin.

The obvious intrusion of human activity in many aspects of the world environment, perhaps seen at its most extreme in the almost complete destruction of the temperate zone forests to clear the land for agriculture – and the surely irresponsible destruction now proceeding in the equatorial forests, such as the Amazon forest – leads many to assume too readily that any change of the climate must also be, at least indirectly, due to human actions. Others, with their eyes concentrated on other aspects of Africa's difficulties, are too ready to say – as was heard on the BBC in August 1985 – 'The weather has not got worse. It is simply a question of our ability to cope'. The first part of this quoted statement is frankly not true, as will be shown in this chapter.

The impact of climatic changes in countries where cultivable land and other resources are limited may, of course, be much aggravated by population growth and, as in Africa today, by civil wars; and these things may indeed make it much harder to cope.

The desert and tropical forest regions of Africa were not always as they are today. During Europe's last Ice Age, which produced enlarged glaciers in low latitudes also, there were periods thousands of years long when the levels of lakes in Africa were high and other long periods when they were

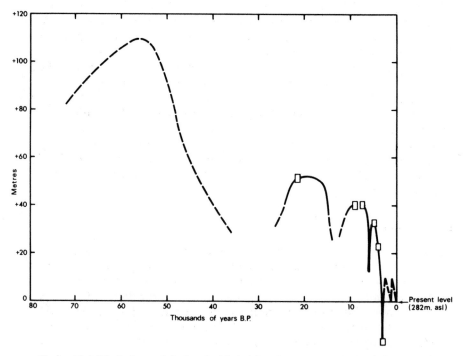

Figure 12.1 Variations of the level of Lake Chad over the last 80,000 years. (The broken curve before 30,000 years ago gives sketchy preliminary indications only.)

quite low (Butzer *et. al.* 1972, in Nicholson and Flohn 1980) (figure 12.1; p. 196). The same is true for the 10,000 years since the last major Ice Age ended (see, for example, Grove and Warren 1968; Grove *et. al.* 1975).

There have been times when sand dunes were formed in some of today's forest regions (Moreau 1969; Sarnthein 1978), and other times when the lakes and oases were more extensive than they are today. At these latter times, the deserts themselves were less extensive and less arid. Eight to ten thousand years ago Lake Chad (figure 12.1), in today's semi-desert region, near 13°N 13°E, stood 30–40 m higher than now, forming an inland sea abundant in fish. According to Maley (1983), there are interesting indications of a change in the prevailing character of the rainfall in the Lake Chad basin: until mid-postglacial times, around 7000 years ago, sedimentation in the water-courses was clayey, presumably corresponding to regular rainfall in which small raindrops (under 2 mm in diameter) predominated. But since about 7000 years ago the sedimentation has been sandy, as would be expected with bigger raindrops in the monsoons and occasional tropical depressions. Between 6000 and 5000 years ago dwellers in the Sahara painted pictures on the walls of caves showing cattle herding and men hunting hippopotamus with canoes on water bodies that do not exist today. Maley writes of a rapid penetration of the Sahel by a spreading savanna with tree species from farther south. At some time after that, there was an arid period in which Lake Chad dropped below its present level (see figure 12.1).

Again in the Middle Ages, the Sahara seems to have been moister than now. There was traffic across the desert between the Mediterranean fringe and Kufra oasis (24°N 22°E), and to Ghana and Mali, which was for some centuries the centre of a considerable empire. It is recorded that in the course of a two-month crossing of the Sahara, in AD 1352, numbers of wild cattle often approached the caravan (Pejml 1962). And two centuries later, early Portuguese explorers experienced summer rains penetrating northwards over the western Sahara as far as 24°N – i.e. well into southern Algeria and Libya (Sarnthein 1978). Maley (1981) has plotted out a record showing that Lake Chad was high in the Middle Ages, probably about 285 m above sea-level from AD 900 or before until around 1200. It may have fallen 2 m temporarily during the thirteenth century, but was again around 284 m from about 1330 to 1400. Important low phases followed around 1450 and 1550, when the lake dropped to 279 m and the local population settled on the former lake-bed, only to be driven away when the water again rose to 285 m throughout the seventeenth century. It is noteworthy that the unlike climatic periods of Europe's warmth in the Middle Ages and the Little Ice Age climax in the seventeenth century both produced moist regimes in the Sahel. Drying out in the eighteenth century was followed by higher levels, up to 283 m, in the nineteenth century, especially in the 1880s. Clearly, the record shows a whole succession of both longer and shorter lasting changes of climate in Africa. And at least some of them took place more or less abruptly.

Climate and climate's mechanisms

What, then, do we understand by climate? It is not just the average weather, but the whole range of weather experienced in a region and its changes, and the extremes and the frequencies of this and that condition, during the period concerned. And we must specify the period that we are referring to, because other periods may have had a somewhat different experience.

The main zones of world climate are shown in figure 12.2 – the prevailing surface winds and the belts of generally high and low barometric pressure, together with the cloud cover and raininess that result from the conver- gences (and consequent upward motion) in the winds' flow and the moisture which the winds transport. These are all statistical results: there are always variations from day to day and over long periods, too. And, in examining the global pattern of rain and snow downput, it must be remembered that the effectiveness of the rainfall in watering the ground is reduced by evapor- ation. The evaporation from wet surfaces is on average about three times greater in latitudes within 20°N or S of the Equator at today's temperatures than in the latitude of the British Isles. Figure 12.3 shows the average rainfall over Africa in recent times and how far the global belts shown in figure 12.2 can be recognized over Africa, despite local variations due to topography, and so on.

Much used to be heard of the constancy of tropical weather. In the early days of trans-Atlantic flying, about 1940, an aircraft captain westbound on

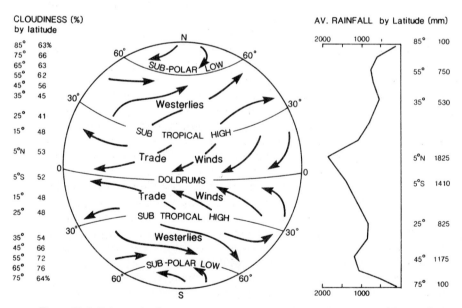

Figure 12.2 Schematic diagram to show the main distribution of prevailing surface winds and average atmospheric pressure, cloudiness and rainfall, by latitude zones.

the old winter route that made use of the Trade Winds, flying from Europe to North America via Africa in 13°N (Bathurst), thence to the West Indies and then north, was shown a forecast weather map with the intertropical front near his route, though there were no observation reports. When he asked about the basis for this front, he was told, 'It has been there for the last 50 years, so there's no reason why it shouldn't be there tonight'.

We are now much more aware that weather varies from day to day, and the activity of fronts and other weather systems varies, even in the tropics. But the net outcome is that the climate also varies and specifications of the averages of meteorological measurements, and the extreme values and frequencies quoted for various phenomena, have no meaning unless the period in question is also specified.

The changes are sometimes a shift of the underlying averages to warmer or colder and wetter or drier, conditions than before, while in other cases it may be rather an increase or decrease in the range of the occasional extremes, a widening or narrowing of the variability presented by any group of years. Some of the changes set in gradually, others seem to take place with surprising suddenness. Statistical analyses of past weather – even recent past weather – are liable to be a poor guide to the future when one of the more abrupt climatic changes takes place.

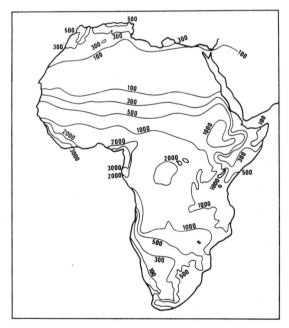

Figure 12.3 Average yearly rainfall over Africa (in millimetres) (present century), adapted from *The Times Atlas*.

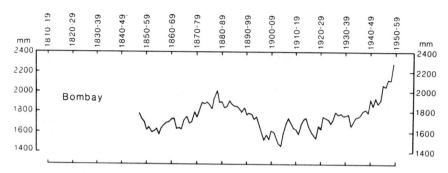

Figure 12.4 Indian monsoon rainfall: 10-year average totals for the period May to September at Bombay, 1840s–1950s.

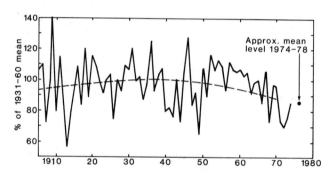

Figure 12.5 Rainfall of the Sahel zone in Africa: yearly values averaged for five places between about 12° and 14°N (Zinder, Niamey, Sokoto, Kano and Maiduguri), 1905–74, and the best fitting smooth, sine curve (broken line); the 5-year average for 1974–8 is shown as a dot which fits neatly to the sine curve. (Adapted from a diagram by Bunting *et. al.* 1976.)

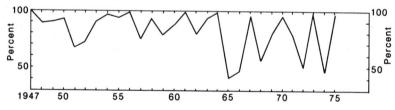

Figure 12.6 Percentage of the area of India not classified as suffering from 'scanty rainfall', year by year, 1947–75, i.e. the proportion of the country over which the monsoon could be considered satisfactory. (Adapted from a diagram by K. R. Saha and D. A. Mooley 1978.)

Records to illustrate this may be taken as well from the Indian monsoon as from the monsoons and other rains of Africa. Examples are provided in figures 12.4, 12.5 and 12.6. Monsoon failures in India were more frequent early in this century, when the average rainfall was relatively low. It was unfortunate that an Indian statistical analysis in 1963, after nearly 40 years of increasing rainfall (figure 12.4) without any significant monsoon failure, suggested that there had been no significant climatic change; for since 1965 the monsoons have again been less reliable (figure 12.6). The same lesson is indicated by the much longer perspective of monsoon changes over the whole span of postglacial time, which can be presented for the area that is now the Rajasthan Desert in north-west India (Bryson 1975; see also Swain *et. al.* 1983) (figure 12.7). It is reflected in the changing levels of Lake Chad and other lakes in Africa. We see that the climate – in this case focusing on the rainfall regime – undergoes changes that may come in quite sharply and are liable to last for decades and sometimes for hundreds, or even thousands, of years.

The world-wide arrangement of climates, associated with the global wind circulation, is set up by the sun's heating and its unequal distribution over the Earth. The inequalities drive the wind circulation like a giant system of convection. So it is the temperature pattern – the gross features of the thermal distribution affecting the atmosphere – that is fundamental, even though in low latitudes we are mostly concerned with the rains and belts of cloudiness. These result from the moisture transport and convergences in

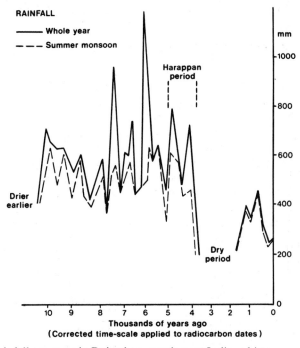

Figure 12.7 Prevailing rainfall averages in Rajasthan, north-west India: a history derived from lake level and botanical (pollen analysis) evidence. (Adapted from work by R. A. Bryson 1975, and G. Singh 1971.)

the wind system. Hence, any changes in the strength of the sun's beam as it penetrates the atmosphere, and any changes in the underlying heating of the atmosphere resulting from changes of the nature of the Earth's surface – e.g. the colour and reflectivity (the 'albedo') of the surface due to vegetation changes, or the warmth of the sea surface due to currents or upwelling and downwelling in the ocean – must affect the pattern of climates.

Among the range of influences that have to be considered are (i) the regular, slow astronomical changes over thousands of years in the Earth's orbit that control the timing and development of ice ages, (ii) the more irregular, but perhaps partly cyclical, variations of volcanic activity loading the atmosphere with ash and aerosol veils and (iii) to some degree the pollution put in by human activity. Some part may also be attributable to human destruction of the vegetation by forest felling, by the grazing of animals and by fire; but these seem to be of somewhat secondary importance, when we view the gross changes that have taken place over postglacial time or even over historical time since the pharaohs of Egypt or the fall of the Roman Empire. The natural climate also produces changes due to secondary effects – e.g. through changes of cloudiness which may be attributable to wind circulation changes having important reactions ('feed-back effects') on the broadest-scale heating pattern. This means that the cloudiness changes may affect the heating of the Earth as a whole.

Recent history of the climate in Africa and its interpretation

A series of papers by Kraus (e.g. 1955, 1958) drew attention to the lowered yield of the equatorial rains between 1895 and 1940 or later (and which ultimately continued to 1960 or thereabouts); this time-span embraced the periods of increased rainfall in other latitudes (a) from the monsoons over northern India (figure 12.4), (b) in the Sahel–Sudan zone across Africa (figure 12.5) and (c) in the prevailing west winds of the temperate zone over the British Isles and western aspects elsewhere in both the northern and southern hemispheres. It coincided with a long period of increased strength of the general wind circulation, as indicated by various measurements of the main currents, such as the westerlies over middle latitudes and the Trade Winds, as seen on mean maps (Lamb and Johnson 1966; Lamb 1977). So it appears that, over those years, the global transport of moisture may have been organized on a grander scale than in the preceding or later years. Both the northern and southern hemispheres seem to have experienced the intensification of the circulation, although not perfectly in step as regards the timing: the peak may have been 20–25 years earlier in the southern hemisphere.

Reversal of these trends seems to have set in around 1955–60, with sharply decreasing rainfall in the Sahel–Sudan zone and in a corresponding zone south of the Equator, over southern Africa, while there was some increase of rainfall close to the Equator, although not to the late-nineteenth-century level. Figure 12.8 illustrates the pattern. What seems to have been happening is that the seasonal march north and south of the equatorial rains system (see figure 12.10), which brings the monsoons, has become more restricted in recent years, and so the rains have remained closer to the equator. This first showed itself in a great rise of the level of Lake Victoria, near the Equator, in 1961 by nearly 2 m (figure 12.11) and similar rises of the other great lakes in east Africa between the equator and about latitude 10°S. The discharge of Lake Victoria into the River Nile doubled, from 20 to about 41 km^3 a year, and the higher level and discharge rate of the lake have been largely sustained into the 1980s (Fohn 1983). Despite this, there seems to have been some overall decline of the rain yield over Africa. The equatorial rain belt seems also to have acquired a mean position on balance a few degrees of latitude south of before.

In South America, some parallel changes have been reported, with a rise of the Bolivian Lake Titicaca, high up on the Altiplano in latitude 16°S, since about 1970 (Kessler 1984).

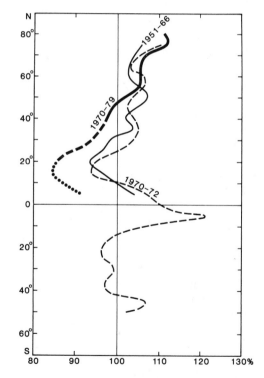

Figure 12.8 Changes in the world distribution of rainfall (strictly the total downput of rain and snow converted to rain equivalent) by latitude; the curves show the average downput in different periods of recent years, by latitude, as a percentage of that experienced from 1931 to 1960.

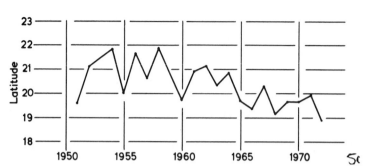

Figure 12.9 Northward limit of the west African monsoon, year by year since 1951. (Diagram by Professor R. A. Bryson of Wisconsin, prepared for a conference at the Rockefeller Foundation's Study Center, Bellagio, Italy, 1974.)

The increase of rainfall close to the Equator in Africa seems to represent 'the other side of the medal' from the droughts in the Sahel/Sudan/Ethiopia and in a corresponding zone over Africa south of the Equator. An important element in the East African part of this post-1960 regime, according to Flohn, has been maintenance of the level of Lake Victoria by increased night thunderstorm activity over the lake. A study by Newell and Kidson (1984), however, shows that in the main the changes are related to changes in the great global wind systems. As such, the changes, including the changes of rainfall, are unlikely to be attributable to any human actions or mismanagement of resources in Africa apart from any activity which might intensify the local consequences (see Postscript at end of chapter). These investigators compared conditions during years of good rains in the Sahel (1958–62) with those in dry years (1970–3), when the long drought reached its first climax. The southern hemisphere's winter jet-stream winds, the westerlies at around 12 km height, were stronger in the Sahel's dry years than in the wetter years. And between the Earth's surface and heights up to 2–3 km there was more northward flow of air across the Equator in the Sahel's wetter years, and it reached farther north than in the dry years. The northern hemisphere subtropical high pressure, at the 3-km height, over Africa was farther south in the dry years, farther north in the wetter period.

What these changes have meant for rainfall in the Sahel year by year from the beginning of the century to 1984 is seen in figure 12.11. The data were collected and analysed by Dennett *et. al.* (1985) and Nicholson (1983), and updated by Folland (personal communications, July 1985 and August 1986). The resulting diagram represents the rainfall in the Sahel by averaging the measurements at five west African stations between latitudes 12° and 14°N from longitude 21° to 14°E. (The places were Zinder, Niamey, Sokoto, Kano and Maiduguri; data from Maiduguri and Sokoto failed after 1980, but the representativeness of the tendency presented is not in doubt.) We see an almost unbroken tale of deficits below the previous average since the mid-1960s. The deficit in 1985 (after the end of the graph) was about half as great as in 1984. In 1986 reports from *Die Witterung in Übersee*, published by the Deutscher Wetterdienst Seewetteramt in Hamburg and from the UK Meteorological Office, again indicate substantial deficiencies of rain in the Sahel–Ethiopia zone of between 20 and 60 per cent and continuing dryness in the corresponding zone over southern Africa. And since April 1987 the relief agencies (Oxfam and others) have alerted the world to a renewed worsening of the long-continued drought in Somalia and elsewhere. (Better crops are attributed to relief measures, particularly water from wells – ultimately a not inexhaustible resource.)

Much of the discussion of the situation, even in meteorological circles (e.g. Bunting *et. al.* 1976) after the first severe phase of the drought had eased in 1974 and 1975, can now be seen to have assumed on the slenderest evidence that the event was essentially over and the rainfall beginning to revert to 'normal'. But the present writer and others have consistently pointed to the drought period's continuance (e.g. H. H. Lamb 1976, 1982,

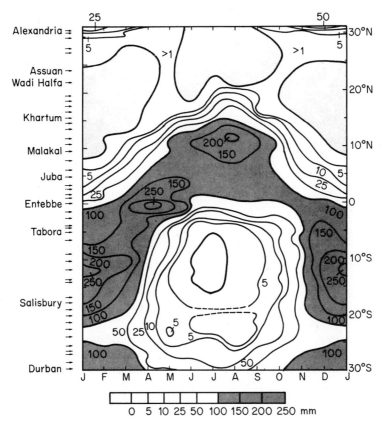

Figure 12.10 Rainfall at different latitudes, month by month, through the year over Africa near 32°E. (From Lockwood 1979, based on the work by H. Flohn.)

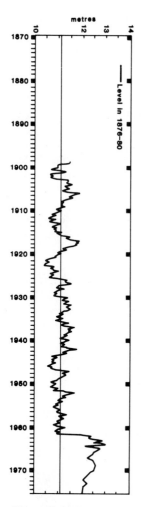

Figure 12.11 Variations in the level of Lake Victoria in eastern equatorial Africa.

p. 265; P. J. Lamb 1982) and attention was repeatedly drawn to this in the *Wetterkarte des Deutschen Wetterdienstes* (on 15 April 1976 and 23 March 1978, for instance).

An important discovery to set beside this prolonged drought is reported in recent, so far unpublished, work by Folland and his colleagues in the Meteorological Office. Changes in the prevailing surface temperatures in the oceans of the world have accompanied the changes in the world's wind circulation, which we have alluded to. From the beginning of the survey in the 1850s to 1870s until around 1950 to 1960 the sea-surface temperatures in the northern and southern hemispheres underwent similar trends; these are illustrated as regards the Atlantic sector, where the changes were biggest, in figure 12.13. But since 1960, the experience of the two hemispheres has parted company, with cooling on balance in the northern hemisphere oceans and some warming in the oceans south of the equator. This has displaced the 'thermal equator' (i.e. the zone of highest temperature) in the ocean surface somewhat south of its previous position (figure 12.14). Such a change must be reflected in the development of the atmospheric circulation. It is not surprising that the zones of the general wind circulation also show some displacement to the south, and there is evidence of this in both hemispheres. But the circulation patterns developing from day to day, and from season to season, show more complications than can be contained in such a bald statement.

The southward displacement of the basic wind zones, which is most marked in and near the Atlantic sector, allows more room for the occasional development of blocking anticyclones in temperate and sub-Arctic latitudes to take place. This may account for the more frequent development of blocking patterns and, in particular, for the occurrences in recent years of more frequent anticyclones and high pressure between 50° and 70°N in the Greenland/east Atlantic/Europe sector as well as occasional almost stationary low-pressure systems in the same region, which includes the British Isles. Also, this may be why there appears to be some positive correlation between the frequency of the west winds over the British Isles (and their decline since the 1950s) and the seasonal rainfall in the Sahel (as noted by Winstanley 1973; Folland et. al. 1985). With this, there seems to be a general parallelism between recent variations of the summer rainfall over England and Wales and over the Sahel (Folland et. al.)

The correlation coefficient connecting the number of days with general west winds over the whole of the British Isles ('westerly days') and rainfall in the Sahel zone 10° to 20°N was +0.56 (statistically significant at the 1 per cent level) for yearly values over the period 1900 to 1973 (see table 11.2, p. 181) appeared to be higher (about +0.8) when 20-year blocks of years

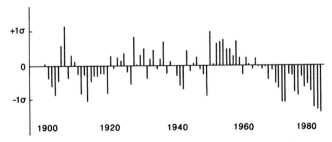

Figure 12.12 Rainfall year by year in the Sahel, 1901–84; mainly as in figure 14.5, but the yearly anomalies being here standardized: i.e. departures from the overall average divided by the standard deviation. (From work by S. Nicholson 1983, and Dennett et. al. 1985, as presented by C. F. Folland: reproduced by his kind permission.)

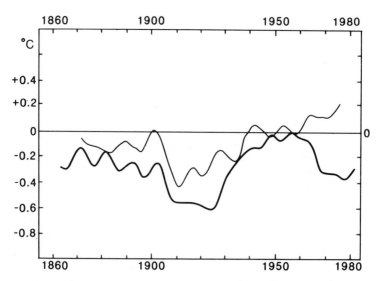

Figure 12.13 History of sea-surface temperatures, 1856–1984, over the North Atlantic, and South Atlantic; the values are departures from the 1951–60 average. (Adapted from two diagrams by Folland *et. al.* (unpublished): reproduced by kind permission.)

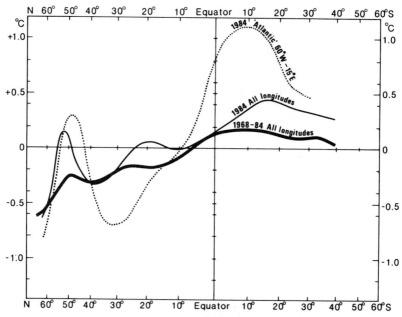

Figure 12.14 Changes of prevailing sea-surface temperatures, shown as departures from 1951–60 averages at different latitudes. (Adapted from a diagram by Folland *et. al.* (unpublished): reproduced by kind permission.)

were compared (Winstanley 1974; Lamb and Mörth 1978). In the present state of scientific knowledge, however, it has to be admitted that there may be other additional reasons why blocking has become more frequent and the frequency of days with general westerly winds over the whole of the British Isles has declined from an average of almost 95 days/year between 1860 and 1960 (109 days in the 1920s) to 73 days in the 1970s and an average of 67 days from 1980 to 1985.

Suspicions that one or more other causes may be important as well are aroused by indications of a cyclic element of about 200 years length in various meteorological and related series (see Chapter 13 and Lamb 1972, p. 242), including for instance variations in the frequency of the westerly and southwesterly winds over the British Isles. The analysis of the Sahel zone rainfall, presented in figure 12.5, indicates a best-fitting sine curve corresponding to a 'cycle' of about 200 years length (H. H. Lamb 1982, p. 264). An oscillation of this length was also found in the variations in the amount of radioactive carbon in CO_2 in the Earth's atmosphere (Stuiver 1961), an association which suggests a solar cause. This indication of a recurring fluctuation in the sun was traced back through many repeats – in part as a 200-year, in part as a 400-year, cycle – over 7000 years. (It was in the earlier part of the record in BC time, that the cycle was substituted by one of just double the length, about 400 years.) Other, shorter-period harmonics may also be traceable in weather data, notably one close to 100 years, by which for instance hot summers and cold winters in Britain and Europe seem to occur most frequently in and around the eighties of each century, often interspersed by swings to the opposite extreme; this behaviour also has to do with the varying incidence of blocking patterns and so may affect the Sahel indirectly, too. In blocking situations, the main belt of westerly winds over middle latitudes is transferred to higher and/or lower latitudes, while northerly, easterly and southerly winds become more frequent in the latitude of Britain and Europe.

Less can be said about the associations of the increased drought tendency in the last 20 years in parts of Africa south of the equator. The southern hemisphere's wind zones seem to have undergone some southward displacement, at least in and near the Atlantic and New Zealand sectors. Nevertheless, blocking situations have apparently been more frequent there, too, than they were in this century before 1950. Blocking in middle latitudes, when it occurs, pushes what is left of the subtropical anticyclones to lower latitudes. So this may be involved in the droughts that have affected Zimbabwe and the Transvaal albeit less severely than the Sahel/Sudan/Ethiopian zone.

Conclusion

Our survey has shown that Africa has a long history of rainfall regime changes, which sometimes set in rather quickly but have lasted over long periods. They bear an intelligible relationship to changes in the global wind circulation and, at least in the recent case, to prevailing sea-surface temperatures. The fundamental causes undoubtedly lie in the effective heating of the Earth, and operate through the distribution and redistribution of the heat received from the sun. Although no precedent for the current length of drought in parts of Africa exists within the period of meteorological instrument measurements in that continent, there are some indications of its relationship to global fluctuations repeating at about 200-year intervals, which may have their origin in the sun. It has not been necessary to mention the increase of carbon dioxide due to human activities or its supposed warming effect in this analysis. There is beginning to be renewed debate as to whether the increase of particulate matter polluting the upper atmosphere (i.e. the stratosphere and high troposphere), attibutable to human activity and to other factors such as the increase of volcanic activity since about 1960, may be cooling the Earth and effectively masking any carbon dioxide warming effect. What is certain is that there will be a continuance of climatic variations on many time-scales and, almost equally certainly, the development of another ice age around 5000 years from now, due to the regularly recurring, slow astronomical changes which affect the Earth's orbit (Berger 1980).

Africa's place in the world pattern of rainfall changes from the situation of 1930–60 to the present time (1985) is rather like that shown in figure 12.15

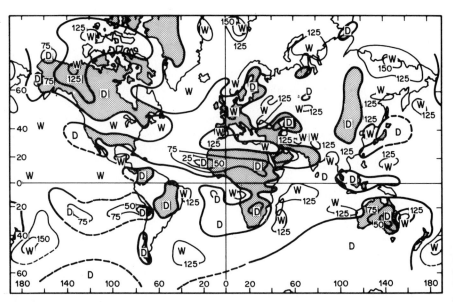

Figure 12.15 World map of rainfall 1970–2, expressed as percentages of the 1931–60 average; letters D and W mark the main drier and wetter areas. (Originally published in Lamb 1974.)

whether we look at maps for the 1960s, 1970s or the 1980s so far. The plain implication of these studies is that there is no basis for expecting an early, or sustained, return to conditions as they were with the greater rainfalls in the Sahel between 1930 and 1960. It seems more likely that the rains will continue for some decades more at the reduced level of recent years, but not necessarily in every decade or in every year. The average may even go lower – as is also the independent conclusion, from somewhat different approaches, of my colleagues in the Climatic Research Unit, G. Farmer and T. M. L. Wigley, reporting to the British government's Overseas Development Administration (unpublished).

Yet this cannot be taken as a firm forecast, to which some specifiable degree of confidence can be attached. The state of our scientific knowledge of climate is not yet ready for such specific climate forecasting over years and decades ahead. And for a long time to come, it will be necessary to treat with scepticism any forecasts, issued from any quarter whatever, which claim a firmly defined probability level. Such a practice would be a scientific ideal, but could be very misleading before the science is ready. The probabilities stated in such forecast presentations must be related to some proposed model of the climate mechanisms responsible and to the frequencies of this or that development when the supposed circumstances were recognizable in the past. The real probabilities could be quite different, if the climatic mechanism were wrongly or incompletely identified, or if some additional influence were to have a strong impact. At this time, many meteorologists are thinking of the alternative prospects of a drastic future warming of world climates, attributable to human activities, or a drastic cooling resulting from enhanced volcanic 'dust' in the atmosphere or, God forbid, from nuclear war.

So the most that can be said about the future is that, on the basis of our present understanding of the climate that has been developing since 1965–70, the drought is likely generally to continue for several decades. Unless some new influence supervenes, the deterioration from the situation experienced between 1930 and 1960 may even increase.

Postscript

We have examples of how human activities can worsen the consequences of situations brought about by climatic change. Evidence of overcropping and over-use of resources has been left by many past civilizations, and the same could be said of the 'dust-bowl' disaster in the US Middle West in the dry years in the 1930s. Similarly, the increased exploitation of the Sahel for agriculture in the moister period in the 1950s paved the way for greater distress when the long drought came.

The natural vegetation in such areas as the Sahel and other parts of both northern and southern Africa is often in a precarious, even 'sub-fossil', condition, because the climate has become drier and more unfavourable for regeneration of woodland, scrub, etc. than in the time when these had established themselves. Destruction of the existing vegetation by human activity in these circumstances may lead to a situation in which no vegetation, neither crops nor pasture, can be maintained. That is truly a human-made desert. Tickell (1986) has attributed to destructive human practices most, though not all, of the loss of the grain growing areas in north Africa that once supported the Roman Empire.

There is thought to be an important secondary effect which follows destruction of the vegetation in areas near the desert margin and compounds the damage. The albedo (reflectivity) of the surface is increased. Hence less incoming energy is absorbed, convection is discouraged and downward motion prevails oftener than before in the overlying air. And so cloud formation and rainfall becomes even more unlikely than before.

It would be a mistake, however, to put down the loss of all the once-fertile lands which once stretched from the Mediterranean to northern India to human activity. This is well shown by Singh (1971, 1977) for the areas where the Indus Valley culture flourished from about 2500 to 1700 BC. He presents a picture, undoubtedly paralleled elsewhere and in other times, in lands near the arid margin of cultivation and settlement possibilities. From pollen studies and comparisons with the evidence of archaeological work in old salt-lake deposits in western Rajasthan, north-west India, and from land-forms and stratigraphy in the area, it is seen that after a severely arid period in early postglacial times, with shifting wind-blown sands, the lake basins were filled with fresh water during the main period of climatic warming in the northern hemisphere from about 8000 BC onward for 6000 years thereafter. A vegetation with trees became established and was increasing during the expansion of Neolithic cultures in the region. This was a very favourable climatic period there, a 'rainfall optimum', illustrated in figure 12.7. Swamp vegetation species were increasing too. Then came a much drier time, particularly between 1800 and 1500 BC. The moisture-demanding vegetation was disappearing from 2000 BC onwards. There was some recovery from about 1500 to 1000 BC, but then a further drought which ended in the lakes drying up. Thus, it seems that the collapse of the Indus civilization coincided with a stage in the development of a drastic deterioration of climate – the retreat of the monsoon after the postglacial warmest times. This was a climatic change of global scale attributable only to natural causes.

References

Berger, A. (1980) 'The Milankovitch astonomical theory of paleoclimates – a modern review', *Vistas in Astronomy*, 24, 103–122, Oxford, Pergamon.

Bryson, R. A. (1975) 'The lessons of climatic history', *Environmental Conservation*, 2(3), 163–179.

Bunting, A. H., Dennett, M. D., Elston, J., and Milford, J. R. (1976) 'Rainfall trends in the west African Sahel', *Quarterly J., Royal Meteorological Society*, 102, 59–64.

Butzer, K. W., Isaac, G. L., Richardson, J. L., and Washbourn-Kamau, C. (1972) 'Radiocarbon dating of East African lake levels', *Science*, 175(4027), 1069–1076.

Dennett, M. D., Elston, J., and Rodgers, J. A. (1985) 'A reappraisal of rainfall trends in the Sahel', *J. Climatology*, 5(4), 353–361.

Farmer, G. and Wigley, T. M. L. (1985) 'Climatic trends for tropical Africa', Research report for the Overseas Development Administration, Norwich Climate Research Unit, University of East Anglia.

Flohn, H. (1983) 'Der Katastrophenregen (1961/2) und die Wasserbilanz des Viktoria-See-Gebietes', *Wiss. Berichte des Met. Inst. Univ. Karlsruhe*, Nr. 4 (*Prof. Max Diem 70. Geburtstag*) 17–34.

Folland, C. K., Parker, D. E., and Palmer, T. N. (1985) 'Sahel drought and worldwide sea surfaces temperatures' (Meteorological Office, Bracknell, unpublished).

Grove, A. T. and Warren, A. (1968) 'Quaternary landforms and climate on the south side of the Sahara', *Geographical J.*, 134(2), 194–208, London.

Grove, A. T., Street, F. A., and Goudie, A. S. (1975) 'Former lake levels and climatic change in the Rift Valley of Southern Ethiopia', *Geographical J.*, 141(2), 177–202.

Kessler, A. (1984) 'The palaeohydrology of the Late Pleistocene Lake Tauca on the Bolivian Altiplano and recent climatic fluctuations', in *Late Cainozoic climates of the southern hemisphere*, J. C. Vogel (ed.), Rotterdam and Boston, Balkema, 115–122.

Kraus, E. B. (1955) 'Secular changes in tropical rainfall regimes', *Quarterly J., Royal Meteorological Society*, 81, 198–210.

Kraus, E. B. (1958) 'Recent climatic changes', *Nature*, 181, 666–8, 8 March, London.

Lamb, H. H. (1972) *Climate: Present, Past and Future. Vol. 1, Fundamentals and Climate Now*, London, Methuen.

Lamb, H. H. (1976) 'Understanding climatic change and its relevance to the world food problem', *Climatic Research Unit Research Publication*, CRURP No. 5, Norwich, University of East Anglia.

Lamb, H. H. (1977) *Climate: Present, Past and Future. Vol. 2, Climatic History and the Future*, London, Methuen.

Lamb, H. H. (1982) *Climate, History and the Modern World*, London, Methuen.

Lamb, H. H. and Johnson, A. I. (1966) 'Secular variations of the atmospheric circulation since 1750', *Geophysical Memoir*, 110, London, HMSO for Meteorological Office.

Lamb, H. H. and Mörth, H. T. (1978) 'Arctic ice, atmospheric circulation and world climate', *Geographical Journal*, 144(1), 1–22, London.

Lamb, H. H. and Weiss, I. (1979) 'On recent changes of the wind and wave regime of the North Sea and the outlook', *Fachliche Mitteilungen*, Nr. 194 (Geophys. BDBW-FM 194). Traben-Trarbach (Amt für Wehrgeophysik).

Lamb, P. J. (1982) 'Persistence of sub-Saharan drought', *Nature*, 299, 46–48. London, 2 September.

Lockwood, J. G. (1979) *Causes of Climate*, London, Edward Arnold.

Maley, J. (1977) 'Palaeoclimates of the central Sahara during the early Holocene', *Nature*, 269, 573–575.

Maley, J. (1981) 'Travaux et documents', *ORSTOM*, No. 129, Paris, Centre National de la Recherche Scientifique.

Maley, J. (1983) 'Histoire de la vegetation et du climat de l'Afrique nord-tropicale au Quaternaire recent', *Bothalia*, 14(3 and 4), 377–389, Pretoria.

Moreau, R. E. (1969) 'Climatic changes and the distribution of forest vertebrates in West Africa', *J. Zoology*, 158, 39–61, London.

Newell, R. E. and Kidson, J. W. (1984) 'African mean wind changes between Sahelian wet and dry periods', *J. Climatology*, 4, 27–33, London, Roy. Meteorol. Soc.

Nicholson, S. E. (1983) *Report of the Expert Group Meeting on the Climatic Situation and Drought in Africa, Geneva 6–7 October 1983*, Geneva, World Meteorological Organization, 32.

Nicholson, S. E. and Flohn, H. (1980) 'African environmental and climatic changes and the general atmospheric circulation in Late Pleistocene and Holocene', *Climatic Change*, 2, 313–348, Dordrecht, Reidel.

Pejml, K. (1962) 'A contribution to the historical climatology of Morocco and Mauretania', *Studia geophysica et geodetica*, 6, 257–259, Prague.

Saha, K. R. and Mooley, D. A. (1978) 'Fluctuations of monsoon rainfall and crop production', in *Climatic Change and Food Production*, K. Takahashi and M. M. Yoshino (eds), Tokyo, Tokyo University Press, 73–80.

Sarnthein, M. (1978) 'Sand desert during glacial maximum and climatic optimum', *Nature*, 272, 43–46, London, 2 March.

Singh, G. (1971) 'The Indus Valley Culture', *Archaeology and Physical Anthropology in Oceania*, 6(2), 177–189, Canberra.

Singh, G. (1977) 'Climatic changes in the Indian desert', ch. 4 in *Desertification and its Control*, 25–30, New Delhi, Indian Council of Agricultural Research for UN Conference on Desertification, Nairobi, 29 August–9 September.

Street-Perrott, A. and Roberts, N. (1983) 'Fluctuations in closed basin lakes as an indicator of past atmospheric circulation patterns', in *Variations in the Global Water Budget*, A. Street-Perrott (ed.), Dordrecht, Reidel, 331–345.

Stuiver, M. (1961) 'Variations in radiocarbon concentration and sunspot activity', *J. Geophysical Research*, 66(1), 273–276, Washington, DC.

Swain, A. M., Kutzbach, J. E., and Hastenrath, S. (1983) 'Estimates of Holocene precipitation for Rajasthan, India, based on pollen and lake-level data', *Quaternary Research*, 19, 1–17, Seattle.

Tickell, C. (1986) *Climatic Change and World Affairs*, Lanham, Maryland and London, University Press of America, Inc. for President and Fellows of Harvard College.

Wetterkarte des Deutschen Wetterdienstses (1976, 1978) Reports in the *Amtsblatt des Seewetteramtes*, Nr. 75, 1976, and Nr. 59, 1978.

Winstanley, D. (1973) 'Rainfall patterns and general atmospheric circulation', *Nature*, 245, 190–194, London, 28 September.

Winstanley, D. (1974) 'Rainfall and river discharges in the sub-Saharan zone (10°–20°N) of Africa', *Background paper for the United Nations Food and Agriculture Organization's mission on inland fisheries in West Africa*.

Part II

Climate and
weather changes,
causes and
mechanisms

13

Causes and time-scales of climatic change

This chapter is taken from an invited paper given at the 21st Inter-University Geological Congress at Birmingham on 2–4 January 1974 and published under the title 'Changes of climate: the perspective of time-scales and a particular examination of recent changes' in the book of that Congress, in honour of Professor L. J. Wills and F. W. Shotton FRS, Ice Ages: Ancient and Modern, *A. E. Wright and F. Moseley (eds), Liverpool, Seel House Press, 1975.*

In the excerpt printed here a few details of the text have been updated, whereas most of the section on recent changes has been omitted to avoid duplication of the content of other chapters of the present book.

Introduction

Changes in the world's climate and their distribution have to be considered in separate groups according to the time-scales involved, which entail differences in the main causes underlying the changes observed as well as differences in the types of analysis that are appropriate.

There is also an important distinction to be made between, on the one hand, the mean conditions of successive climatic regimes, which may be deduced by considering the atmospheric circulation characteristics that would be in equilibrium with the environmental conditions prevailing (the extent of persistent ice and snow, and the known sea temperatures, etc.) and, on the other hand, the conditions which bring about the changes from one regime to another. The latter will have to be derived by quite different techniques and from different types of data.

Future effort is needed to extend the methods of analysis already used to some further climatic periods of great interest – e.g. the so-called 'Upton Warren' warm stage in the middle of the last (Weichselian) ice age – for which so far the field data have been inadequate. Fresh attention must, however, also be focused on a more detailed analysis of periods of rapid change, including those quite recent periods for which most information is available.

Because of the premature conclusion, about the beginning of the century, that climatic changes belonged only to the geological past, and that for the last 2500 years at least climate had been constant apart from largely random short-term fluctuations, the subject was for many decades neglected by meteorology apart from isolated figures such as C. E. P. Brooks, C. C. Clayton, W. Köppen, M. Milankovitch, G. C. Simpson and A. Ångström, most of whom were not strictly meteorologists. Investigation of the facts of past climate – and, indeed, their interpretation – was left to the field sciences: there was little contact between meteorologists and geologists, geomorphologists, botanists or marine biologists, and none at all with historians. Scientific research in meteorology over the last 35 years has concentrated too largely on developing mathematical models of the physical and dynamical processes of the atmosphere aimed at numerical methods of daily, and now longer-range, weather forecasting. The models have been based essentially on physical theory and the performance of the atmosphere during recent years, particularly since the International Geophysical Year in 1957–8 and the introduction of artificial satellites in 1960, when the regular daily observational coverage has for the first time extended to the whole Earth and to great heights above the Earth. One may say that the theoreticians of meteorology have, in seeking to verify our understanding of the development of climate, preferred global completeness of observation coverage to the long run of observations needed to survey long processes.

In 1969 the World Meteorological Organization abolished its Commission for Climatology. Nevertheless, awareness of significant climatic shifts occurring within our own century had been growing in many quarters for some years; and the last session of the Commission for Climatology established a working group on climatic fluctuations to report on the new demand for climatic forecasts and for advice on any undesired effects of human activities on the global climate. These demands arise from those concerned with forward planning in agriculture, industry, tourism, trade and government, and particularly in connection with provision for the energy and water needed by the increasing population and the anxieties about pollution. Thus, in the 1970 volume of the WMO Bulletin itself we read: 'From the viewpoint of human welfare, climatology will soon become the most essential branch of the atmospheric sciences' (Flohn 1970). This prediction began to be fulfilled when, in 1979, the WMO proclaimed a 'World Climate Programme' of research, to which all member states were urged to give priority.

The view in meteorology, which is possibly too readily accepted by planners in the computer age as applicable now, is that the problems of prediction of climate will be met by computation from suitably adapted mathematical models of atmospheric processes. There is some danger that the present inadequacies of knowledge will be overlooked by many because of the brilliance of the theoretical constructions. The doubling time of errors in the forecast evolution of atmospheric patterns, stemming from inaccuracies of specification of the initial situation, is found by experiment (Lorenz 1969a, 1969b; Washington 1972) to be of the order of 3–8 days; this probably sets an ultimate limit of about 7–15 days to the range of forecasts which attempt to specify the positions of individual cyclones and anticyclones. Moreover, computation costs become prohibitive once really long periods are attempted, e.g. 1200 hours of computer time at the ESSA Geophysical Fluid Dynamics Laboratory, Princeton were used by an experiment by Manabe and Bryan (1969) with an elaborate model of atmospheric processes, in the usual time-steps on a scale of minutes, integrated over one year and combined with a model of the processes of 100 years in the ocean.

Climatic studies seem likely therefore to be mainly pursued with the much simpler models of climatic mean state, which explore the relationships between the average budgets of radiation and of heat transported by the mean winds and ocean currents, treating the effects of convection and turbulence also in terms of averages. Probably the most reliable estimates of the manner of operation, and the net warming effect, of the human-made increase of carbon dioxide on world temperature have already been achieved in this way (Manabe and Wetherald 1967). Doubling the atmosphere's carbon dioxide content should apparently raise the overall mean temperature at the surface of the Earth by 1.9 degC. Similar treatment of other pollutants, particularly the increase of particulate matter in the atmosphere (Rasool and Schneider 1971), must modify this conclusion somewhat (e.g. Mitchell 1972) and it may even be that the various effects of man's activities as regards warming or cooling of world climate at present tend to cancel one another out; though a net warming has to be expected to take over in future centuries if human energy production – particularly the waste heat of industrial processes – goes on increasing.

Improving the reliability and completeness of the theoretical models will be a long process. The most fundamental need, however, in the quest for climatic – and even for seasonal – forecasts is for improvement of the observational base of the subject. Our knowledge of the most basic facts of the mechanism of climate, let alone of the long record of its behaviour and the multiplicity of influences upon it, is still insufficient. The accepted estimates of global albedo have been revised downwards in successive stages over the past 15–20 years from 43 to 29 per cent, but it is now reported (J. Charney, personal communication, 9 January 1973) that the albedo of the great sandy deserts has now been revised upwards from about 30 to 40 per cent.

Satellite observations have recently shown (G. J. and H. J. Kukla, unpublished) that the total area of snow and ice is subject to much greater variations over periods of a few years (of up to 12 per cent over the northern

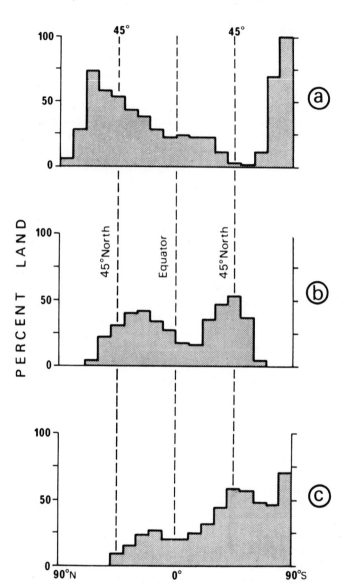

Figure 13.1 Percentage of land by 10° latitude zones: (a) present Earth; (b) with the North Pole in mid-Pacific but with the continents in their present relationships; and, (c) with the supposed geography of 400 million years ago: South Pole in Gondwanaland. (Adapted from Cox 1968.)

hemisphere) than had previously been suspected. Another great uncertainty is in the overall moisture exchanges, owing to the unsolved problem of measuring precipitation over the oceans which constitute 71 per cent of the Earth's surface.

The greatest need, however, must be the provision of the longest possible observation record of past climate, in as much detail and over as much of the Earth as the availability and types of data allow. Without this data base, there can be no prospect of identifying any recurrent processes of climatic fluctuation which operate over periods longer than a few years nor any related causative phenomena in the terrestrial or extraterrestrial environment. This is where the field sciences obviously have much to contribute. Meteorology and climatology greatly need this contribution. Because of past neglect, and the dire problems for humanity (such as the global cooling since around 1950 and the currently increasing droughts in Africa) which demand understanding of climatic trends, rapid steps forward must be made along several new roads:

(i) in observational climatology, monitoring the present and reconstructing the past;
(ii) in synoptic climatology, developing methods of handling the data of the long climatic record and current climatic processes;
(iii) in theoretical understanding, a physical (really a geo/astrophysical) and dynamical climatology;
(iv) in mathematical modelling of the long-term processes in the atmosphere and oceans.

The Climatic Research Unit at the University of East Anglia was set up in 1972 with these aims.

The time-scales of climatic variation

The different time-scales involved in the climatic record correspond to different processes and different main underlying causes. They may be conveniently listed as follows:

(i) Changes over geological time, say, 10^6–10^9 years.
(ii) Changes on time-scales of the same order as the astronomical variations of the Earth's orbital arrangements, 10^4–10^5 years.
(iii) Recurrent changes, many of them apparently cyclic, which seem to be associated with fluctuations of solar output or solar disturbance, combined tidal force of the sun and moon, volcanic activity, etc., over periods of from 1–2 years up to 10^3 years.
(iv) Shorter-term changes plainly associated with variations in the circulation and heat economy of the oceans and atmosphere, but which may in some cases also be related to an external impulse, e.g. a solar flare or particle invasions of solar or cosmic origin.

Climatic changes on the geological time-scale

The climatic changes which geologists' evidence shows have taken place over millions of years, and longer, including both (a) the prevalence of warm climates with no permanent ice through most of the course of the Earth's history and (b) occurrences of glaciation (or alternations between glacial and interglacial climates) over a few million years, apparently at rather evenly spaced intervals of about 250–300 million years, may be interpreted largely in the following terms:

  (i) palaeogeographic changes, namely:
    (a) polar wandering and changes of latitude of the main land-masses;
    (b) drift of the continental plates, changing their positions relative to one another;
    (c) uplift and erosion, changing the general relief and the disposition of mountain barriers;
 (ii) possible changes in the total mass and chemical composition of the Earth's atmosphere and oceans (the latter would be important if they changed the overall radiation balance);
(iii) change in the heat flux from the Earth's interior;
(iv) changes of the energy output of the sun.

Only the changes listed under (i) and (iv) have been generally thought to be of sufficient magnitude to have affected global climate significantly during the last 500 million years. Budyko (1982) has, however, argued that the changes of climate through the geological past can be largely explained in terms of the carbon dioxide and water supplied by volcanic activity, which has gradually changed the character of the Earth's atmosphere. The Earth's climate is some 40 degC warmer than would be expected at this distance from the sun, thanks to the extent to which the Earth's out-radiation is intercepted by the carbon dioxide, water vapour, and so on, in the atmosphere. Budyko presents the changes of climate through the geological past largely as a response to the varying amount of carbon dioxide in the atmosphere. Decline of this CO_2 set in as the volcanism eased off and as vegetation appeared and increased. The vegetation removed some of the carbon dioxide and converted it into oxygen. This, according to Budyko, was accompanied by a lowering of temperature and the appearance of ice around the poles. Other, shorter-term temperature drops have – as Budyko agrees – undoubtedly been associated with periods of specially active volcanoes and frequent loading of the atmosphere with veils of dust and aerosol. Not all geologists, however, agree that the long record of carbon dioxide in the atmosphere fits so well with past variations of prevailing temperature level.

   Polar wandering and continental drift – jointly considered as 'crustal wandering', to use F. W. Shotton's term – are likely to produce changes in the distribution of temperature with latitude, since they make great changes in the land fraction at different latitudes (figure 13.1). That this is clearly to be expected can be seen from empirical formulae such as that of Spitaler (1886), quoted by Brooks (1949), expressing the present mean temperature of a latitude zone as a function just of latitude and of the land fraction. Such formulae cannot, however, be expected to give wholly trustworthy results if applied to gross changes from the present geography, since the effects of consequential complicated changes of heat transport by the ocean currents and atmospheric circulation are not provided for. Nevertheless, in the extreme case of the pole lying in the middle of a great ocean it may reasonably be assumed that ice and snow could not accumulate. Thus, a geography with a large land fraction in high latitudes may be seen as a necessary precondition for glaciation, though it is known that glaciation has not invariably occurred when such a situation existed. Hence, other conditions must also be needed.

High mountainous relief may be conducive to glaciation in two ways: (a) by deforming the flow of the prevailing upper westerly winds in middle latitudes, and so promoting mean meridional (north–south) components of flow and extensions of the polar climatic regime in certain longitudes; and (b) by providing a high terrain on which summer snow-melt is more liable to fail and ice-accumulation to begin.

According to Cox (1968), there can be little doubt that as the composition of a star like the sun changes its luminosity increases. The rate of the increase is not known, but Öpik (1965) has suggested that during the pre-Quaternary ice ages the solar output may have been less than 88 per cent of the present value and that glaciation should then extend to the Equator. Such complete coverage of the Earth would probably not be consistent with the evidence of occurrences of glaciation and its distribution between Devonian and Permian times, 400 to 250 million years ago; but the suggestion may provide an explanation of severe Precambrian ice ages more than 500 million years ago (Harland 1964). It is not clear, however, whether or how the Earth could recover from 'complete glaciation' – or an ice cover over all the land areas – because of the enormous (approx. 80 per cent) loss of the solar radiation falling on a snow or ice surface due to its reflectivity (albedo). Öpik (1958) regards it as certain that the sun's luminosity also undergoes a slow 'flickering', and regards the occurrence of terrestrial ice ages at fairly uniform intervals of 250–300 million years as evidence of this. Umbgrove (1947) first suggested, and Steiner (1967) developed the theory, that flickering on this time-scale may be ascribed to the gravitational disturbance due to the regular motions of the other systems within the galaxy.

It is a remarkable fact that the variations of climate on the Earth have never been great enough to extinguish all life on the planet – if not altogether since life first appeared over 3000 million years ago, then at any rate not for many hundreds of millions of years past. Nevertheless, some episodes when there was a massive extinction of species, as with the dinosaurs about 65–70 million years ago, may be attributable to severe climatic upsets. There were other cases of such disasters, when reptiles or amphibian species died out, at the end of the Triassic and the Permian, respectively about 180 and 230 million years ago. Evidence has been cited of exceptional volcanic activity and dust deposits at these times (e.g. Kennett and Watkins 1970), making a temporary drastic lowering of world surface temperature likely. One suggestion has been the impact of some body from outer space, as cause of some of these disturbances.

Climatic changes associated with variation of the Earth's orbital elements

The cyclic changes of solar radiation at the top of the atmosphere which must occur due to the variations of:

   (i)   obliquity (tilt of the Earth's rotation axis) – period about 40,000 years,
   (ii) ellipticity of the Earth's orbit – period nearly 100,000 years,
   (iii) precession of the equinoxes – period about 21,000 years,

first reliably calculated by Milankovitch (1930), have latterly been recalculated by Vernekar (1972), giving the radiation totals at each latitude for each of the caloric halves of the year at 1000-year intervals from 2,000,000 years in the past to 110,000 years into the future.

These are variations about which we can be certain, and which must occur even with constant energy output from the sun (which is assumed in the calculations). When allowance is made for consequential changes of the Earth's albedo due to changes in the accumulation of snow and ice, and for probable adjustments of the global atmospheric circulation, they appear likely to produce temperature changes of about the same magnitude as those deduced from observation of the differences between glacial and interglacial periods in the Quaternary and, presumably, likewise in other periods of the Earth's history which had a suitable geography for ice ages to develop. They probably do not, however, account for the rapidity of some of the climatic changes, which points to the contributory effect of other influences, such as solar fluctuations, variations of tidal force or of volcanic dust load in the atmosphere, all on various shorter time-scales.

The timing of the last glaciation (the only one for which radiometric dating methods give some key points with sufficient accuracy) seems to fit reasonably well with the radiation variations due to the orbital variables, subject to a lag of about 5000 years required for the melting of the great ice-sheets. There is some interest, therefore, in the reproduction of Vernekar's results here in figure 13.2 for comparing the present climatic situation with corresponding points in previous interglacials and as a frame of reference for probable future climatic development.

Figure 13.2 is from a presentation by Mitchell (1972) of the radiation totals for each latitude in the summer and winter half-years calculated by Vernekar (1972), the upper panels of the diagram covering the time from

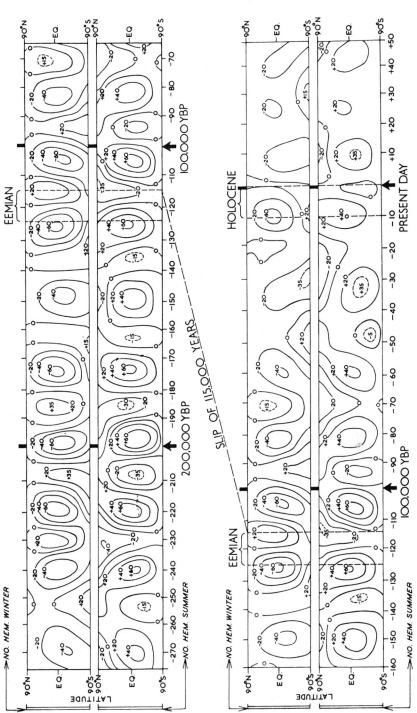

Figure 13.2 Distribution of solar radiation reaching the top of the Earth's atmosphere in the two halves of the year, by latitude, from pole to pole (upper panel) from 275,000 years ago to 65,000 years ago, and (lower panel) from 160,000 years ago to 50,000 years in the future. Units: calories/cm²/day, expressed as departures from present values at each latitude.

Table 13.1 Periodicities most frequently and prominently indicated in meteorological and related data

Item analysed	Period lengths (years)								
Sunspot numbers (direct observations since c.1700, indirect data from antiquity)		c.5½		10–12		c.22–23 'Hale cycle'	80–90	c.170–200	(400?)
Radioactive carbon in the Earth's atmosphere (clear indication since 200 BC)								c.200	
Temperatures in Europe (1760+)	c.2.2								
Temperatures in England (c.1100+)								170–200	
Rainfall in Europe (c.1800+)	2–2.5								
Rainfall in England (c.1100+)							c.100	170–200	
SW winds in England (1340–1965)								c.200	
Baltic ice (1900–1950)	3	5–6	8	11–14		21–24			
Severity of European winters (1215–1905)							89.5		
W. and E. Greenland and Barents Sea ice (1820+)						c.71–77			
Greenland ice-cap, thickness of annual layers (AD 1200+)							78	181	
Night cloudiness in China (2300 BC+)									c.400
Pressure differences in January (1850+) Madeira–Iceland and Siberia–Iceland	3	6		11–12	15	18–24			
Pressure differences in January and July (1750+)								(c.190?)	
for W'ly winds over Britain						18–23			
for S'ly winds over the North Sea						18–23	55–80		
Latitude of the anticyclone belt:									
(a) over Siberia in January (1836+)							80–85		
(b) over the N. Atlantic in July (1865+)							85–110		
Zonal index at 500 mb measured right round the northern hemisphere between 35° and 55° N in January (1943+)	2.2								
Meridional circulation pattern frequency over Europe (1881–1964)	2.1–2.2								
Pleistocene ice-age varves	2–3		7–8	10–12	15–20				
Tertiary strata (e.g. Miocene, Oligocene and Eocene, various strata in Russia and USA)	2–3	5–6		11–12		23			
Miocene (Sarmatian anhydrite in Sicily)				9–10	15	21 and 32–33			
Mesozoic strata (e.g. Cretaceous, Jurassic)	c.2½	6	8	10–13				170–180	
Palaeozoic strata (including Permian, Carboniferous, Devonian)	c.2½	c.5½	7–8	10–14					
Upper Permian (Zechstein) salt deposits in Germany		5–6		10–12		23	85–105	170–210	400
Precambrian	c.2½	c.5½		11–12					

Full references to sources given in Lamb (1972).

275,000 to 65,000 years ago and the lower panels from 160,000 years ago to 50,000 years into the future. The presentation is designed to show how far the present situation (right-hand arrow on the lower panels) is similar to that 115,000 years ago (placed immediately above it in the upper panels). That was in the last (Eemian) interglacial, and full interglacial conditions ended some time within the following 5000 years. There is no doubt about the qualitative resemblance between this and our present situation, but the radiation changes that followed were greater in amplitude than any expected in the next several thousand years: this is because the ellipticity of the Earth's orbit has remained slight over the past 100,000 years and is expected to continue slight for the next 100,000 years, a situation which last occurred between 1.8 and 1.6 million years ago. On this ground, Mitchell doubts whether the climatic oscillations that lie ahead will be so severe as those which marked the last several interglacial–glacial oscillations. It is worth remarking, however, that the severest developments of the last glaciation seem to have set in within the last 70,000 years, particularly some time between about 60,000–70,000 years ago and from about 30,000 years ago onwards, a 40,000-year spacing which suggests association with the obliquity variations. Figure 13.2 shows that there is some quantitative as well as qualitative resemblance between our present situation and that about 68,000 and 42,000 years ago.

Other workers (e.g. Kukla and Matthews 1972) have concluded from the parallel course of vegetation history in postglacial times and previous interglacials that the present situation is that of an interglacial in its late temperate stage – the third of the four zones of a typical interglacial defined in West's scheme (1968, p. 303) – and that successive future climatic oscillations on time-scales from centuries to millennia superposed on the declining tendency of the radiation regime will bring step-wise advances of the natural vegetation boundaries equatorward. There are also parallels in the last two interglacials for the mid-postglacial elm decline, which appear to support this analogy. Sample interglacial and postglacial pollen diagrams for the British Isles will be found in Sparks and West (1972).

Since this section was written, Berger (1980) and Kukla and Berger (1979) have recalculated the radiation available at each latitude of the Earth, month by month and season by season, at thousand-year intervals over the past 400,000 years and more than 100,000 years into the future. Their work confirms the agreement of timing between the glacial periods of the last 400,000 years (see also Imbrie and Imbrie 1979) and the variations of the available energy budget due to the astronomically predictable variations in the Earth's orbit and attitude or tilt to the plane of the orbit. The calculations also support the expectation of another glaciation, with its first, relatively modest climax about 5000 years from now and more severe stages later, so that around 60,000 years in the future conditions should be not much different from the climax of the last two Ice Ages.

Climatic changes during and since the last glaciation

There is enough observational evidence, from palaeobotany, marine biology, geology, glaciology and isotope work, etc., to permit reconstructions of the temperature fields and prevailing patterns of the atmospheric and ocean circulations of the last main climax of the last glaciation

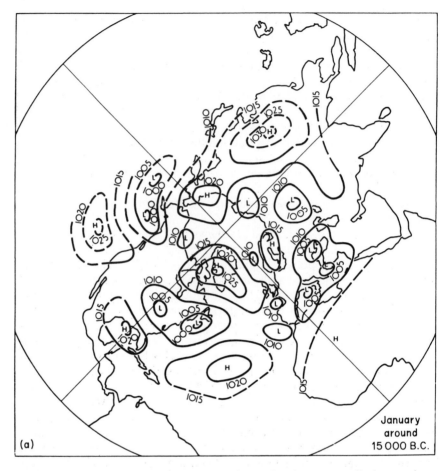

January
around
15 000 B.C.

(a)

and different well-marked climatic regimes that have prevailed at various stages since.

Examples of the results of this work for the glacial climax about 17,000 to 20,000 years ago are illustrated here in figure 13.3 and in figures 9.5 and 9.6 (pp. 150–1). Already interesting patterns are coming to light which tend to verify each other – e.g. the ocean current implied by the transport of volcanic debris on sea-ice in the North Atlantic (figure 13.4) fits with the prevalence of northerly winds suggested by the atmospheric pressure distribution maps reproduced here (figure 13.3) and others for the late glacial phases published elsewhere (Lamb and Woodroffe 1970, Lamb 1977). It may also be possible to verify the circulation pattern by fossil evidence of the winds that prevailed, e.g. in sand dunes or loess of relevant age, and by evidence of the distributions of precipitation.

Corresponding maps for various postglacial stages have been given by Lamb, Lewis, and Woodroffe (1966) and, with dates corrected by the

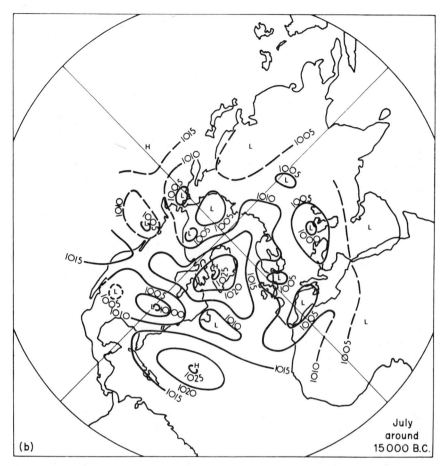

Figure 13.3 Prevailing distribution of barometric pressure at mean sea-level, and implied winds, about the time of the last maximum of glaciation: (a) January; and (b) July. The winds blow counterclockwise around regions of low pressure, and clockwise around regions of high pressure, in the northern hemisphere. (As derived by Lamb and Wright 1974.) The present-day geography of the base map has not been adjusted to show ice-age features in this preliminary presentation.

subsequent calibration of radiocarbon measurements, in Lamb (1977). It must be hoped that similar derivations may be made for earlier stages of the last glaciation, when the available field evidence (particularly of prevailing surface temperatures) provides sufficient coverage of the hemisphere. Such reconstructions will hardly be possible, however, for times beyond the limit of radiocarbon dating with an error margin of at most a few thousand years. Other, more elaborate methods of meteorological reconstruction have been applied (e.g. Williams *et. al.* 1973), so far with results not too different. In the present writer's opinion, whatever method involves the least input of

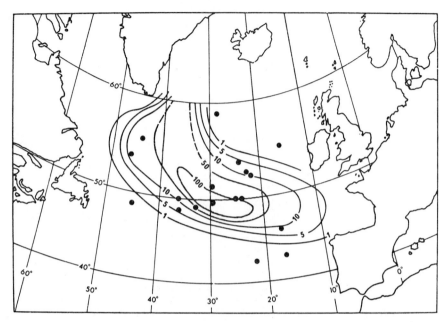

Figure 13.4 Distribution in the Atlantic ocean-bed sediment of Iceland and Jan Mayen volcanic debris of sand-size particles, too large to be carried by the winds and presumed therefore to have travelled on floating sea-ice. The position in the cores indicates that the time of greatest accumulation of this debris was about 10,000 years ago. (After Ruddiman and Glover 1974.)

theory, and fewest assumptions about parameters for which direct indications are lacking, must give results that are closest to the facts (even if those results do not cover at all many items that might be of interest). His own reconstructions have therefore made use only of well-known and understood physical laws governing temperature changes with height in the atmosphere and the dynamics of the wind flow which generates cyclones and anticyclones.

It must be understood that all derivations of the kinds here discussed concern average conditions which prevailed over long periods of time. The derivations of the atmospheric circulation, in principle, indicate the circulation regime that would be in equilibrium with the given surface distribution of snow and ice and ocean temperature.

To understand what conditions brought about the changes from one climatic regime to another, we must reconstruct the events that occurred over much shorter spans of time. This is almost a virgin field waiting to be investigated. It is of great importance for what it may tell of the rapidity, and evolution and processes, of climatic changes liable to occur even at the present time. And, by the same token, examination of the most rapid climatic shifts observable in historical or recent times may throw some light upon the sharper changes at the times of onset and ending of glaciation. For work on this, great care must be exercised in identifying lags in the response of the vegetation, or other evidence used, to changes of weather and climate. Particular value attaches to those types of evidence, such as tree-rings, varves and ice-sheet stratigraphy, as well as documents of the historical period, which give evidence of the events of individual years and make possible some study of their sequence.

Modern methods of spectral analysis of time series, made possible by computers, have renewed interest in searching for quasi-periodic components in long series of meteorological and other weather-related observations (table 13.1). This table, which includes analyses of data from the modern era of observations with meteorological instruments and from eras in the remote geological past, seems to confirm the existence of some more or less cyclic, or oscillatory, tendencies in climate which have continued to operate over great stretches of the Earth's history.

The most prominent quasi-periodicities seen in this table, from the greatest diversity of data and times, are a somewhat variable 2- to 3-year oscillation and others with time-scales of about 5½, 11, 19, 23, 90–100 and 170–200 years.

Some of these quasi-periodicities have also been identified in the data on solar fluctuations. The 19-year period is probably tidal. Symptoms of a 400-year oscillation, also seen in the data in the table, have been reported in the long record (over 7000 years) of variations of atmospheric radiocarbon as well; this may therefore be another period of solar origin. Bray (1971) has also reported prominent oscillations, indicated in glaciological data (e.g. from Dansgaard's work on the Greenland ice core at 77°N 56°W), on periods of about 1000–1300 and 2000–2600 years. Much of the evidence suggests that the amplitude of the oscillations on time scales of about a century and longer is greater than that characterizing some of the shorter-term fluctuations. Presumably, therefore, the operation of these, superposed on the still longer-term trends, is liable to produce the sharpest changes of climate. Any recurring oscillation in climatic time series must correspond to some process which characteristically requires a certain time for its completion and which it should be the aim of future research to identify.

References

Berger, A. (1980) 'The Milankovitch theory of paleoclimates – a modern review', *Vistas in Astronomy*, 24, 103–122, Oxford, Pergamon.

Bray, J. R. (1971) 'Solar–climate relationships in the post-Pleistocene, *Science*, 171, 1242–1243.

Budyko, M. I. (1982) *The Earth's Climate: Past and Future*, New York and London, Academic Press. (Originally published in Russian, Leningrad, Gidrometeo. Izdat., 1980.)

Cox, A. (1968) 'Polar wandering, continental drift, and the onset of Quaternary glaciation', *Met. Monogr.*, 8, 112–125.

Flohn, H. (1970) 'Climatology – descriptive or physical science?', *WMO Bull.*, 19, 223–229.

Harland,W. B. (1964) 'Evidence of Late Precambrian glaciation and its significance', in A. E. M. Nairn (ed.), *Problems in Palaeoclimatology*, London, Interscience, 19–149, 179–184.

Imbrie, J. and Imbrie, K. P. (1979) *Ice Ages. Solving the Mystery*, London, Macmillan.

Kennett, J. P. and Watkins, N. D. (1970) 'Geomagnetic polarity change, volcanic maxima and faunal extinction in the South Pacific, *Nature*, 227(5261), 930–934, London, 29 August.

Kukla, G. J. and Matthews, R. K. (1972) 'When will the present interglacial end?', *Science*, 178, 190–191.

Kukla, G. and Berger, A. (1979) 'The astronomic climate index', paper presented in the International Scientific Assembly, 'The Life and Work of Milutin Milankovitch', Belgrade, Serbian Academy of Sciences and Arts, 9–11 October.

Lamb, H. H. (1966) *The Changing Climate*, London, Methuen.

Lamb, H. H. (1967) 'Britain's changing climate', *Geogr. J.*, 133, 445–468.

Lamb, H. H. (1972) *Climate: Present, Past and Future. Vol. 1, Fundamentals and Climate Now*, London, Methuen.

Lamb, H. H. (1977) *Climate: Present, Past and Future. Vol. 2, Climatic History and the Future*, London, Methuen.

Lamb, H. H., Lewis, R. P. W., and Woodroffe, A. (1966) 'Atmospheric circulation and the main climatic variables between 8000 and 0 BC: meteorological evidence', in J. S. Sawyer (ed.), *World Climate from 8000 to 0 BC*, London, Proc. Internat. Sympos., R. Met. Soc., 174–217.

Lamb, H. H. and Woodroffe, A. (1970) 'Atmospheric circulation during the last ice age', *Quaternary Res.*, 1, 29–58.

Lamb, H. H. and Wright, P. B. (1974) 'A second approximation to the circulation patterns prevailing at the time of the last glacial maximum', in *Proceedings of the International Conference on Mapping the Atmospheric and Oceanic Circulations and Other Climatic Parameters at the Time of the Last Glacial Maximum about 17000 Years Ago*, Norwich, University of East Anglia, Climatic Research Unit, CRU RP2.

Lorenz, E. N. (1969a) 'The future of weather forecasting', *New Scient.*, 42, 290–291.

Lorenz, E. N. (1969b) 'The predictability of a flow which possesses many scales of motion', *Tellus*, 21, 289–307.

Manabe, S. and Bryan, K. (1969) 'Climate calculations with a combined ocean-atmosphere model', *J. Atmos. Sci.*, 26, 786–789.

Manabe, S. and Wetherald, R. T. (1967) 'Thermal equilibrium of the atmosphere with a given distribution of relative humidity', *J. Atmos. Sci.*, 24, 241–259.

McIntyre, A. (1974) 'The CLIMAP 17000 years BP North Atlantic map', in *Proceedings of the International Conference on Mapping the Atmospheric and Oceanic Circulations and Other Climatic Parameters at the Time of the Last Glacial Maximum about 17000 Years Ago*, Norwich, University of East Anglia, Climatic Research Unit, CRU RP2.

Milankovitch, M. (1930) 'Mathematische Klimalehre und astronomische Theorie der Klimaschwankungen', in Köppen, W. and Geiger, A. (eds), *Handbuch der Klimatologie*, Berlin, Teil A, Borntraeger.

Mitchell, J. M. (1972) 'The natural breakdown of the present interglacial and its possible intervention by human activities', *Quaternary Res.*, 2, 436–445.

Öpik, E. J. (1958) 'Solar variability and palaeoclimatic changes', *Ir. Astr. J.*, 5, 97–109.

Öpik, E. J. (1965) 'Climatic change in cosmic perspective', *Icarus NY*, 4, 289–307.

Rasool, S. I. and Schneider, S. H. (1971) 'Atmospheric carbon dioxide and aerosols: effects of large increases on global climate', *Science*, 173, 138–141.

Ruddiman, W. F. and Glover, L. K. (1974) 'Counterclockwise circulation in the North Atlantic subpolar gyre during the Quaternary', in *Proceedings of the International Conference on Mapping the Atmospheric and Oceanic Circulations and Other Climatic Parameters at the Time of the Last Glacial Maximum about 17000 Years Ago*, University of East Anglia, Climatic Research Unit, CRU RP2, Norwich.

Sparks, B. W. and West, R. G. (1972) *The Ice Age in Britain*, Methuen, London.

Spitaler, R. (1886) 'Die Wärmevertheilung auf der Erdoberfläche, *Denkschr. Akad. Wiss., Wien*, 51, Abt. 2, 1–20.

Steiner, J. (1967) 'The sequence of geological events and the dynamics of the Milky Way galaxy', *J. geol. Soc. Aust.*, 14, 99–131.

Umbgrove, J. H. F. (1947) *The Pulse of the Earth*, The Hague, Martinus Nijhoff.

Vernekar, A. D. (1972) 'Long-period global variations of incoming solar radiation', *Met. Mongr.*, 12(34).

Washington, W. M. (1972) 'Numerical climatic-change experiments: the effect of man's production of thermal energy', *J. Appl. Meteorol.*, 11, 768–772.

West, R. G. (1968) *Pleistocene geology and biology*, London, Longman.

Williams, J., Barry, R. G., and Washington, W. M. (1973) 'Simulation of the climate at the last glacial maximum using the NCAR global circulation model', *Occ. Pap. Inst. arct. alp. Res. Boulder, Colorado*, 5.

14

An outline of the world's wind circulation and weather systems, and their part in the variations of weather and climate, with notes of some relationships to birds and the living world

The following brief sketch of the development of the general wind circulation and individual weather systems is based on a lecture given to members of the Royal Society for the Protection of Birds and others at the 1975 Joint Conference of the British Trust for Ornithology and Climatic Research Unit, at Swanwick, Derbyshire, on 15 February 1975, and originally published in Bird Study, *22(3), 1975, pp. 121–141. Some attention is given to the effects on birds, insects, pollen transport, etc.*

The climate of any place on the Earth is broadly determined by:

(1) the radiation conditions – the balance of gain and loss, of incoming solar and outgoing terrestrial radiation and the exchanges within the atmosphere;
(2) the heat and moisture transported (to and from) by the winds;
(3) the heat transported and the moisture made available by the ocean currents (which are wind-driven);
(4) influences of the locality:

> the nature of the surface, its albedo (i.e. its property of reflecting away some part of the incident radiation), the specific heat, thermal conductivity and water content of the soil, urbanization (e.g. paved surfaces and artificial drainage), the effects of buildings, trees and hills upon the winds, and the aspect or direction of the slope with respect to the incoming solar beam.

Different surfaces differ greatly in these regards. New snow reflects away about 90 per cent of the radiant energy falling on it, most other surfaces only 10 to 30 per cent. The specific heat of water is 1, that of most rocks only 0.2–0.3. Friction reduces the speed of the wind over most land surfaces to on average about half its speed over the sea and a third or less of that in the free air at a height of a few hundred metres.

The flux of heat from the Earth's interior is generally insignificant, only about 0.1 g cal/cm^2/day compared with a global average of 720 g cal/cm^2/day that would be supplied by the sun to a surface normal (i.e. at right angles to) to the solar beam in the absence of any interference by the atmosphere.

The radiation available is, of course, graduated according to latitude and season. And owing to cyclic changes in the Earth's orbital conditions over tens of thousands of years, small but significant changes take place in the budget over those periods. The amount of radiation absorbed, and hence, the surface temperatures developed, change very significantly with changes in the nature of the surface – most of all when the surface becomes snow covered.

It is the inequalities of heating of different parts of the Earth which set the atmosphere in motion. Suppose for a moment that the air is everywhere at rest on the Earth's surface. The atmospheric pressure would have to be everywhere the same at sea-level, for where a pressure gradient exists, there is a force to move the air and cause a wind. The air must, however, become less dense, and expand vertically, over the warmer regions than over the places where it is cold and vertical columns of air contract. Hence, if we consider any particular height above the Earth's surface, there will be more of the atmosphere above this height over the warm regions of the Earth than over the cold regions. So a pressure gradient will be bound to exist in the upper air, from high pressure over the tropics to low pressure over the poles.

These pressure gradients are just the pattern we do find prevailing through a great depth of the atmosphere from about 2000–3000 m above sea-level up to 15–20 km. Pressure throughout this range of heights is generally high over the tropics and low over the polar regions. It is also relatively low over the colder regions in each latitude zone. This is illustrated by figure 14.1(a) and (b), which show the average height at which the pressure is 500 millibars (about half the pressure of the atmosphere at sea-level) over the northern and southern hemispheres. We see that the prevailing pressure patterns at this sample height of 5–6 km are very simple and consist of a single low-pressure region in high latitudes, surrounded by increasing pressures as far as the tropics. We also see troughs of low pressure over cold regions like north-east Canada, and ridges over the warmer waters of the north-eastern parts of the North Atlantic and Pacific Oceans. The gradients are much stronger over the southern hemisphere (particularly over the great ocean in middle latitudes) than over the northern hemisphere, as shown by the closer spacing of the contours. This is because all the land in the highest southern latitudes is still in an ice-age situation.

The differences of pressure represent a force which starts the air moving. But a state of balance is soon approached between this force and others, such as centrifugal effects and the deviating force of Earth rotation, which act upon the air once it gets under way. The result is that the wind blows

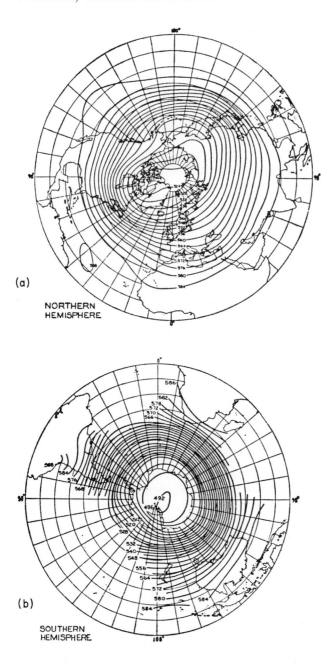

Figure 14.1 Contours of the average height (in metres) of the 500-mb pressure level: (a) northern hemisphere, (b) southern hemisphere. (All seasons of the year averaged; data approximately 1950–4.)

nearly along the lines of equal pressure ('isobars') – or along the contours of a constant pressure surface, as shown in figure 14.1(a) and (b) – counter-clockwise around regions of low pressure in the northern hemisphere, and clockwise in the southern hemisphere. Its speed increases with the steepness of the gradient. Hence, as these maps indicate, the main wind current aloft is a single great vortex or whirl of more or less westerly winds blowing around the low-pressure centre over the polar regions in each hemisphere – the circumpolar vortex. This basic circulation is stronger in winter, when the temperature contrasts between tropics and poles are greatest, than it is in summer.

This wind system which we find through a great depth of the atmosphere between about 2–20 km above the Earth's surface is, in fact, the main flow which carries most of the momentum. As the upper winds circle the Earth, they pass from regions of weak to regions of strong pressure gradients and vice versa. Hence, the air undergoes acceleration in one part of the map and has to slow down in another. The temporary disequilibrium between the forces acting upon the air in either case causes some departure from strict flow along the pressure lines (contours or isobars). This is the origin of those shifts of mass which create the anticyclones and cyclones, or depressions, of the surface weather map. These familiar features develop and decay in response to changes in the upper wind pattern over them and are carried along with the general direction of the massive flow of the mainstream of the upper winds.

High pressure at the surface is the dominant development along the warm flank (equatorward side) of the strongest upper flow, and produces a belt of anticyclones there. Low pressure is produced by the dynamical effects prevailing along the cold (poleward) side of the main flow. The strongest upper flow at any given moment is represented by great concentrations of high wind speeds, i.e. by 'jet-streams', in parts of the strong wind belt called 'frontal zones', where the temperature contrast is particularly strong. Speeds of 50–75 m/s (100–150 knots) are fairly commonplace in the stronger jet-streams at heights of 5–10 km in winter, and 100 m/s (200 knots) is sometimes exceeded at 10–12 km. An opposite distribution of surface pressure developments, with high pressure on the cold side and low pressure on the warm side of the mainstream of the upper winds, tends to occur over limited sectors of each hemisphere near the confluences to the strongest parts of the upper windstream and in some special configurations of the flow, when and where greater than usual deviations of the flow extend the polar regime towards lower latitudes.

We have already met the circumpolar vortex in chapter 9, where the principal types of variation of its pattern were illustrated in figure 9.1 (p. 143). These show sometimes great deviations of the flow, which may amount to complete blocking of the westerlies in parts of the latitude zone where these winds usually prevail and the surface weather systems then become almost stationary (or may even move in a westward direction) in

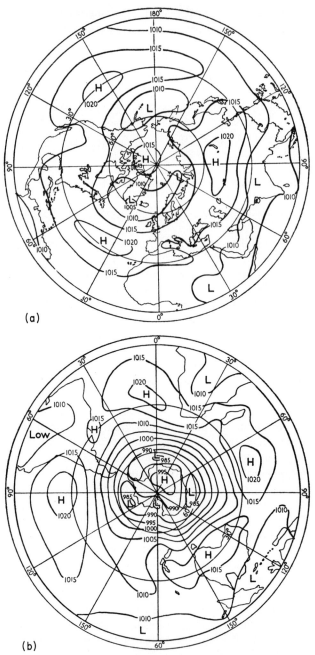

Figure 14.2 Average barometric pressure distribution at sea-level: (a) northern hemisphere 1900-39, (b) southern hemisphere 1900–50 approx. (All seasons of the year averaged.)

that part of the map. The full list of variations of the circumpolar vortex includes variations of the strength and of the latitude of the mainstream of the upper winds, changes in the spacing (or wave-length) between one meander and the next and changes in the amplitude or latitude-range, and distortion of these excursions of the flow. Changes in these items occur both seasonally and irregularly over shorter periods, as well as with the longer-term changes of climate. Finally, there are times when the whole pattern becomes eccentric – i.e. centred fairly far away from the geographical pole – as must have been a persistent feature of the situation during the last ice ages: at those times the Arctic cold surface spread far towards the south over North America and Europe, and over the Atlantic Ocean in between them, while there was much less change in the Pacific Ocean. There is a similar but much smaller eccentricity of the Antarctic ice and snow surface, and of the prevailing patterns of the large-scale wind circulation, over the southern hemisphere today.

The principal features of the prevailing distribution of surface barometric pressure which results are seen in figure 14.2(a) and (b). We may distinguish the following:

(i) a subtropical belt of high pressure over either hemisphere, divided into separate anticyclone cells over the oceans (and over central Asia in winter);

(ii) subpolar belts of low pressure, which are most marked over the oceans, though in the northern hemisphere in summer the regions of lowest mean pressure are over the land;

(iii) in summer the continent of Asia is dominated by the monsoon low pressure centred near 30°N. This is actually part – a displaced and invigorated part – of the equatorial low-pressure zone, which is also present in our winter, though farther south. Over the oceans the equatorial low-pressure zone is known as the Doldrums belt, the winds being light except in thunder-squalls;

(iv) regions of rather high average pressure near the poles and some extensions of them over regions of cold surface, especially over parts of the northern continents in winter. These extensions sometimes occur as separate anticyclones right in the subpolar zone of prevailing low pressure, where they are known as 'blocking anticyclones', because they appear to block, or reverse for a time, the prevailing westerly winds in middle latitudes. Their development is related to the confluent parts of the pattern of jet-streams aloft, in the manner mentioned, and to extensions of the polar cold regime to lower than usual latitudes. On the western sides of these blocking anticyclones (which are often stationary), however, warm air streams far towards higher latitudes.

An idealized scheme of the global arrangement of prevailing surface winds which the general pressure distribution produces, is shown in figure 14.3. The prevailing wind belts in each hemisphere – the polar easterlies,

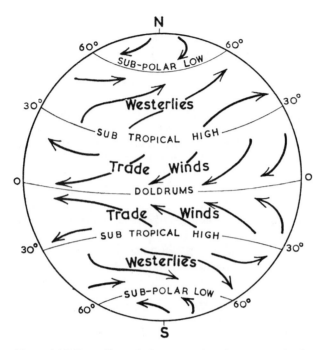

Figure 14.3 Prevailing wind zones and surface atmospheric pressure over an idealized Earth (with no geographical complications.)

middle-latitude westerlies and the easterly Trade Winds in low latitudes, as well as the belt of equatorial calms, or Doldrums, where the wind systems from the two hemispheres meet (the Intertropical Convergence Zone) – move north and south with the seasonal changes of the heat input; but this seasonal shift of the wind zones is much smaller than the range of the zenith sun. It is also more complex, being distorted by the geography of oceans, continents and great mountains. And it takes place in a series of erratic movements, some of which greatly exceed the total seasonal range and are followed by shifts in the reverse direction.

The detail of prevailing surface winds in January and July, as observed in the first half of the present century, is indicated by figure 14.4(a) and (b). The world distribution of surface ocean currents, which are largely wind driven, has been shown in figure 9.2 (p. 144). Their pattern resembles that of the average winds.

The flight of birds is influenced by the weather and wind-flow of each passing moment, and their ranges and migration patterns have obviously been affected by the changes of climate from ice age to interglacial and even by the smaller changes within the present century. Many bird species extended their ranges northwards during the warming period in the first 50–60 years of the twentieth century, and evidently learnt to adapt their migration patterns to the somewhat modified wind circulation patterns, occurrences of anticyclones, storm tracks, etc. that went with the warming. Swallows (*Hirundo rustica*) began to breed in the Faeroe Islands and in Iceland in the 1930s, starlings (*Sturnus vulgaris*) in Iceland from 1941, and fieldfares (*Turdus pilaris*) in Greenland and on Jan Mayen Island (71°N) after 1937. Some Arctic-nesting northern birds which had disappeared from

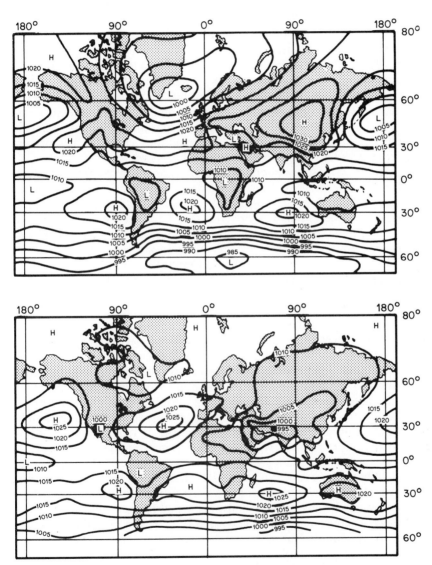

Figure 14.4 Prevailing barometric pressures at sea-level over the Earth; (a) January, (b) July. (Data 1900–50 approx.)

Britain during the warming period have returned to Scotland since the 1960s: examples are the snowy owl (*Nyctea scandiaca*) and the great northern diver (*Gavia immer*). Similarly, birds for which England is near the northern limit of their range, such as the wryneck (*Jynx torquilla*) and the woodlark (*Lullula arborea*), have been seen less frequently since about 1960–1970. (More examples and more discussion of this and similar matters are given in Lamb, 1977.) It is also known from Konrad von Gesner's *Historia animalium* (Zurich 1551–8) that the hermit ibis (*Geronticus eremita*), which today is confined to North Africa, Arabia and Asia Minor, where it nests on cliffs, bred as far north as central Europe in the warmer climates of the Middle Ages. As late as 1550 it was a familiar yearly migrant to Germany, where it was known as the *Waldrapp*, a German name which presumably indicates that its croak (*Rapp*) had long been familiar. But it is reported to have become extinct in Europe in the seventeenth-century cold climate period. Similarly, it is reported (Fredskild 1985) that Lapland buntings were living in northernmost Greenland, well north of their present limit, in the warmest postglacial times 5000 years ago.

It is clear from these considerations that the migration patterns and routes of various bird species, even those now well established, have largely been developed during the last few thousand years since the European and North American ice-sheets melted. Moreover, they have been subject to modifications even in the smaller climatic changes of our own day – and this despite the 'leeway' apparent in the birds' reactions to the weather. The late Sir Landsborough Thomson wrote (1964) of migration and the whole annual cycle of a bird's life alternating between the reproductive and non-reproductive states, adapted to seasonal fluctuations in the food supply, with necessary migration and periods of moulting as appropriate accessory adaptations: 'Migration from the breeding area often begins before the need is apparent, proceeds farther than the minimum necessity requires, and displays greater regularity than do the seasonal changes.'

The migrational behaviour of birds is thus evidently governed by internal (endogenous) compulsions, which commonly manifest themselves in increasing restlessness as the time for migration approaches. The time of actual take-off seems in many cases to be influenced by relatively high or rising temperature in spring and relatively low or falling temperature in autumn and, more particularly, by a period of good (anticyclonic) weather in the area where the birds are. Settled weather and clear skies along the route evidently help towards successful achievement of the migration. Dense fog tends to bring movements to a standstill. Storms, and particularly very strong headwinds or side-winds, are liable to cause disaster to many birds. Such weather – and prolonged winds from an abnormal direction, e.g. from easterly points in Europe – leads to 'accidentals' and the arrival of rare visitors. Thus, for ornithologists, as well as for the general purposes of this book, we must also turn our attention to the individual, passing weather systems.

The changing patterns of the wind circulation are at least equally important in connection with variations in the occurrence of insects, pollen transport and all kinds of atmospheric pollution and airborne diseases. Fredskild (1984, 1985) has found from study of lake-bed deposits that there was a long period in and about the warmest postglacial times between about 5000 and 2500 years ago when exotic pollens were reaching western Greenland, evidently borne by much more frequent southwesterly winds than before or since. Similarly, we have noticed in Chapter 4 how occurrences of locusts in central Europe, which were rather frequently reported in the late thirteenth and fourteenth centuries seem to be evidence of a heightened frequency of long-track southerly and southeasterly winds in that period. Red dust from the Sahara was deposited in Finland one day in 1947, and after the passage of a cold front, black dust and ash from a volcano erupting in Iceland was deposited there on the following day. The lethal deposits of nuclear material carried by southerly and southeasterly winds from the Chernobyl power station accident site in the Ukraine in 1986 first over the Baltic states to Finland and Scandinavia, and then veering about with the winds to reach other parts of Europe as far away as Wales and western Scotland before returning over the middle of the continent to revisit European Russia, are a reminder of the importance of such long-distance wind transports also to the human and animal populations of our planet.

Much interesting work has also been done, notably by R. C. Rainey, over recent decades by the internationally supported Centre for Overseas Pest Research – formerly the Anti-Locust Survey – in London, showing how locust swarms are affected by the vertical as well as horizontal components of air movement in the Trade Wind zone and, particularly, near the intertropical convergence of the winds. This has resulted in great advances in control of the swarms. Similarly, use of radar has made possible the monitoring of bird and insect migrations (e.g. over the North Sea). The late Sir Alister Hardy pioneered a systematic study of the insect populations of different air-masses and the windstreams that bring them, paralleling his earlier studies of plankton in the seas.

So we must now describe the structure and air motions involved in the individual travelling cyclones (or depressions) and in typical anticyclones and their development. These weather systems are the major eddies that continually occur within the general wind circulation. They are generated at certain points by favourable features of the upper wind pattern and are steered by the main flow of the upper winds. Three stages in the development of a typical frontal cyclone with its winds, clouds and weather are shown in figure 14.5. Most cyclones in middle and higher latitudes are of this type. They develop from a wave-like disturbance on the front, which marks the convergence of warm and cold wind-streams (the so-called 'polar front') that have originated in widely different latitudes. The disturbance travels more or less eastward along the front, steered in the general direction of the massive flow of the upper winds (jet-stream) – which more or less

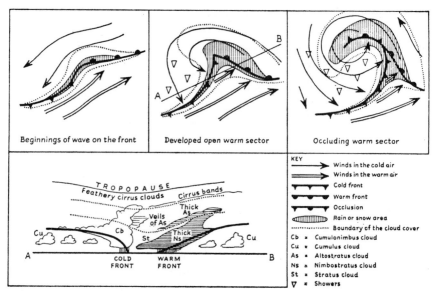

Figure 14.5 Three stages in the development of a frontal cyclone and vertical section through typical cloud systems and frontal surfaces along the line AB. (Invert the maps for the southern hemisphere.)

accords with the direction of the low-level winds in the warm sector of the depression – and tends ultimately to the cold side of the jet-stream. The system gains kinetic energy, as the frontal pattern evolves in such a way that the colder, denser air-masses spread underneath, and lift, the warmer air-stream: a process that lowers the centre of gravity of the system. Typical frontal cyclones attain a diameter of 1500–2000, occasionally 3000 km at maximum development.

Figure 14.6 displays the characteristic relationships of a frontal cyclone (the full lines are isobars, i.e. lines of equal surface pressure) – with a new wave disturbance beginning to develop on the front – and of an anticyclone to the pattern of the upper winds (broken lines) and temperature distribution: the isotherms nearly parallel the flow-lines of the upper winds. (For southern hemisphere situations, both figures 14.5 and 14.6 should be inverted; the winds in that case blow clockwise round the low-pressure area and anticlockwise round the high-pressure area.) The anticyclone in these common cases is at the warm side of the jet-stream. A vertical section through the air motions and cloud distribution in a typical warm anticyclone is displayed in figure 14.7. The gentle subsiding motion of the air in anticyclones, and wherever anticyclonic development occurs, suppresses the upward growth of clouds. The diameter of such a typical anticyclone is commonly 3000–4000 km along the long axis, and its breadth is of the order of 1000–1500 km.

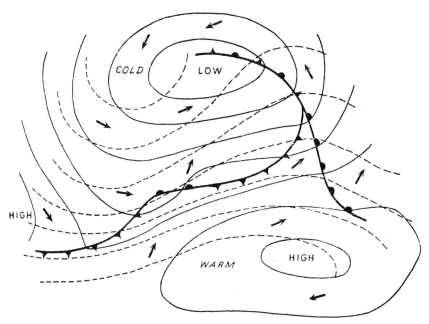

Figure 14.6 Characteristic relationships of: (i) an eastward-moving cyclone, (ii) a nearly stationary warm anticyclone, (iii) a small wave on the cold front and (iv) a newly developing anticyclone (left edge of the map) to the thermal pattern, shown by the broken lines (isopleths of thickness of the layer between the 1000 and 500 mb pressure levels), which indicate the flow of the jet-stream. Arrows show the directions of the surface winds.

Anticyclones over high latitudes and over the continents in winter are sometimes of this type, although very low temperatures develop in the surface air layer through ground radiation cooling under clear skies. This is the vertical structure of typical blocking anticyclones in either middle or high latitudes. But other high-pressure systems over high latitudes and over the continents in winter exist only in the surface layers and the high pressure in them is entirely due to the density of the very cold air there, while above them are various wind patterns which, if vigorous, may cause the anticyclone to be rapidly displaced or decay.

The occasional tropical cyclones (typhoons, hurricanes) of low latitudes, which develop more or less in the Trade Wind zone and sometimes produce tremendous winds and seas, derive their energy from convection intensified by the latent heat release accompanying phase-changes of the moisture content in the air over the warmest seas (where the surface water temperature is over 27 degC). They are also associated with the inter-tropical convergence of air-streams from either hemisphere, at times when this has advanced farther than usual from the geographical equator into the hemisphere where it is summer. Although the energy developed is

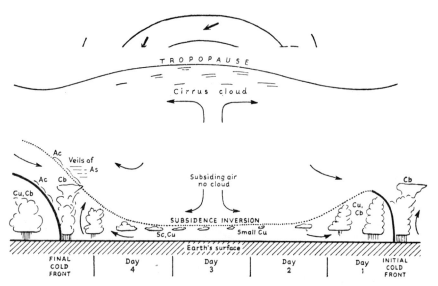

Figure 14.7 Vertical cross-section through the atmosphere – with clouds, frontal surfaces and subsidence inversion – passing over one place during the passage of a typical anticyclone.

impressive, and destructive, tropical cyclones are commonly only 300–500 km in diameter, unless and until they move into higher latitudes and become involved with the polar front. In the latitudes of their origin, tropical cyclones generally drift westwards, but if they encounter an extended trough in the upper westerly wind-flow of the circumpolar vortex, they then recurve and are steered with the upper westerlies and ultimately to higher latitudes.

Many aspects of the normal seasonal change of the wind circulation are explained by the changes in the position and strength of the heating gradients, as the insolation declines and a snow and ice surface spreads over the higher latitudes and continental interiors, while the temperature of the ocean surfaces in the lower latitudes changes hardly at all. This process is paralleled by an intensification of the wind circulation and a shift of the main circulation activity – particularly the points or origin of the disturbances – towards lower latitudes.

If we examine the changes of climate in the past, and even from decade to decade in the present century, we find evidence of similar, though generally smaller, changes of overall warmth, paralleled by changes in the prevailing intensity and shifts of latitude of the main features of the wind circulation. Figure 14.8 shows the calculated changes of prevailing surface temperature, averaged over the whole Earth, by 5-year periods from 1870–4 to 1965–9. Notice how similar to this is the graph of average barometric pressure gradient over the North Atlantic Ocean in winter (figure 14.9) and the

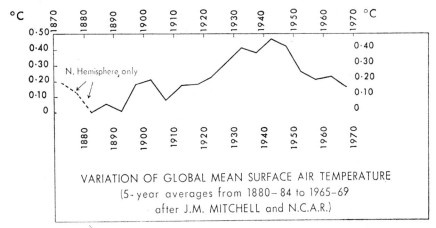

Figure 14.8 Global surface air temperature: successive 5-year averages from 1870–4 to 1965–9, expressed as departures (°C) from the level in 1880–4.

frequency of the (commonly vigorous) prevailing west-wind situations over the British Isles (figure 6.5, p. 95). These changes of vigour of the global wind circulation affect the amount of rain (or the total downput of rain and snow) penetrating far east to the heart of Europe and Asia, and into the Arctic and Antarctic, as well as, obviously, the downput on the west-facing hill and mountain slopes in Britain's wet western districts (figure 14.10).

The secular (i.e. long-term) increases and decreases of temperature, and still more of rainfall, are far from being uniform over the whole Earth. They each have an interesting geography, which is of importance to many aspects of the human economy and environment – to the possibilities of successfully growing grain and fruit crops, to the natural vegetation and, of course, to the insect-life, birds and other land fauna. Corresponding changes in the wind-driven ocean currents bring about quick responses also in the ocean, its microfauna and flora and in the distribution of the fish-stocks which feed on them.

The waxing strength of the global wind circulation in the early part of this century had its greatest effect in warming those parts of the Arctic near the Norwegian Sea and Barents Sea, which are penetrated by the warm North

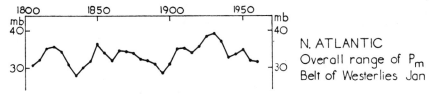

Figure 14.9 Pressure difference measured across the middle latitudes belt of prevailing westerly winds over the North atlantic in January: 10-year averages, from 1800–9 to 1950–9, plotted at 5-year intervals.

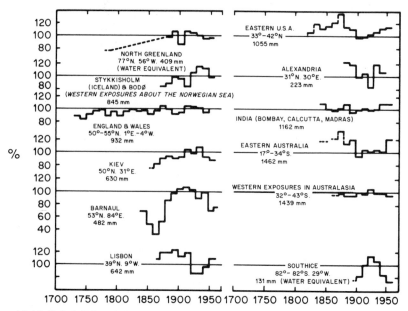

Figure 14.10 Rainfall (total downput of rain and snow) histories in selected parts of the world. Successive decade values as percentages of the 1900–39 averages quoted.

Atlantic Drift water, which in those years was impelled farther north and produced a great recession of the Arctic sea-ice.

World mean surface air temperature rose just 0.1 degC between the 1910–19 decade and 1920–9, this being about one-fifth of the total rise between the 1880s and 1940s. The rise between those same two decades, 1910–19 and 1920–9, in Iceland averaged 1.0 degC; and in Spitsbergen and Franz Josefs Land, near the receding edge of the Arctic pack-ice, about 80°N, the rise was over 2.5 degC. The temperature fall from the peak level of the 1940s to the decade centred on 1960 averaged 0.2 degC for the whole Earth, 0.6 degC in Iceland and 2.4 degC in Franz Josefs Land. Up to that point the Canadian Arctic had hardly begun to feel the cooling trend, and some areas there were even warmer than before; but the rest of the Arctic was generally colder, and an area centred over Franz Josefs Land was 5 degC colder than in several preceding decades. This area of Arctic cooling extended over northern Europe and Iceland, including many of the breeding areas of northern birds. Figure 14.11 shows the recovery of the Arctic sea ice to its most forward position for 50 years in May 1968.

There are lags in the responses of birds which we have mentioned to the possibilities opened up to them by a climatic warming. The early-twentieth-century warming of the Arctic was accompanied by northward extensions of range. It may be significant that the Black-headed gull *Larus ridibundus* extended its breeding range to Iceland quite early, by 1911, because the responses of fish-stocks following the transport of their food by the ocean currents seems to be, at least in some cases, almost immediate. Some bird species were still extending their range northward when the warmth was long past its peak; the lapwing *Vanellus vanellus*, for example, began breeding in Iceland about 1960. By contrast, one must suppose that deteriorating climatic conditions bring disasters to birds. The individual disasters are probably always somewhat haphazard in their incidence, but the effect on the broad scale is presumably to impose a quicker reaction to climatic cooling or increasing aridity than to warming and to conditions which improve the availability of water and vegetation.

The climate of the Arctic in the Iceland and Greenland sectors has continued into the 1980s to be colder than in the period before 1960, as to a much lesser degree has the average temperature of the entire northern hemisphere. By contrast, the overall average temperature of the southern hemisphere has increased, partly through the greater warmth in the largely ocean zone in the southern tropics (which we have noticed in Chapter 12), and partly through a remarkable warming of the Antarctic, which seems to have continued over the decades since about 1930 though there have been some indications of a levelling off or reversal in recent years.

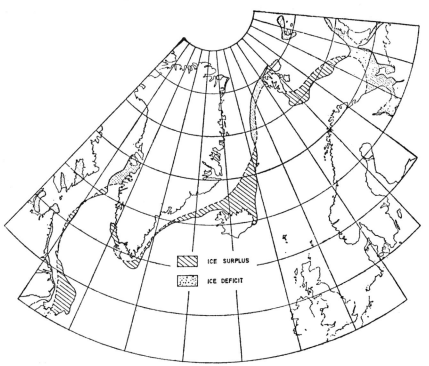

Figure 14.11 Ice extent (outlined by solid line) on the Arctic seas on 8 May 1968, compared with the 1911–56 average (broken line) for that time of year.

References and recommended reading

Blüthgen, J. (1966) *Allgemeine Klimageographie* (2nd edn), Berlin, Walter de Gruyter: *Lehrbuch der Allgemeinen Geographie*, Band II.

Flohn, H. (1969) *Climate and Weather*, London, Weidenfeld & Nicolson, World University Library.

Fredskild, B. (1984) 'Holocene palaeo-winds and climatic changes in West Greenland as indicated by long-distance transported and local pollen in lake sediments', in *Climatic changes on a Yearly to a Millennial Basis* (N.-A. Mörner and W. Karlén (eds), Dordrecht, Reidel.

Fredskild, B. (1985) 'The Holocene vegetational development of Tugtuligssuaq and Qeqertat, northwest Greenland', *Meddelelser om Grønland, Geoscience*, 14, Copenhagen.

Johnson, C. G. and L. P. Smith (eds) (1965) *The Biological Significance of Climatic Changes in Britain*, London, Inst. Biology and Academic Press.

Lamb, H. H. (1964) *The English Climate*, London, English Universities Press.

Lamb, H. H. (1966) *The Changing Climate*, London, Methuen.

Lamb, H. H. (1972) *Climate: Present, Past and Future. Vol. 1, Fundamentals and Climate Now*, London, Methuen.

Lamb, H. H. (1977) *Climate: Present, Past and Future. Vol. 2, Climatic History and the Future*, London, Methuen.

Manley, G. (1952) *Climate and the British Scene*, London, Collins, New Naturalist series.

Thomson, A. L. (1964) *A New Dictionary of Birds*, London and Edinburgh, Nelson.

15

Christmas to New Year weather as an indicator of the tendency of the large-scale wind circulation and climate

Variations of the wind circulation come over as wide a range of frequencies and time-scales as those of the weather and climate. The most important patterns of variation are evidently those of the largest scales and which also show a noteworthy tendency to persistence. We are not concerned here just with the extremes of ice age and the warmest interglacial behaviour of the circulation, patterns which are stabilized by the inertia of the ice and snow surfaces on the one hand, and ocean temperatures, land vegetation and soil surface conditions on the other.

There are two behaviour patterns which show important persistence and frequency variations even in times like the present. One is of global extent in its ramifications, the so-called Southern Oscillation and associated three-dimensional Walker Circulation across the Pacific Ocean east and west in low latitudes, named after its discoverer in the early years of this century, Sir Gilbert Walker, FRS. The maps of this phenomenon (figure 15.1(a) and (b)) indicate a 'see-saw' of prevailing barometric pressure level in the tropics between the Indian Ocean and the eastern South Pacific. Pressure tends to be relatively high over one of these regions for months at a time and low over the other. Figure 15.2 shows the record of pressure at Batavia (now Jakarta) in the first of these regions over 100 years. The range of between 1 and 2 mb in the six-month average pressure is very significant in relation to pressure gradients and winds in the tropics. Striking changes of weather, surface temperatures and winds in the upper air are associated. The maps in figure 15.1 also indicate that changes in the occurrence of blocking anticyclones in high northern latitudes and in the normal zone of travelling cyclonic storm depressions over the North Atlantic and Greenland or northern Eurasia, and over Canada and the North Pacific, as well as in sub-Antarctic, have some relationship to the Southern Oscillation. The switch between cyclonic storm sequences with vigorous westerly winds in middle latitudes and the

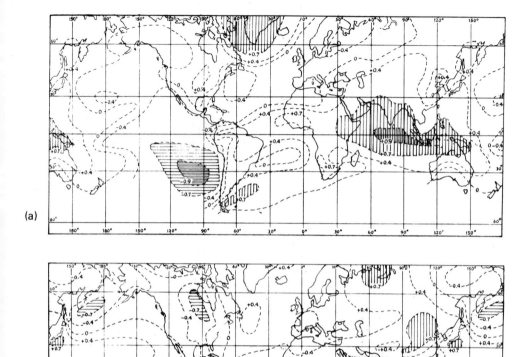

Figure 15.1 The patterns of the Southern Oscillation and its associations: (a) in 1931–8, (b) July 1949–July 1957. Correlation coefficients connecting 12-month average atmospheric pressure anomalies at each place with the pressure at Jakarta (6°S 107°E). Vertical hatching where the correlation is positive. Horizontal hatching where the correlation is negative. (After Berlage 1961.)

occurrence of so-called blocked patterns, with an anticyclone somewhere in the usual low-pressure zone, and easterly, northerly and southerly winds prevalent in some parts of the usual zone of west winds, is the other easily recognized main variation of the large-scale wind and climate regime.

So far as there are hints of regularities in the recurrence of these types of variation, both show evidence of a repeat tendency at intervals of two to three years (average about 2.2 years). Blocking, at least in the northern

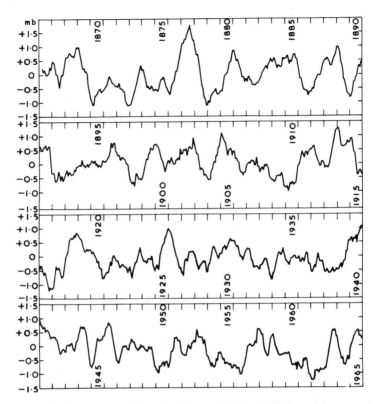

Figure 15.2 Atmospheric pressure at Batavia/Jakarta 1866 to 1965: 6-month means
– as departures from the overall average. (After Berlage 1966.)

hemisphere seems to be more frequent around times about 23 years apart, as
first noticed by Brezowsky *et al.* (1951). Occasionally as many as 5–9 years
go by between the obvious repeats of the Southern Oscillation. And it is
clear that there are some much longer-term tendencies in the prevalence of
'blocking' situations, at least in the European sector of the northern
hemisphere. Analysis of recorded history of the seasons in Europe over the
last 1000 years seems to indicate that blocking has been more frequent in and
about the thirties and eighties of most centuries than in the other decades,
implying something of a 50-year cyclical tendency (Lamb 1964a), but with
some characteristic differences that would produce a tendency for a 100-year
interval between the recurrences of some situations. Longer-term fluctuations than even this may also be suggested by the data, e.g. weather and
wind direction tendencies in the years since about 1960 (up to 1986) which
were apparently better matched between about 1760–80 and 1820, and
between 1560 and 1615–20, than at any other time for which we have suitable
wind or weather data in Europe for analysis. This variation was first noticed
by Brooks and Hunt (1933). These may be symptoms of 200- and 400-year

periodicities such as were seen in other data, including, perhaps significantly, indices of solar variation, in table 13.1 (p. 224). Thus, periods
stretching over many decades with either notably heightened or much
reduced frequency of 'blocking' and prominent easterly, northerly and
southerly winds are liable to occur in Europe.

Variations of the large-scale wind circulation such as we are concerned
with, in this and the previous chapter, are with little doubt responses to
variations in the energy available to drive the winds and ocean currents or in
the way in which this energy supply is fed into the system. They are thus a
fundamental indicator of the state of the climatic system. As symptoms of
this, statistically significant correlations are found between the long-year
frequency of the westerly winds over the British Isles and the weather and
temperature levels prevailing over widely separated parts of the world
(Lamb and Mörth 1978): see table 11.2 (p. 181). The variations in the wind
circulation are probably the most readily observed indicator of the state of
the world climate system and are of wide interest for this reason.

Before proceeding to look at the relevance of the weather around
Christmas and New Year in this connection, we must first look briefly at the
natural seasonal structure of the year apparent in Europe. An examination
(Lamb 1950) of the frequency over a 50-year period (1898–1947) of long
spells of this or that set weather type in the British Isles revealed that more
than two-thirds (70 per cent) of the summers were marked by a long spell of
more than 25 days (with no interruptions of more than three days) of one
weather type or another at some time between 18 June and 9 September: the
types were mostly either cyclonic, westerly or anticyclonic, sometimes with a
good deal of northwesterly weather. There was another time when long
spells were almost equally common, between late September and early
November, and less marked peaks in December and February to early
March. The different peaks of frequency of set weather occurrence were
associated with characteristically different patterns, the December spells
being mostly westerly and the late-winter spells often with anticyclonic or
easterly weather in the British Isles. The tendency for a thorough change in
the prevailing wind and weather pattern was very marked about mid-June,
early September, mid-November and the turn of the year. In the period
from late March to early June, long spells were so infrequent (except in years
with much northerly wind) that considerable changes of weather from one
week to the next were more to be regarded as the norm, and this could be
said of most of the month of November.

Besides the natural seasons:

   (i) high summer – July and August;
  (ii) autumn – centred on October;
 (iii) forewinter or early winter – December;
 (iv) late winter (in some years early spring) – centred on February–
       March;
  (v) spring – April to early June.

Table 15.1 The most frequent singularities (*Regelfälle*) in the weather of central Europe 1881–1947 (after Flohn and Hess 1949)

Dates	Type of weather	% of years	Mean duration
9–18 June	Cyclonic, European summer monsoon	89	7.3 days
21–31 July	Cyclonic, European summer monsoon	89	7.2 days
1–10 August	Cyclonic, European summer monsoon	84	7.2 days
3–12 September	Anticyclonic, late summer fine weather	79	6.5 days
21 September–2 October	Anticyclonic, early autumn fine weather ('Old Wives' Summer')	76	6.3 days
28 October–6 November	Anticyclonic, mid-autumn fine weather	69	6.3 days
11–22 November	Anticyclonic, late autumn weather	72	6.4 days
1–10 December	Cyclonic, oceanic, mild westerly types	81	5.9 days
14–25 December	Anticyclonic, European winter monsoon	67	7.4 days
23 December–1 January	Cyclonic, oceanic, mild westerly types (the 'Christmas Thaw' and mid-winter mildening)	72	6.2 days
15–26 January	Anticyclonic, European winter monsoon	78	7.4 days
3–12 February	Anticyclonic, European winter monsoon	67	6.1 days
14–25 March	Anticyclonic, early spring fine weather	69	6.2 days
22 May–2 June	Anticyclonic, late spring fine weather	80	6.4 days

many have noticed fairly regularly recurring shorter seasonal episodes or singularities (Lamb ibid). Attention was perhaps first drawn to these by Pilgram (1788) and Dove (1856) in central Europe, and by Alexander Buchan (1867) in Scotland. Perhaps the most useful definition has been given by Flohn and Hess (1949), listing periods of the year of 10-12 days duration in which some specific type of wind circulation pattern occurred over Europe on at least three successive days in more than two-thirds of the years 1881–1947. The list of these *Regelfälle,* or regular recurrence tendencies, in central Europe was as given in table 15.1.

For the British Isles, the same list broadly applies, this area commonly being, however, near the western edge of the anticyclones, especially in late September and early November. Indeed, at those times, especially around 23 September, Scotland is liable to be swept by Atlantic gales from a southerly or southwesterly point. Anomalies in the wind circulation behaviour when compared with this apparently normal natural calendar are liable to reveal longer-lasting tendencies of the climatic system.

The remainder of this chapter is taken from the relevant section of a paper on 'Contributions to historical climatology', which appeared in Klimatologische Forschung, *[Climatic Research], the* Festschrift *for the sixtieth birthday of Professor Hermann Flohn,* Bonner Meteorologische Abhandlungen, *Heft 17, Bonn, Dümmler Verlag, 1974, pp. 549–67.*

Christmas weather and the tendency of the general circulation

A curious feature noticed by Flohn (e.g. 1954, p. 119) in his studies of the varying behaviour of the general circulation connected with climatic changes, is that many items of the singularity calendar (e.g. Flohn and Hess 1949; Lamb 1964b), which superposes itself on the broad seasonal development of the northern hemisphere atmospheric circulation as it affects Europe, remain recognisable, for example, in the Zurich observations of 1546-76 as in the twentieth century (Flohn 1949). This is doubtless due to the apparent nature of the singularities, largely as an approximately 30-day wave phenomenon, reported by Flohn in 1947 and lately found again by Sawyer (1970) in a new type of analysis. Nevertheless, as noted by Rudloff (1967a, p. 209), some singularities appear most strongly in periods of strongly developed zonal circulation and are largely suppressed in meridional periods when blocking is frequent, whereas others are more characteristic of periods of markedly meridional circulation. This applies with particular force to the Christmas thaw, which in south Germany might be described as mainly a phenomenon of the mild climate period of the first half of the twentieth century with its strong zonal circulation (Rudloff 1967a, 1967b). Indeed Trenkle (1951–52) has pointed out that the circulation development and temperatures prevailing in south-west Germany in the pentads 72, 73, 1 and 2* over about the last 100 years shows a similar variation, but with greater amplitude, compared to that over the years as a whole and appears to have a certain limited prognostic value for the character of the winter.

In England snowy Christmases are rare, but a popular legend that the 'old-fashioned Christmases were 'white' persists and is commonly associated with the great English novelist Charles Dickens (1812–70), more particularly because of his vivid description in *A Christmas Carol* (1843), but also because of the wintry weather in London and south-east England described in some of his other novels. The winter snow motif also appears in English

---

* For studies of the seasonal development, the year is commonly divided into 73 five-day periods or 'pentads', Pentads 72 and 73 comprise the last ten days of the year, from 22 to 31 December, and pentads 1 and 2 cover the period from 1 to 10 January. For comparisons of the circulation development in periods of history when the Old Style, Julian, calendar was in use, the dates are always converted to the modern calendar.

Christmas hymns (carols) written in the nineteenth century, as it does in the Flemish and Dutch religious paintings of the seventeenth century, indeed from the time of the great winter of 1564–5. These pictures and snowy scenes of the English countryside in the heyday of long-distance travel by coach, around 1800–30, are probably still the most popular motifs of the English Christmas card industry, which doubtless perpetuates the legend.

As it happens, we can compile from various sources a 312-year record of the Christmas weather in the London area, from 1660 to 1971 inclusive. Most of this record has become reasonably accessible thanks to the labours over many years of Professor Gordon Manley in assembling, and critically examining, daily weather observation registers. Fairly recently, the years 1669 to 1717 were added to the collection by his discovery of manuscript weather diaries in the Bodleian Library, Oxford, the existence of which had long been hidden by a mistake in cataloguing. This means that London has the longest continuous record of its daily weather for any place in the world. And for some purposes, there is enough about the weather in the famous diary of Samual Pepys (Charles II's Secretary to the Navy) to extend our knowledge back to 1660. (The dates given in seventeenth-century diaries and observation registers must be adjusted to the modern calendar by adding 10 days. The modern, Gregorian calendar was introduced in England in September 1752. Thus, 15 December in Pepys and other seventeenth century sources = 25 December New Style. From March 1700 onwards the correction became 11 days, which had to be skipped in 1752 when the calendar was changed.)

The long record of London's Christmas weather makes a meteorologically interesting story. Indeed, as in south-west Germany, it not only reproduces the ups and downs of our winter climate, but in some ways amplifies them. And as it turns out, Dickens had an obvious excuse for his writings which

Table 15.2   Winter weather in England: frequencies per half century

	Snow lying on Christmas morning* in the London area	Christmas day* white with either snow or frost in the London area	Winters (DJF) with average temperature in central England below 3°C (37.4°F)
1650–99	6 to 10	about 20	8 to 11 between 1680 and 1699
1700–49	2	13	8
1750–99	about 7	22	22
1800–49	6 to 8	20 to 22	16
1850–99	5 to 8	21 to 22	16
1900–49	3	7 to 10	6
1950–71	4 in 22 years	9 in 22 years	4 in 21 years

* Adjusted to the modern (Gregorian) calendar throughout.

started the legend of the 'old-fashioned' white Christmas, for of the first nine Christmases of his life, in London, between 1812 and 1820, six were white with either frost or snow. There were eight or nine more in the next 20-odd years before *A Christmas Carol* was written. Then, as now, however, it seems that a persistent frost, often accompanied by a fog which deposits thick white rime and through which the sun may or may not shine palely in the afternoons, was a commoner cause of the English white Christmas than snow. (Snow appears to have covered the ground in just six of these 15 cases in Dickens' young life.)

The change to the great predominance of green Christmases of the early twentieth century (see table 15.2) was remarkable. It was paralleled by a rise from the 1880s to the 1920s of 1 degC in the average winter (DJF) temperatures in inland, low-lying districts of England, where on average there were only 5–7 days a year with snow-covered ground between about 1910 and 1937. Since 1950 there has been a change in the reverse direction: snow lay on average 10–15 days a year in equivalent places in the 1950s and is known to have averaged 15–20 days in some of them in the 1960s, when the central England mean winter temperatures were 1.2 degC below the level of the 1920s and 0.8 degC below the 1900–60 average. Apart from a run of mild winters between about 1972 and 1976, the greater incidence of snow in England has continued at least to 1986: no less than four of these winters between 1979 and 1986 produced totals of 30–40 days with snow lying at the 8.00 hours morning observation in parts of southern England. The remarkable change in the total frequency of snow is, of course, related to the fact that the winter temperatures in England vary around just the critical level for this. The cooling noticeable in other months (except October) in the 1960s has had the effects that were to be expected. Thus, a recent contributor (R. A. Canovan) to *Weather* (London, Royal Meteorological Society, November 1970) remarks that 'until 1962 snow was rare in November although the last few years have seen an increasing tendency towards early snowfalls in the London area' and that these have been associated with more frequent, and colder, northerly winds.

Consideration of the winds brings us nearer the heart of the matter. Southwesterly and westerly winds sweeping as a broad stream across Britain and all northern Europe are the prevailing characteristic of our climate, and represent the commonest individual type of situation, nevertheless, the long history revealed by early weather diaries and observation registers shows that their frequency does vary widely. In the first half of this century, westerly situations were very predominant, averaging 100 days a year and together with hybrids and cyclonic situations having some westerly characteristics accounting for about half the days of the year. Since 1950 the frequency of westerly situations has declined, averaging only 80 days a year in the 1960s and 73 days in the 1970s. In four years between 1969 and 1981, the total was 60 days or less, a frequency only known to have occurred before in 1784 and 1785. There have been compensating increases since 1950 of cyclonic, northwesterly, northerly, easterly and southerly patterns, the latter giving some notable mild interludes in winter (particularly in some Januarys); but the main increase has been in the colder types.

The average frequency of the westerly situations, which bring the most reliable and persistent flow of mild air from the ocean across Britain and much of Europe in winter, seems in recent years to have fallen lower than at any time since before 1850 if we rely on comparison of the frequency of southwesterly surface winds at London (figure 15.3) which can be constructed from the daily observations available since 1669 and on various types of indirect evidence – including actual weather diaries from different places in north-western and central Europe for certain runs of years – back to 1340 (Weiss and Lamb 1970). The shape of this curve suggests a possible 200-year periodicity, of which there are suggestions in other geophysical and parameteorological phenomena (Lamb 1969, Anderson 1961, Johnsen *et al.* 1970) and which may be of some value in connection with any possibility of ultra-long-range weather or climatic forecasting. The decline of the westerlies around 1960 seems to some extent paralleled by earlier cases around 1750 and 1560 (cf. figure 15.3). Indeed, in his analysis of Haller's sixteenth-century observations in Zurich, Flohn and Hess (1949) noticed an increase in the proportion of snow in the winter precipitation observations from 44 per cent in 1550–63 to 63 per cent in 1564–76. That the recent similar change in the frequency of snow lying reported in England is proportionately much greater is due to the fact that the prevailing temperature level in England is much more critical for snow.

Analysis of the wind data and their normal seasonal changes shows why the weather around Christmas and the turn of the year may be a sensitive indicator of the current tendency of the general circulation over the hemisphere and the climatic tendency that goes with that. Westerly type over the British Isles in general corresponds to strong upper westerlies occurring as a well-developed zonal flow over middle latitudes, i.e. with high values of the zonal index* measured at 35–55° or 40–60°N. Seasonally this index increases to its strongest values of the whole year in mid-winter, when thanks partly to the snow and ice distribution over the northern hemisphere the Equator–pole thermal gradient is greatest. When the circulation was particularly strong, as during the first half of the present century, the seasonal maximum of the zonal index and the highest frequency of westerly situations over the British Isles were attained in the 3–4 weeks after the winter solstice, between about Christmas and mid-January. But in other periods, this climax seems to be frustrated by the development of the blocking patterns associated with cold weather and an extended snow cover over the continents and over the Arctic, which throw the mainstream of the atmospheric circulation farther south or confine the main cyclonic activity more narrowly to the oceans and limited longitude sectors elsewhere. At

---

* The zonal index is a measure of the average atmospheric pressure gradient between two chosen parallels of latitude, favouring westerly winds when the gradient has big values.

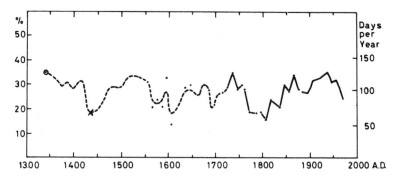

Figure 15.3 Frequency of south westerly surface winds in south-eastern England since AD 1340. Daily observations at London from 1669. Indirect evidence before 1669, including available weather diaries kept in other places in England and elsewhere in central or western Europe.

these times the seasonal maximum of the westerlies is already over by late November to mid-December. Thus, the weather over Christmas and New Year in England in the one type of regime is nearly always mild and makes a strong contribution to the year's total of 'westerly-type days'; in the other type of regime, it is likely to be cold and unlikely to produce westerly days.

Figure 15.4 displays the varying frequency of different surface wind directions as shown by the observations at London on 24, 25 and 26 December in each decade from 1670 to 1970. The main features are the strong maxima of winds from about SW between 1700–1750, 1850–70 and 1910–1960. Secondly, the high frequencies of winds from about N to E between 1750 and 1850 or somewhat after while during much of this time surface winds from W (corresponding presumably to northwesterly gradient wind) were commoner than SW. the frequency of which was low: the 1960s, and to a less extent the 1950s, show a reversion towards this pattern.

Since the winds and weather prevailing at Christmas are clearly a response to a long-term tendency of the general circulation favouring high or low index, zonal or blocked situations, and their frequencies may be regarded as a fairly sensitive index of this tendency, their possible forecasting significance was explored. It was found to be limited, but an interesting pattern of statistically significant relationships was revealed by chi-squared tests, details of which are listed in table 15.3. In all the tests, a significant association was found between cold weather in the London area at Christmas, or in pentads 72, 73, 1 or 2 of the year and cold winters (quintile 1 of the temperature distribution) with the mean temperature for the three months December, January and February <3 degC in central England the same winter. The association was strongest for cold weather just at Christmas and in pentad 2 (6–10 January). But some relationships at or approaching the conventional statistical confidence levels were also indicated between cold weather at Christmas, or in pentads 72 or 73, and cold (quintile 1) winters in central England the next following winter and the second following winter, whereas the third following winter showed (at least in the last 100 years) a marked reversion to normal. In some of the tests, however, there was a distinct suggestion that there was also a minor boost to the chances of a very warm winter, presumably through blocking developments so placed at to give England southwesterly winds. It is thought that these surprising relationships must be connected with the 13– to 14– and 26-month oscillations which have often been reported in meteorological phenomena. If so, it seems likely that other similar long-term forecast indicators could be found at other times of the year.

The relationships here reported, and listed in table 15.3 have been found without making any allowance for the urbanization effect which has progressively raised the Kew temperatures above those of the country outside London over the last 100 years. Until now this difference averages over 1.5 degC in winter. No allowance has been made either for the fact that the long-term mean winter temperatures were lower, and the frequency of cold winters was higher, before 1850 than since that time. The quintile boundaries used throughout this study were those applying in the years 1874–1963.

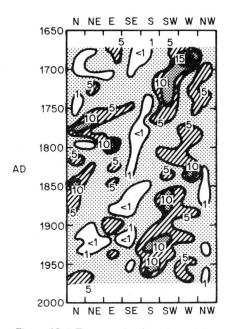

Figure 15.4 Frequencies (number of days per decade) of different surface wind directions in the London area on 24, 25 and 26 December (New Style, i.e. Gregorian calendar).

Table 15.3 Occurrence of winters with mean temperatures in different ranges in central England, following certain preconditions. Manley's (1959, 1961) temperature series was used.

Precondition/indicator	Period of reference	Chi-squared value and significance level	Following winters (D, J, F): numbers of cases in quintile				
			1	2	3	4	5
Christmas day (25 XII) with snow or persistent frost in the London area	1869–1968	15.0 (0.5%) — Same winter	15	7	4	2	3
		3.3 (—) — Next winter after	9	8	6	2	6
		6.7 (c. 12%) — 2nd winter after	10	10	2	4	4
(31 cases)		3.4 (—) — 3rd winter after	3	6	10	6	4
Pentad 72 mean temperature ≤ 2.0°C at London (Kew)	1869–1968	10.2 (3%) — Same winter	12	7	3	1	5
		1.0 (—) — Next winter after	8	6	6	4	4
		6.8 (c. 12%) — 2nd winter after	9	9	3	1	6
(28 cases)		7.9 (10%) — 3rd winter after	4	3	12	6	3
Pentad 1 mean temp. ≤ 2.0°C at London (Kew)	1869–1968	9.2 (5%) — Same winter	9	4	4	2	0
		4.6 (—) — Next winter after	7	6	3	2	1
		1.5 (—) — 2nd winter after	6	4	4	2	3
(19 cases)		3.3 (—) — 3rd winter after	1	7	4	3	4
Pentad 2 mean temperature ≤ 2.0°C at London Kew	1869–1968	21.1 (0.1%) — Same winter	14	6	4	1	0
		3.1 (—) — Next winter after	8	6	2	3	6
		1.2 (—) — 2nd winter after	7	5	6	3	4
(25 cases)		4.9 (—) — 3rd winter after	2	9	6	4	4

Pentad 73 mean temperature ≤ 2.0°C at London (Kew) (22 cases)

Period	Winter	Statistic (%)					
1869–1968	Same winter	15.4 (0.5%)	11	6	5	0	0
	Next winter after	5.2 (c. 14%)	8	6	5	1	2
	2nd winter after	2.0 (—)	5	7	5	2	3
	3rd winter after	1.6 (—)	3	6	6	5	2

Christmas Day (25.XII Gregorian, i.e. New Style calendar ≤ 2.0°C) (60 cases)

Period	Winter	Statistic (%)					
1680–1868	Same winter	35.0 (0.1%)	26	21	5	6	2
	Next winter after	14.1 (0.7%)	20	18	10	3	9
	2nd winter after	8.8 (6%)	17	19	9	9	6
	3rd winter after	16.3 (0.4%)	22	18	8	8	4

Pentads 72 and 73 overall mean temperature ≤ 2.0°C at London (Kew) and in both pentads < 3.0°C (16 cases)

Period	Winter	Statistic (%)					
1869–1968	Same winter	17.1 (0.2%)	10	4	2	0	0
	Next winter after	2.9 (—)	5	5	4	1	1
	2nd winter after	3.9 (—)	5	6	1	1	3
	3rd winter after	2.0 (—)	2	3	6	3	2

Christmas Day (25.XII New Style, i.e. Gregorian calendar ≤ 2.0°C) (91 cases)

Period	Winter	Statistic (%)					
1680–1969	Same winter	51.0 (0.1%)	41	28	9	8	5
	Next winter after	18.4 (0.1%)	29	26	16	5	15
	2nd winter after	16.1 (0.4%)	27	29	12	13	10
	3rd winter after	9.7 (5%)	25	24	18	14	8

Note: The quintile boundaries used are those applying (Murray 1970) to the winters 1874 to 1963: $T_5 > 5.2°$, T_4 4.5 to 5.2°, T_3 4.0 to 4.5°, T_2 3.0 to 4.0°, $T_1 \le 3.0°C$

The material investigated in this chapter indicates further that the character of the weather around Christmas and New Year also shares in the longer-term variations of the climate, with the same sort of relationship as revealed in table 15.3. This longer-term persistence of one tendecny or the other is very clear when the figures for the cold seventeenth century are compared with the milder regime during parts of the period 1700–50, as seen in table 15.1.

Note: This essay has demonstrated in a new way the value of historical climatology, in this case for its potentiality in improving meteorological understanding of the seasonal development and awareness of long-term factors underlying the circulation behaviour.

It is many years since Hellmann (1926, 1927) reviewed, and gave a first list of, the early weather diaries and meteorological observation registers known to exist. The list included some in central Europe from as early as 1490 onwards. Yet apart from Flohn's interesting analysis of the observations by Haller in Zurich from 1546 to 1576, and Lenke's (1960, 1961, 1964, 1968) studies of mostly seventeenth- and early-eighteenth-century data, not much of this early material has yet received detailed meteorological scrutiny. Valuable diaries still come to light from time to time, for instance, the appearance in a Zurich antiquarian bookseller's (Helmut Schumann A/G) catalogue in 1969 of a diary kept by one Frederico Bonaventura in Urbino, northern Italy, in 1591–97. This diary promises to be the more valuable, because the Danish astronomer Tycho Brahe was making daily weather and wind observations on the island Hven in Öresund (The Sound) in the same years. Happily, the Urbino diary has been secured for scientific study by the Swiss Technical Library, Zurich, but this outcome at first seemed doubtful because of the price and was only achieved after long negotiations involving several national meteorological services and other authorities. It may not always be possible to secure such works for serious use when they can command a high price as a collector's item.

References

Anderson, R. Y. (1961) 'Solar–terrestrial climatic patterns in varved sediments', *Annals of the New York Acad. Sci.*, 95, (Art. 1), 424–439.

Brezowski, H., Flohn, H. and Hess, P. (1951) 'Some remarks on the climatology of blocking action', *Tellus*, 3, 191–194, Stockholm.

Brooks, C. E. P. and Hunt, T. M. (1933) 'Variations of wind direction in the British Isles since 1341', *Quart. J., Royal Meteorological Society*, 59, 375–378, London.

Buchan, A. (1867) 'Interruptions in the regular rise and fall of temperature in the course of the year', *J. Scottish Meteorol. Soc.* (n.s.), no. 13, 3–15, Edinburgh, Blackwood.

Dove, H. W. (1856) 'Über Rückfälle der Kälte in Mari', *Abhandlungen der Königlichen Preussischen Akademie der Wissenschaften zu Berlin.*

Flohn, H. (1947) 'Stratophärische Wellenvorgänge als Ursache der Witterungssingularitäten', *Experimentia*, 3(8), 1–12, Basel.

Flohn, H. (1949) 'Klima und Witterungs-Ablauf in Zürich im 16. Jahrhundert', *Vierteljahresschrift der Naturforsch. Gesell. in Zürich*, 95, 28–41.

Flohn, H. (1954) *Witterung und Klima in Mitteleuropa*, Zurich, Hirzel, 27, 96, 117–120.

Flohn, H. (1967) 'Klimaschwankungen in historischer Zeit', in H. von Rudloff, *Die Schwankungen und Pendelungen Klimas in Europa seit dem Beginn der regelmässigen Instrumenten-Beobachtungen (1670)*, Braunschweig (Vieweg), 81–90.

Flohn, H. and Hess P. (1949) 'Großwettersingularitäten im jährlichen Witterungsverlauf Mitteleuropas', *Met. Rundschau*, 2, 258–263.

Hellman, G. (1926) 'Die Entwicklung der meteorologischen Beobachtungen in Deutschland von den ersten Anfängen bis zur Einrichtung staatlicher Beobachtungsnetze', *Abhandl. der Preuss. Akad. Wiss. Physmath, Klasse*, Nr. 1, Berlin.

Hellman, G. (1927) 'Die Entwicklung der meteorologischen Beobachtungen bis zum Ende des XVIII, Jahrhunderts', *Abhandl. der Preuss. Akad. Wiss. Phys-math. Klasse*, Nr. 1, Berlin.

Johnsen, S. J., Dansgaard. W, and Clausen, H. B. (1970) 'Climatic oscillations 1200–2000 AD', *Nature*, 227, 482–483.

Lamb, H. H. (1950) 'Types and spells of weather around the year in the British Isles: annual trends, seasonal structure of the year, singularities', *Quart. J., Royal Meteorological Society*, 76, 393–438, London.

Lamb, H. H. (1964a) 'Atmospheric circulation and climatic changes in Europe since 800 AD', *Report of the VIth International Congress on Quaternary (INQUA), Warsaw 1961. Vol. II, Palaeoclimatological Section*, Lódź, Poland, 291–318 (esp. 307–309).

Lamb, H. H. (1964b) *The English Climate*, London, English Universities Press, 1.

Lamb, H. H. (1969) 'The new look of climatology', *Nature*, 223, 1209–15, London, 20 September.

Lamb, H. H. and Mörth, H. T. (1978) 'Arctic ice, atmospheric circulation and world climate', *Geographical Journal*, 144(1), 1–22, London.

Lenke, W. (1960) 'Klimadaten von 1621–1650 nach Beobachtungen des Landgrafen Hermann V. von Hessen, *Ber. des Deutsch. Wetterd.*, 9, Nr. 63, Offenbach.

Lenke, W, (1961) 'Bestimmung der alten Temperaturreihen von Tübingen und Ulm mit Hilfe von Häufigkeitsverteilungen, *Ber. des Deutsch. Wetterd.*, 10, Nr. 75, Offenbach.

Lenke, W. (1964) 'Untersuchung der ältesten Temperaturmessungen mit Hilfe des strengen Winters 1708–9, *Ber. des Deutsch. Wetterd.*, 13, Nr.92, Offenbach.

Lenke, W. (1968) 'Das Klima Ende des 16. und Anfang des 17. Jahrhunderts nach Beobachtungen von Tycho Brahe auf Hven, Leonhard III. Treutwein in Fürstenfeld und David Fabricius in Ostfriesland', *Ber. des dt. W.*, 15, Nr. 110, Offenbach.

Manley, G. (1959) 'Temperature trends in England, 1698–1957', *Arch. Meteor. Geophys. Bioklim.*, B9, 413–433,

Manley, G. (1961) 'A preliminary note on early meteorological observations in the London region with estimates of monthly mean temperatures 1680–1706', *Meteor. Mag.*, 90, 303–310, London, Meteorological Office.

Murray, R. (1970) 'A note on central England temperatures in quintiles', *Meteor. Mag.*, 99, 292–294, London, Meteorological Office.

Pilgram, A. (1978) *Untersuchungen über das Wahrscheinliche in der Witterungskunde durch vieljährige Beobachtungen*, Vienna.

Rudloff, H., von (1967a) *Die Schwankungen und Pendelungen des Klimas in Europa seit dem Beginn der regelmäßigen Instrumenten-Beobachtungen (1670)*. Braunschweig, Vieweg-Die Wissenschaft, Band 122.

Rudloff, H., von (1967b) 'Die Schwankungen der Grosszirkulation in den Hauptjahreszeiten in Europa', *Annalen der Met.* (n,s,), Nr. 3, 86–95, Offenbach.

Sawyer, J. S. (1970) 'Observational characteristics of atmospheric fluctuations with a time scale of a month', *Quart. J. Roy. Meteor. Soc.*, 96, 610–625, London.

Trenkle, H. (1951–52) 'Eine langperiodische Klimaschwankung um die Jahreswende in Karlsruhe', *Abhandl. des Badischen Landeswetterdienstes.* 185–192, Freiburg, Wetteramt.

Weiss I. and Lamb, H. H. (1970) 'Die Zunahme der Wellenhöhen in den Operationsgebieten der Bundesmarine, ihre vermutlichen Ursachen und ihre voraussichtliche Entwicklung, *Fachl. Mitteilungen,* Nr. 160, Porz-Wahn, (Geophys., BDBw).

16

Fronts and their life-history

The discovery of fronts, the more or less sharp boundaries between unlike wind-streams, with their associated cloud-belts and weather, and their role in the development of cyclones (barometric depressions) was made by the Norwegian school of meteorologists under Professor Vilhelm Bjerknes at Bergen in 1917 and expounded in a series of scientific papers, now regarded as classics of the subject, over the years that followed (e.g. J. Bjerknes 1919, 1922; V. Bjerknes *et. al.* 1933). Some of the related atmospheric structures showing convergence and uplift of one of the two air currents meeting at a boundary had been spotted much earlier, notably by Loomis in 1840–1 in the USA, and by Shaw and Lempfert in Britain in 1905–6, but their full significance had not come out.

This chapter reprints the substance of a three-part essay on the life-history of fronts which the author wrote for the *Meteorological Magazine*, 80, London, 1951, pp. 35–46, 65–71, 97–106. It was remarked then that 'The theoretical treatment of frontogenesis and frontolysis (the formation and decay of fronts) due to Bergeron (1928) and Petterssen (1936) has not led to such systematic use . . . in weather analysis and forecasting as the importance of the subject might suggest'. Fronts are better known now to a wider public and recognized by all weather-watchers; yet, surprisingly, the statement quoted is still partly true. Around 1950, advances in understanding of the role of the thermal gradient concentrated in a frontal zone in the upper air, and the associated jet-stream pattern of flow, in the development of depressions and anticyclones (Sutcliffe 1947) switched attention somewhat away from the importance of the precise location and structure of the front near the Earth's surface for the local weather and sometimes subtle changes of surface wind. Incorporation of the newer understanding of developing and decaying weather systems opened up by Sutcliffe, and by Rossby and others' jet-stream studies, has led on to great advances in the quality and reliability of forecast weather maps over several days ahead, which are now produced in the main international meteorological centres. Nevertheless, presentation of the weather analysis to a wide public on television services remains sometimes crude and unconvincing.

In Britain, in particular, where the main forecasting offices are in the most urbanized parts of south-eastern England and southern Scotland, the broadcasting authority has for decades past allotted so little time to presentation of the general weather forecast as to suggest a belief that the public does not really want to know about the weather. On main channel television presentations, fronts are often broken off into short lengths, or omitted altogether, in order to simplify the map, when significant vestiges of their activity still survive. A worse fault is that they are commonly drawn in awkward – straight or angular – shapes, which the fluid atmosphere could never produce, or are swung round a low-pressure centre like the spokes of a wheel with the outer regions (where the winds are weakest) apparently showing the fastest progress. Wrongly placed fronts lead to mistaken expectations of when their weather will arrive.

Hence, there is still a need today to present an account of the life-history of fronts, and the evolution of their geographical patterns, as well as their activity. In what follows, the essay is slightly reworded for the general reader, and updated by a contribution from the era of satellites, though the original examples are still used. The only disadvantages of this are the appearance of old units such as degF which are still widely understood, and the antiquated way of referring to the times of the weather maps as if they were precise to the minute instead of just to the hour. For scientific usage, a front may be defined as essentially a line or a narrow zone of discontinuity in the horizontal distribution of temperature, with an associated discontinuity in the barometric pressure and wind fields. The actual criterion for entry or omission of a front on the weather map must be somewhat arbitrary and governed by usefulness. But, though simplification of a complicated map may sometimes legitimately be achieved by omission of weak fronts, it can never be right to achieve it by distortion to conform to some preconceived model. And attention given to the frontal shapes and satellite-surveyed cloud-patterns arising with various flow-patterns and distributions of wind

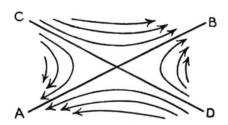

Figure 16.1 Deformation Field (Stream-lines).

speed will help greatly in interpreting isolated or vestigial weather phenomena and shifts or discontinuities of surface wind.

Petterssen pictured a typical 'deformation field', as reproduced in figure 16.1, and defined frontogenesis and frontolysis by reference to it. An entity such as temperature and its distribution could be represented by isotherms drawn on top of this flow-map. Frontogenesis will be taking place if the isotherms are brought closer together, as time goes on, by the wind-flow. In this case, the temperature gradient will be becoming sharper. Frontolysis will be occurring if the wind-flow takes the isotherms farther apart, weakening the temperature gradient.

In figure 16.1, AB and CD are axes of confluence (inflow) and diffluence (outflow) respectively. The diagram properly represents stream-lines of the wind-flow; and if we use the isobars on a pressure map as a rough approximation to the wind pattern, ageostrophic motions (i.e. departures from the ideal isobaric wind), convergence and divergence, must be allowed for. Isotherms initially parallel to AB will be brought progressively closer together along the axis AB, strengthening the temperature gradient. Convergent (ageostrophic) motion is, however, needed to produce an actual discontinuity within the narrowing frontal zone: surface friction normally ensures that such convergence does occur, and upgliding of one airmass or undercutting by the other consequently develops. Isotherms initially parallel to CD would be drawn apart towards A and B, so that CD is the axis of a frontolytic zone.

Isotherms initially having other orientations will be progressively swung into alignment with AB or CD, depending on the initial angles between these axes and between the isotherms and the inflow axis (AB): the 'col' in the barometric pressure and wind pattern has a frontogenetic or frontolytic influence accordingly. A specially strong thermal gradient may be produced when air from an Arctic or Antarctic cold source at C is brought directly towards air from a warm, tropical source at D. A strong thermal wind then develops in the upper air in the same direction as the surface winds in the right-hand branch of the confluence, near B in the diagram. In a zone along this axis, jet-streams are generated in the upper troposphere, though the strongest jet may culminate perhaps 2000–5000 km down-wind as cyclonic systems develop from wave disturbances on the sharpened front and travel in that direction.

Nevertheless, frontogenetic cols seldom stay long enough in one place to produce a fully fledged front where none was before. What happens far more often is that already existing weak fronts, old occluded fronts, and frontal belts containing more than one old frontal system, arrive in an area with such a flow-pattern and become sharpened into essentially one strong simple front that is once more capable of major activity. This is how nature simplifies the weather map.

New fronts also commonly appear on our maps through the occlusion process (figure 16.2), repeatedly reproducing the patterns first drawn by

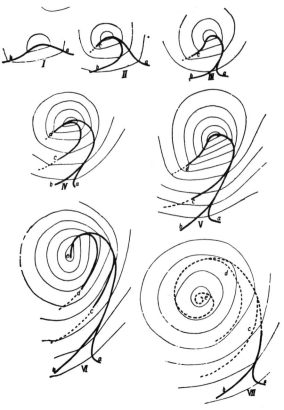

Figure 16.2 The life-history of a big frontal cyclone (after Refsdal 1930).

Refsdal (1930) and plainly revealed in nature when satellite cloud surveys became available (figure 16.3). In the occlusion process, the lighter warm air is lifted off the Earth's surface as the cold front overtakes the warm front and narrows the warm sector of the developing cyclone. Refsdal explained the sequence as follows (my translation):

> The wave disturbances which, by subsequent twisting up (*Verwirbelung*), develop into the common type of cyclone are relatively small disturbances in more southerly latitudes [I in figure 16.2]. They are not at that stage ready to grow into big cyclones. The occlusion process [II and III in figure 16.2] and the slipping of the centre along the occluded front [III and IV in the diagram] have the effect of producing a new, pseudo-warm sector. And if the air masses in this new warm sector have a store of potential energy by virtue of vertical instability (depending on the lapse of temperature with height), as is generally the case in winter over the ocean, the cyclone may deepen (intensify) further. After the occlusion of this

new warm sector . . . the process can repeat itself [V, VI, and VII in figure 16.2]. An ordinary wave disturbance may in this way grow into a large cyclone.

The ability to recognize, long before the days of satellite cloud pictures, the shape of things produced by twisting up of the frontal systems in the ways commonly brought about by the distribution of wind strengths with distance from the centre in a normal cyclonic depression seems to have been a rare gift. Refsdal and V. and J. Bjerknes, Palmén and notably Bergeron (1930) deserve credit for first indicating in numerous writings the reasonable drawing of the types of frontal pattern which evolve in the fluid atmosphere. The commonest situation with light winds near the outskirts of a depression, and also lighter winds near the centre than in the intermediate zone, twists the frontal system forward in the region of strongest winds and back near the edge of the storm. Nowadays, with the guidance of daily satellite cloud surveys, there can no longer be any doubt about the characteristic shapes of fronts becoming gradually coiled around the centre of cyclonic depressions as the system becomes older and more fully developed.

The trailing front outside the limits of the storm commonly lies in an orientation close to the inflow axis of the neighbouring col. Hence, the front becomes sharpened and more wave disturbances arise, so giving birth to a whole sequence or 'family' of depressions. These travel in the same direction, occluding and deepening, until they are commonly 'fed' into the circulation of the previous, ageing storm, which they rejuvenate and maintain for some days.

A frontolytic orientation of the trailing front in the col rarely arises. Frontolysis in practice, as we shall see, comes about in other ways.

Frontogenesis

A frontogenetic col is most effective if the axis of inflow lies in a geographical belt where a steep gradient of temperature of the Earth's surface induces a corresponding zone of quick transition in the overlying air.

The most important geographical belts favouring frontogenesis, following lines first mapped out by Bergeron (1930), are:

(a) In the North Atlantic Ocean off the east coasts of North America and Greenland, near the boundary of the polar ice/Labrador Current and Gulf Stream waters. The cold surface west of the boundary is far more extensive in winter, when the American continent is cold, than in summer.

(b) The corresponding belt in the North Pacific Ocean off the coast of Asia to the Aleutians, to which similar remarks apply.

(c) From late summer through autumn the thermal gradient around the fringe of the permanent ice in the Polar Basin along the north coast of Asia and North America rivals (a) and (b) in importance.

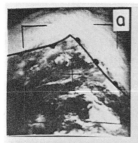

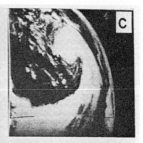

| TIROS IV, 807/806, fr 18 | TIROS IV, 880-direct, fr 26 | TIROS III, 778/777, fr 16 |
| 1619 GMT, APRIL 5, 1962 | 2012 GMT, APRIL 10, 1962 | 1357 GMT, SEPT 4, 1961 |

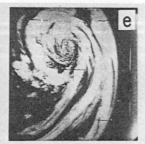

| TIROS III, 679/678, fr 12 | TIROS III, 793/792, fr 15 | TIROS IV, 1005/1004, fr 8 |
| 1625 GMT, AUG 28, 1961 | 1506 GMT, SEPT 5, 1961 | 1123 GMT, APRIL 19, 1962 |

Figure 16.3 Stages in the life-cycle of North Atlantic cyclones, from satellite surveys: reproduced by kind permission from *Weather*, vol. **21**, 1966, p. 175.

(a) open wave (52°N 37°W, 1,004 mb), 5 April 1962;
(b) wave cyclone beginning to occlude (46°N 59°W, 996 mb), 10 April 1962;
(c) cyclone 12–18 hours after occlusion began (48°N 45°W, 1,006 mb), 4 September 1961;
(d) deep occluded cyclone near maximum intensity, 30–36 hours after occlusion began (51°N 24°W, 978 mb), 28 August 1961;
(e) weakening occluded cyclone, 36–48 hours after occlusion began (50°N 45°W, 1,006 mb), 5 September 1961;
(f) dissipating cyclone, more than 48 hours after occlusion began (49°N 4°W, 1,009 mb), 19 April 1962;

(Photographs kindly supplied by ESSA, National Weather Satellite Center, Maryland).

The following likenesses to Refsdal's diagram sequence in figure 16.2 (1930) stand out:

Tiros *satellite* *picture*	*Compare* *Refsdal's frontal* *diagrams on p. 268*
a	I
b	Intermediate between I and II
c	II, though without the back-bent occlusion
d	Intermediate between III and IV,
	With the back-bent occlusion this time
	The pseudo-warm sector is beginning to occlude, a process repeated in V.
e	V and VI
f	VII

(d) In late autumn through winter, as the continents cool down particularly in their eastern portions, (c) tends to be replaced by two belts dipping south-eastwards into the heart of the great continents:

  (i) through Alberta along the flank of the northern Rockies, where mild air from the Pacific Coast meets the snow-covered interior of the continent;

  (ii) a rather less definite north-west to south-east line near the White Sea and north Urals where mild air from the Atlantic and land-locked seas of Europe reaches the main continental interior with its winter snow cover. This line of thermal contrast is more fitfully maintained than (i), being itself considerably dependent on the patterns of air circulation prevailing. Occasionally advection around the Asiatic anticyclone carries warm air east along the north coast and restores belt (c) even in the depth of winter. In some winters, however, the Eurasian snow-cover ultimately becomes firmly established as far west as the Harz Mountains in northern Germany, or even as far as the British Isles, and the frontogenetic belt is near this boundary, orientated from NW to SE or from N to S.

(e) In the spring and autumn transitional periods, (i) the Baltic Sea and (ii) the St Lawrence Valley mark important zones of contrast between cold largely snow-covered regions to the north and warmer lands to the south; (ii) has also some importance at other seasons.

(f) In winter and early spring, a belt of strong thermal gradient runs from the coast of north-west Africa through the Mediterranean across Turkey and the Persian highlands finally becoming weaker and more variable in position towards Mongolia and Tibet. The corresponding feature of deep winter in the western hemisphere is a prolongation of belt (a) into the Gulf of Mexico.

(g) In the southern hemisphere, the main belt is that which surrounds the Antarctic continent and, in winter, the edge of the pack-ice.

(h) Subsidiary frontogenetic zones occur near the coasts of the warmer southern continents at suitable seasons of the year.

Climatological charts show that belt (a) in the western Atlantic corresponds roughly to the inflow axis of the col in the mean winter pressure distribution between the Azores–Bermuda high and the north Canadian anticyclone and between the Iceland low and low pressure in the Mexican Gulf. This means that the frontogenetic pattern repeatedly occurs when cold, continental winter anticyclones move south-east towards the Atlantic seaboard of North America behind the occluded fronts of depressions which have passed out on to the Atlantic. These old occluded fronts are then sharpened and transformed by the frontogenetic deformation field into the so-called Atlantic polar front, on which new depression families arise. Similar sequences take place off the east coast of Asia and less regularly in other parts of the world.

If one watches what happens when a frontogenetic col appears in a single air mass over one of these geographical belts favourable to frontogenesis, embryonic frontal phenomena begin to appear. Sheets of thin altostratus or altocumulus cloud start to form as the convergence in the horizontal wind field forces upgliding at any minor surfaces of discontinuity, even former subsidence surfaces, existing aloft. Stratus sheets may be similarly formed in the lower levels. Slight rain or drizzle sometimes results. All this may happen before the surface temperatures and humidities have developed any sharp discontinuity. And then, in most cases, the movement of distant pressure systems causes the col to shift; the pattern of flow ceases to be favourable for front formation, and the frontal phenomena, that have been appearing, dissipate without having come to anything. Even in the most likely geographical regions, the frontogenetic circulation pattern must

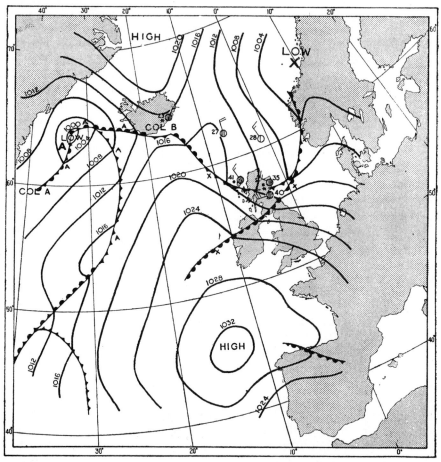

Figure 16.4 Synoptic chart, 0600 GMT, 22 February 1946.

persist for many hours if a front is to be formed. The length of time required is a variable quantity depending on wind components along the axis of inflow in the opposing air streams as well as on the physical properties of the air involved.

Not many fronts would be formed in the atmosphere if the process described by the theoretical deformation-field diagram were the only one capable of creating a front. Too great a coincidence of favourable circumstances is required. The number of fronts shown on the daily weather maps of every forecasting office compels us to consider what alternative processes may lead to front formation.

The most prolific process giving new discontinuity lines in the horizontal pattern of flow seems to be that diagnosed by Refsdal: bent-back occlusion and re-occlusion of the pseudo-warm sector (figures 16.2 and 16.3),

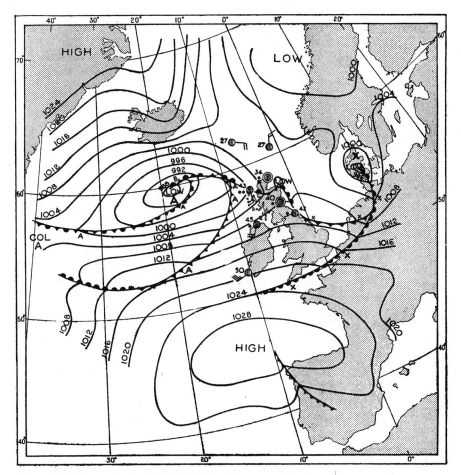

Figure 16.5 Synoptic chart, 1800 GMT, 22 February 1946.

sometimes repeated several times in the life of one depression. The successive bent-back occlusions may be treated as secondary cold fronts at quite an early stage in their history. And these may later become sharpened, or bunched together, into a single frontal zone of major importance if they come into a frontogenetic col situation. Occlusions, however, which are followed by warmer air coming round the rear of the depression obviously tend to weaken if they move on farther south, although it may be a long time before they become so weak as to disappear. After some days of northerly wind bringing a vigorous outpouring of air from the polar basin as far south as the British Isles, it is not uncommon for the next frontal system to be a warm-type front or occlusion that has crossed the North Pole and so is followed by less cold air from the north. In such cases, the cloud systems of the front may reach the latitude of Europe intact though seldom very active.

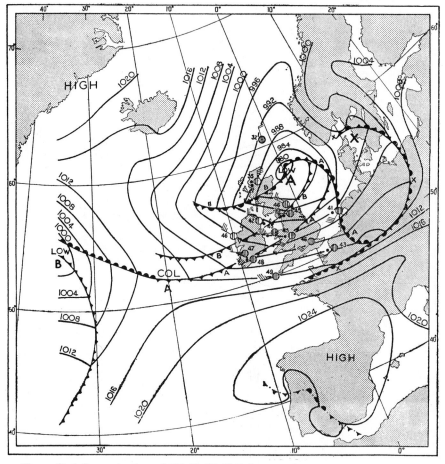

Figure 16.6 Synoptic chart, 0600 GMT, 23 February 1946.

Conventional analysis tends to pick upon the major fronts and ignore the minor discontinuities produced by Refdal's process. Difficulties arise with some old depressions in which all the fronts are rather minor, and all have a more or less equal claim to recognition.

A case of this kind arose on 22–25 February 1946 (figures 16.4–16.7). Conventional analysis confined to a single warm front, cold front and occlusion entirely failed to explain the patterns of wind and weather in depression A as it developed and moved across the British Isles. Considerable areas experienced continued warm-sector-like conditions after one or two wind veers and rising pressure had set in. The depression might reasonably have been treated as more or less non-frontal, with intermittent rain and drizzle and local clearances over its entire area. Individual stations where detail was important might nevertheless trace three or four successive

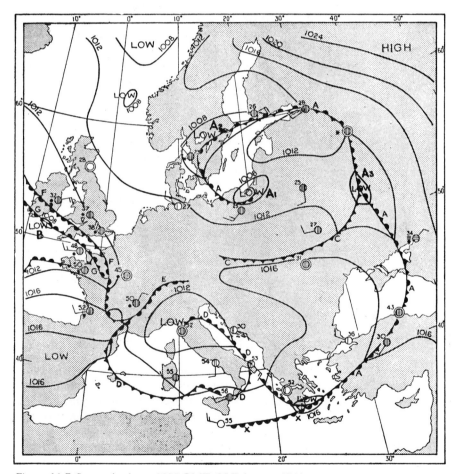

Figure 16.7 Synoptic chart, 1800 GMT, 25 February 1946.

weak fronts passing and bringing a final clearance only when the veering wind at last reached a point giving some orographic shelter.

Low A is seen, already partly occluded and with a depth of 1000 mb, west of Iceland on the morning of the 22nd (figure 16.4). In the following 24 hours, the centre moved quickly east-south-east into the North Sea, the original occlusion A became secondary cold front B on the map for the 23rd (figure 16.6), and in the course of re-occlusion the centre deepened to 980 mb, producing a long new frontal line, occlusion B, which already showed signs of repeating the process. The centre deepened further to 968 mb when it approached Denmark on the evening of the 23rd and as the second pseudo-warm sector occluded. Ultimately, occlusion X of the former Scandinavian low was swept around the rear of low A as a further secondary cold front over Denmark when the deep centre moved into the Baltic. With this final development completed, the centre began to fill up, and by the evening of the 25th it was represented only by several shallow centres of 1008 mb along a single occlusion coiled up over the Baltic States, south Sweden and central Russia (figure 16.7).

Figures 16.4 and 16.5 further show how occlusion A was not only bent back, but necessarily rotated into alignment with the inflow axis of col A over the Atlantic south-east of Greenland and thereby came under the influence of the frontogenetic effect of the deformation field, which seems to have prolonged the front to the westward as time went on. Figure 16.6 shows the primary cold front A likewise brought into line with the frontogenetic axis of the col as a natural consequence of the circulation pattern. We see too how a normal development of the circulation is quite unlikely to produce a frontolytic orientation of the front in the col.

Figure 16.7, marking the end of the sequence, is an interesting example of the fertility of the occlusion process at different points along the subtropical subpolar airmass boundary in producing a welter of new, offshoot fronts, any one of which may later be subjected to the sharpening process of a frontogenetic deformation field. The occlusion marked A on this map is already the product of closing together (consolidation) of a bunch of occlusions (A, B, and X) developed in the earlier stages of the life-history of depression A. Consolidations of this kind are frequent, but complete amalgamation of the successive fronts requires ageostrophic motion and is not always achieved. Cold front C was formerly an extraneous system drawn into the circulation of the rear side of low A from Scandinavia and now helps the development of a new depression over Russia with a warm sector formed between the old warm occlusion A and the cold front C; this system moved south-east towards the Caspian Sea with renewed intensity of precipitation. Occlusions D, E, F, and G are new derivatives of the activity along the same parent front, and the occlusions F and G are in process of being closed together into a single, sharp front by the pattern of winds over northern France and southern England. Ageostrophic components are evident. The subsequent activity on the sharpened front between the mild Atlantic air-masses to the south and the Arctic air-mass with freezing temperatures to the north of the confluence gave one of the heaviest snowfalls in Kent.

The spotting of this situation, which will squeeze together parallel fronts (simplifying the map) or sharpen and energize the frontal belt, is an important use of the deformation-field pattern (figure 16.1). The building up of potential energy and quickening of upgliding motions with increased rainfall commonly take 24–48 hours to mature. The warm-front wave disturbance forming south-east of Iceland on figure 16.4, rather as a break-away running ahead of the depression A and travelling south-east to Scotland during the 22nd, may be looked upon as an example of this action in the case of the frontogenetic col B. The south-westerly air-stream (maritime tropical air) which gave a temperature of 43 degF at Vestman-naeyjar, south Iceland, and the Arctic air-stream with temperatures around 23 degF met from directly opposite directions at this section of the front. The warm-front wave, which appeared, travelled quickly south-east in the strong thermal gradient, bringing rain and sleet far ahead of the main depression.

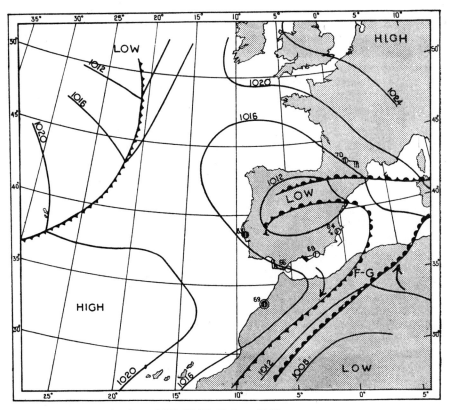

Figure 16.8 Synoptic chart, 0000 GMT, 19 June 1945.

Breakaway depressions forming at the point of occlusion and running ahead of the parent low, first studied by Sawyer (1950) commonly produce one long new frontal line upon our maps by prolonging the occlusion. This process is, therefore, another source of new fronts.

Another interesting frontogenetic sequence occurred over north-west Africa and the regions to the north on 19 June 1945. The chart for 0000 on the 19th (figure 16.8) shows a rather complex situation with parallel fronts over north Africa associated with the very old depression over Spain. Pressure gradients are weak but a frontogenetic col is marked on the chart near Algiers, and the temperature contrasts between the air masses approaching one another at this point are great. By 1200 on the same day (figure 16.9) the situation in the frontogenetic belt has been reduced to a single sharp front. The resulting simple warm sector over the western Mediterranean continues northwards aloft as a broad expanse of alto-cumulus castellatus clouds over France; and its trough-line, the upper cold front over the Bay of Biscay, is producing a sharp trough in the isobars over the Bay. Such lines of an upper front have been usefully given the name

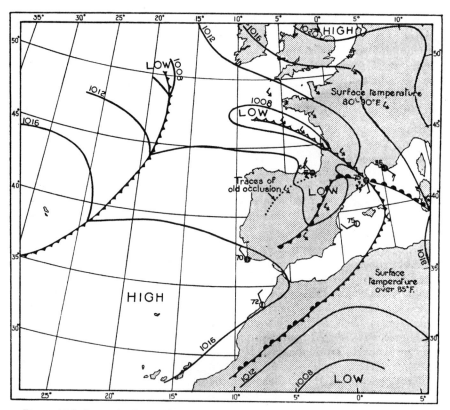

Figure 16.9 Synoptic chart, 1200 GMT, 19 June 1945.

'trowal' in Canadian meteorology: this is the line of the 'valley bottom' formed by the cold front overrunning the warm front surface in the free air, ultimately thousands of feet above the ground. Aircraft reports indicated conditions in the lifted air mass getting progressively more stable westwards to tail out as a thin sheet of altocumulus, or high stratocumulus, about the 8000-ft level in 9°W over the Bay. Other aircraft flying along the Atlas Mountains *en route* from Malta to Rabat reported severe thunderstorms in frontal middle-level cloud in the region where the sharpened front is shown, while the general increase in activity of the depression system was demonstrated by the much greater number of reports of lightning and atmospherics shown at 1200 GMT on the 19th than at the previous midnight (cf. figures 16.8 and 16.9). Recent thunderstorms were reported on both charts but none at the hour of observation at the stations used. The transformation of the frontal pattern went one stage further during the afternoon and evening of the 19th. The westerly breezes developing behind the pressure trough associated with the upper cold front over the Bay of Biscay brought in cool Atlantic air to south-west France. The trough now corresponded to the main temperature contrast on the surface map between the cool oceanic air (maritime polar) and the strongly heated European and African airmasses with high surface temperatures (continental polar and continental tropical). This situation is already apparent in figure 16.9. As a result, the upper front quickly extended itself down to the surface as a normal cold front, which together with the wide belt of high-level instability cloud and thunderstorms ahead of it, became the dominant feature of the map, leaving merely traces of the old occlusions over Spain as relics of little consequence.

Rapid changes of the isobar pattern, under the influence of considerable pressure falls probably associated with the release of instability aloft superimposed on a previously flat situation, made the stages of this quick development (figures 16.8 and 16.9) hard to follow in detail. Nevertheless, this conversion of an upper front to a surface front is by no means unique. British meteorologists stationed in the Mediterranean speak of a fair number of occasions on which fronts appeared to 'build downwards' and some instances have been noted in higher latitudes.

A further type of frontogenesis to which C. K. M. Douglas drew attention is liable to occur in southerly or south-westerly sweeps of returning maritime polar air over the North Atlantic. When the southernmost extremity of a tongue of polar air has become very shallow and considerably modified, the effective warm front may be transferred north to the limit of deeper cold air. This is most likely to happen where an old, weak front, such as one of the trailing secondary cold fronts of depression A (figures 16.4–16.7) already exists and may be revived. In many cases of this kind, however, vestiges of both fronts remain, unless a well-marked frontogenetic convergence and deformation pattern occurs to simplify and sharpen the situation, or the geographical factors are particularly favourable.

Circulations in the vertical plane

Subsidence is commonly considered as a frontolytic agency. Frontal lines of discontinuity in the horizontal plane tend to disappear in the accompanying (horizontal) divergence, and clear skies and sunshine favour modification of the shallow layer of cold air in the lowest levels by surface heating.

Nevertheless, convergence in the vertical plane tends to sharpen existing discontinuity (e.g. frontal) surfaces aloft wherever air subsiding from higher levels meets either vertical up-currents or air-masses rising at a slanting angle. The same effect occurs where subsiding air descends upon another body of air which is sinking less rapidly, as the air layers next to the Earth's surface are normally compelled to do in regions of active subsidence. Namias (1934) has given diagrammatic illustration of these points (see also Chromow 1940). Conversely, frontal surfaces associated with general upward motion (as at normal, active fronts with horizontal convergence in the lowest layers) tend to weaken and become diffuse aloft.

It is a curious fact that when horizontal divergence supplied by downward motions in an anticyclonic situation weakens a front on a surface map, subsidence tends to maintain and sharpen the discontinuity surface aloft. The vertical analysis will sometimes show these sharpened discontinuity surfaces maintaining their identity aloft throughout the life of an anticyclone and eventually becoming involved in renewed horizontal convergence and uplift when the anticyclone breaks down. Upgliding cloud-sheets of thin stratus, altostratus or cirrostratus, then appearing in the region of general fall of pressure far ahead of any surface fronts, may give rainfall, sometimes well within the decadent anticyclone. Such cloud systems tend to be located at the surviving discontinuity surfaces aloft, themselves beginning to be deformed by the convergence and vertical expansion of the lower air layers.

An example of the maintenance and sharpening of discontinuities aloft in the vertical convergence associated with subsidence may be followed through all stages of the process (figure 16.10), which shows a vertical analysis (time cross-section) of the lower atmosphere over Downham Market, Norfolk, during the life of an anticyclone centred over England on 17–20 April 1945. The time cross-section has been given a time-scale that reads from right to left, so that the motion of winds and frontal surfaces from west to east may be read in normal fashion from left to right across the paper. The anticyclonic centre first appeared on the 17th in the mouth of the Bristol Channel after the decadent remains of cold front P, which had been well enough marked over Scotland and Ireland the day before, had passed east-south-east. This front was so weak as to be virtually impossible to follow across East Anglia on the surface maps of the 17th. The isotherms on figure 16.10 show it from the temperatures observed on successive soundings between the 950-mb and 650-mb levels. It was also shown by a wind shift which was sharpest near the ground, though somewhat blurred by a thermal trough in the isobars ahead of the front on the 16th. Between the 800-mb and

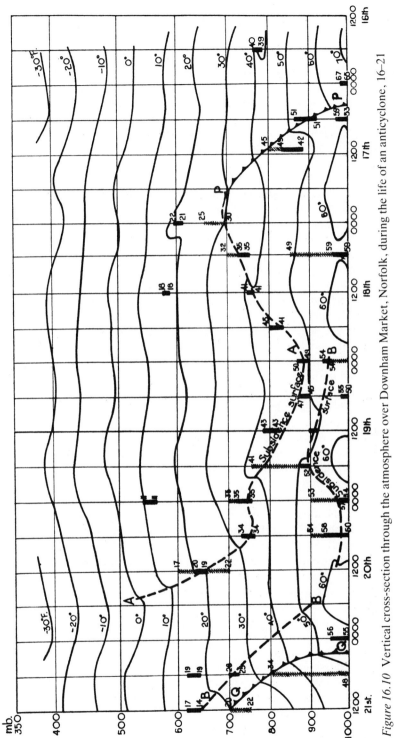

Figure 16.10 Vertical cross-section through the atmosphere over Downham Market, Norfolk, during the life of an anticyclone, 16–21 April, 1945.

Vertical lines correspond to successive soundings. Observed temperatures are shown at the top and bottom of stable layers only. Inversions and isothermal layers are marked with thick black lines, stable layers with hatching.

700-mb levels, the incursion of the colder air-mass was short-lived: the temperatures at each level returned in 24 hours or less to the earlier, higher values, rising approximately 3 degF as the frontal surface sank on the 18th and 19th. The frontogenetic sharpening process working on the subsidence surface A, which is the continuation of the cold-front surface P, eventually produced an inversion layer on 18–19 April where only an isothermal state had existed in the cold-front upgliding stage on the 16th and 17th. Temperatures rose 2–4 degF at each height with the downward passage of the subsidence surface, and later they fell 3–7 degF when the same sharpened discontinuity surface was lifted again on the 19th and 20th. From the 19th onwards, subsidence sharpening seems to have made a distinct discontinuity, designated subsidence surface B, out of the top of the ground turbulence layer; and this too can be traced through the uplift phase as the anticyclone collapsed before the advance of the cold front Q from the north-north-west. An interruption of surface B on the afternoon of the 20th is attributed to convective turbulence resulting from the high ground temperatures in the afternoon at Downham Market. Analysis of the aircraft-ascent readings at the neighbouring coastal station of Langham, where less ground heating occurred, shows surface B without interruption. The anticyclone collapsed on the 20th, pressures at East Anglian stations falling about 15 mb in the 24 hours before the arrival of cold front Q, which is clearly traced in the abrupt temperature falls between the ground and the 700-mb level on the 21st in figure 16.10. The new air-mass brought temperatures 10–20 degF lower than had previously been observed. A very interesting feature of the analysis is the cooling, at various levels aloft, starting 24 hours before cold front Q arrived. At the 700–750-mb level, three stages in the cooling were traced, and at 650 mb there were two stages in the cooling from 29 degF on the 19th, ultimately to 15 degF on the 21st. That this pre-frontal cooling aloft was not just a gradual process is seen from the observed inversions and isothermal layers at successively higher levels on successive ascents, most obviously in the case of the subsidence surface A in the ascents of the 19–20 April. The lifting of the subsidence surfaces with cooling of several degrees Fahrenheit in each case explains the successive upper cold fronts often noticed ahead of major cold fronts at the ground, and is attributed to the setting in of renewed horizontal convergence accompanied by vertical expansion (spreading upward) of the lower air layers ahead of the active cold front.

The increase in depth or re-gathering of a preliminary cold air-mass ahead of cold front Q, actually considered as the surface air-mass which had become shallow in the central regions of the anticyclone during its lifetime, may imply of course considerable horizontal advection. In the case of the English anticyclone analysed in the vertical diagram, figure 16.10, C. K. M. Douglas found that the air concerned was drawn north from a cold pool over the Azores in which a deep reservoir of the cold airmass had accumulated, presumably after the passage of cold front P. Similar effects were revealed by an analysis of the anticyclones of the end of March 1946 over the British Isles and indeed appear to be very usual.

The same deduction of sudden stages of cooling in the upper levels ahead of cold front Q on figure 16.10 follows from the appearance of altocumulus castellatus and cirrostratus bands in the previously clear sky ahead of the front on the 20th. These bands of middle level and high cloud were orientated parallel to the surface cold front which they preceded, indicating the controlling influence of this front, ahead of which scattered thunderstorms occurred in the Midlands during the evening of the 20th. There was a thunderstorm at Downham Market at midnight, also an hour or two in advance of the front Q. The pre-frontal cooling aloft was of course responsible for introducing the latent instability, indicated by the crowding of the isotherms on figure 16.10 on the 20th. The same developing instability in the uplifted layers of air between former subsidence surfaces always favours turbulence in these layers, and this may produce turbulence cloud, whilst upgliding motions at the former subsidence surfaces themselves may lead to the appearance of stratus layers or thin veils of stratiform cloud. This is clearly one way in which the frequently observed multilayered cloud at major cold fronts can develop. Observers on the ground report varying types of altocumulus according to the number of different layers and the degree of instability apparent in the most unstable of them.

Similar circulation patterns and development of frontogentic deformation fields in the vertical plane are considered to explain the re-forming of the tropopause, sometimes at a different level from before, after the previously existing tropopause has become diffuse.

Cloud structure at warm and cold fronts

Bergeron (1937a, 1937b) gave models of the cloud structure at simple warm and cold fronts and in occluded depressions viewed in vertical cross-section. These models have been widely quoted in meteorological literature. Nevertheless, observation does not always confirm the standard sequence of cirrus clouds joining with, and giving place to, the middle level and then the lower cloud-sheets of the lowering frontal surface as one approaches the front, but rather suggests that in very many cases, outside the regions of moderate and heavy rainfall, the cirrus and medium cloud-sheets constitute separate systems one above the other and both more or less separate from the lower cloud masses. Aviation forecasters and aviators may usefully look for clear 'lanes' or interstices between the upper and lower cloud masses, indicated by stable layers in the upper air soundings, whenever these may be found.

Modification or refinement of the accepted model of frontal cloud structure is needed in some situations, and obviously the charting of these clear 'lanes' between the upper and lower cloud masses at a front may be of great importance to aviation. We have already made an approach to the problem in relation to the cold front Q in the vertical analysis of figure 16.10, which may be taken as an example; and this has suggested an explanation of the development of multilayered cloud ahead of some cold fronts. The vertical cross-section offers the most promising technique for following the trend of any one stable layer aloft from one sounding to the next, and therefore of correctly diagnosing the course of the clear 'lanes' when these exist between cloud layers. This, and the *a priori* likelihood of common occurrence of incomplete stages of frontogenesis and frontolysis, seems to imply that there are in reality few simple warm and cold fronts, and that most pronounced fronts move in association with the decadent remains of older systems

C. K. M. Douglas (1934) gave descriptions from early flying experience of some of the more complex cloud structures which occur, perhaps often merely as a result of complicated vertical interchanges of humidity. Conditions which, however, favour deep cloud development at a front and therefore tend to make the upper and lower cloud decks merge into a simpler, so to speak, 'solid' system are:

(a) frontogenetic effects, leading (i) to the supply of fresh warm and humid air for uplift at the frontal surface, and (ii) to increased potential energy available for release as kinetic energy in the upgliding motion;
(b) instability in the uplifted air-mass which, in extreme cases, produces altocumulus castellatus as the frontal cloud and causes such turbulence as to destroy minor discontinuities surviving in the higher levels (partial destruction is liable to result in frontal cloud in multiple, broken, layers, these chaotic skies being specially common in fronts approaching the British Isles from Spain);
(c) high moisture content in the uplifted airmass;
(d) moderate or heavy rainfall raising the humidity of the layers it falls through.

In addition, deep cold airmasses (or cold pools) bounding the warm sector, resulting in rather steep frontal surfaces proceeding to high levels, also favour the realization of Bergeron's models of frontal cloud structure; although the existence of vestiges of an inner warm sector with additional surfaces of upgliding and duplicate cloud systems is not precluded even here. All such cases involve some unavoidable blind flying for aircraft travelling across the fronts, except possibly near ground level. The reverse state of affairs, where frontolytic influences or subsidence are at work, where upgliding is replaced by downward motion and instability replaced by stability and high humidity by low humidity, causes the cloud layers to thin and break and may lead to complete clearance. The nearest cloud layer to the ground and its humidity sources is naturally the one most likely to persist (except in arid regions) and is widely separated by deep layers of clear air from whatever vestiges of thin altocumulus or cirrus may survive at higher discontinuity surfaces above the subsidence zone.

Observers on the ground can corroborate the upper cloud-sheets ahead of a cold front encroaching upon a decaying anticyclone in which there has been no low cloud. Banded cirrostratus, ranged parallel with the advancing cold front, sometimes occurs 250 miles ahead. Later, multilayered cloud and rain may appear. On some occasions all the cloud, and more often all the higher cloud, clears as the front passes, implying that the cloud systems are on these occasions entirely ahead of the front. Figure 16.11 gives a common model of frontal cloud structure which fits these observations.

Apart from the main frontal surfaces and the tropopause (both indicated by full lines), two residual discontinuity surfaces (marked by dotted and dashed lines) appear in the model. In origin, these residual surfaces, which may be only weakly stable layers when upward motions are general (i.e. under conditions which reduce the main frontal surface to a deep transition layer), belong to older systems which have been imperfectly frontolyzed. Inner warm sectors are fairly commonly noticed on the surface maps, and, with careful scrutiny, some traces of such earlier frontal discontinuities might often be discerned.

Such discontinuities are most readily preserved in air-masses characterized by vertical stability and are, therefore, most commonly found as complicating factors in warm air-masses and lifted warm sectors.

Upgliding at weak discontinuity surfaces in the free air produces stratiform cloud, either in thin veils or thicker sheets. Turbulence cloud types may form in the intervening layers, given suitable depth of the layer and lapse-rate and humidity conditions. In as much as the depth of the layers is controlled by the main fronts and associated convergence at the ground, all these cloud systems are liable to show a banded structure parallel to the controlling front. Very fine cirrus or cirrostratus, sometimes only revealed by optical phenomena, may also form at or just below the tropopause.

The narrower, inner warm sectors or residual warm tongues aloft are able to give cloud sheets associated with upgliding and turbulence ahead of the cold front, and in the narrowest ones the cloud system may become continuous across the entire warm sector. It is not surprising, therefore, that

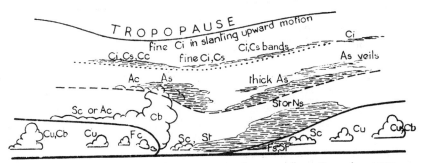

Figure 16.11 Section showing cloud structure of common type through a warm sector.

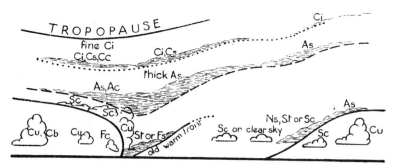

Figure 16.12 Section showing cloud structure through a warm sector of more complex origin.

cirrostratus and sometimes altostratus are often continuous across narrow warm sectors, and probably across most warm sectors near the tip.

A slightly different picture is presented by figure 16.12, in which the cold front may be described as a complex front. This arises where the cold front has formerly been an occlusion or is preceded by relics of the warm front of a nearly occluded warm sector having practically the same orientation as the cold front. Such surviving traces of occlusion character were perhaps first noticed in Australia, but are common in Europe and elsewhere. In these cases, there is a tendency for the warmest and most humid air, occasionally with the only traces of really low cloud in the whole warm sector, to be experienced just before the cold front passes; whereas the simpler case represented by figure 16.11 is liable to show some slight cooling in advance of the cold front.

Frontolysis reconsidered, with examples

Frontolysis has tended to be dismissed in meteorological literature as a mere counterpart of frontogenesis, to be explained in terms of the same deformation field. The subject needs separate study. We have already noticed that the frontolytic col seldom arises in practice.

Figure 16.13, which shows a front, A, lying quasi-stationary between the twin centres of an intense anticyclone over England and France on 14 February 1946, probably typifies an unusually close approach to the frontolytic orientation of a front in the col. It was not at once clear whether the situation would prove frontolytic, but the intensely anticyclonic conditions ensured that the front would be quiescent. At the time of the chart shown in figure 16.13, the situation had already existed substantially unchanged for more than 24 hours, yet the front A is still marked by a sharp wind discontinuity. Eight hours later, a small movement of the front west of Hurn Airport brought a wind veer from NE to S and eventually to SW, and a sharp deterioration of visibility as the smoke from the towns nearby,

Bournemouth and Poole, was transported in the direction of the airfield. At the same time, the low stratus lifted from about 500 to 1000 ft with the arrival of the drier continental air-mass. Dew points over France were mainly between 20 degF and 30 degF, and the dry-bulb temperatures show the effect of radiation cooling in the area of clear skies in figure 16.13, but approached the level of the warm-air temperatures in the strip of territory near the front covered by a frontal cloud sheet in the form of 10 tenths stratocumulus. Dew points in the warm air-mass ranged from 42 degF to 48 degF and there was much fog in this air and at the front.

On the 15th, the two high centres coalesced and it became impossible to trace the front in the central region where the surface winds fell calm and fog was widespread. Nevertheless, the difference between the air-masses remained, and when the continental airmass reached Brest, the dew point there fell below 40 degF on the morning of the 15th. A study of the upper air temperatures over Brussels, Paris, Bordeaux and Toulouse revealed the continued existence of the frontal surface aloft throughout the life of the anticyclonic system, details on the 14th at 0500–0600 being:

Brussels: inversion of 6 degF, base 2700 ft;
Paris: inversion of 2 degF, base 2000 ft;
Bordeaux: isothermal layer, base 2000 ft.

It is worth noticing that the discontinuity was sharpest at Brussels in the centre of the high, where subsidence had sharpened the discontinuity surface aloft.

The surviving airmass difference again became noticeable further west as a weak frontal line on the 16th, when the anticyclonic centre drifted away south-westwards before the main Atlantic cold front advancing from the north. Figure 16.14 shows the situation at 0900 hours on the 16th, when drier and cooler air had reached the Scilly Isles from the west around the centre of the high, and the weak frontal line in the mouth of the English Channel marked the edge of the stratus-covered skies in the warm, maritime tropical air-mass still streaming eastwards through a narrowing corridor over England and Ireland towards the Continent. RMS *Queen Elizabeth* in 49–50°N and 18–19°W was also in the drier air-mass affecting Scilly and had only broken cloud above 5000 ft with light westerly breezes. At the same time, the weak warm front was pressing south and south-westwards over France with a belt of particularly low ceilings (300–600 ft) and poor visibility (less than a mile) nearing Bordeaux, a major consideration for aircraft approaching that area.

Nevertheless, in spite of the way the front over England and neighbouring countries in this example was partially regenerated after surviving several days of what have generally been considered ideal frontolytic conditions, fronts commonly disappear from the charts. Some of them do so through a process that some American workers have called consolidation of several fronts into one. This is in reality a frontogenetic process in the sense that the resulting single front is sharpened by it. Secondary cold fronts almost habitually tend to catch up and become consolidated with the leading cold front or occlusion. Experience of Atlantic Ocean weather analysis and forecasting shows that many of the secondary cold fronts which pass eastwards over Labrador and Newfoundland never reach Europe or the Azores as separate frontal systems, and consolidation is undoubtedly the reason for this in the vast majority of cases in winter time; in summer, it is probable that a number of secondary fronts that are little more than internal convection lines within the cold air-mass coming from the north Canadian territories where ground heating is at work at this season die away completely when the air-mass is stabilized over the cool sea. The frequent occurrence of consolidation comes about through the stronger pressure gradients which are more often seen across the rearward members of a succession of frontal systems than across the foremost members, a situation

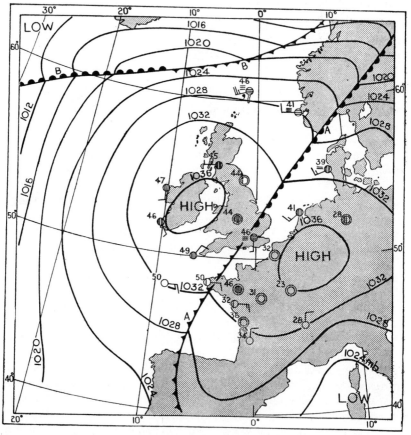

Figure 16.13 Synoptic chart, 0600 GMT, 14 February 1946.

of which figure 16.6 gave a typical example. Consolidation is, however, something quite different from frontolysis.

The fundamental condition for frontolysis is air-mass assimilation. If the air-masses on either side of a given front become alike, the front disappears. Moreover, this process is final: there is no special likelihood of a front ever being formed again in the same region; no transitional zone remains to favour such regeneration.

Frontolysis through air-mass assimilation will be brought about by any of the following circumstances:

(a) stagnation of the air-masses concerned, side by side over a single terrain, where both are subjected to like influences from the under surface and from direct radiation;

(b) movement of the air-masses side by side along parallel tracks without any important difference of velocity between them, so that both undergo like influences;

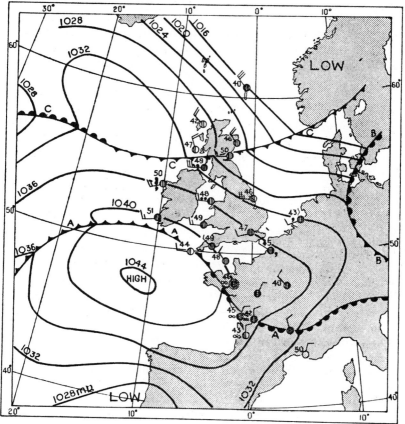

Figure 16.14 Synoptic chart, 0900 GMT, 16 February 1946.

(c) movement of the air-masses one after the other over the same track without any important difference of velocity or shear – with a transverse orientation of the front;

(d) arrival of the front in a geographical region where the sense of the ground-temperature gradient is unfavourable for its continued exist- ence (this occurs where the air-mass fed into the cold side of the front is as warm as, or warmer than, the air on the warm side; and vice-versa when the warm-sector airmass encounters still warmer air ahead of it).

Cases (a), (b) and (c) are very similar in nature. Most rapid assimilation takes place where an adiabatic lapse rate of temperature with increasing height is established. Consider figure 16.15, in which the curve of state PQRS represents a sounding through a shallow layer of cold air PQ underlying a warmer air-mass RS at no very great distance horizontally from the front. This is a typical upper air temperature sounding for a north African station after the passage of a cold front bringing in a polar air-mass from the Atlantic Ocean or from Europe, normally quite shallow in such latitudes. The effect of intense ground heating is already seen in the superadiabatic lapse rate between P and Q. At first, there may have been a frontal cloud-sheet of altocumulus or stratocumulus aloft, but once enough rifts were made in this cloud to let the African sun get to work, the hot ground would soon change conditions in the lower layers of air. In a day or two, surface temperatures of 85 degF might occur in place of the 64 degF recorded at P; and the curve P'RS would be substituted for PQRS, the former shallow polar air-mass being obliterated by its entire transformation into African tropical air. The second curve, LMNO, in figure 16.15, is plotted from an upper air temperature sounding for Lisbon, Portugal (surface temperature 50 degF), after a cold front had come in from the Atlantic, and corresponds exactly to the African curve PQRS just con- sidered. Surface heating over the Iberian peninsula to 79 degF would establish a dry adiabatic lapse rate from L' to N and destroy the polar air-mass. Finally, EFGH is the curve of a sounding made in a similar synoptic situation at Stornoway in the Hebrides. Here the prevailing humidity makes it probable that considerable amounts of convection cloud will accompany surface heating, and that the establishment of a saturated adiabatic lapse rate from E' to G will follow from a warming of only 5 degF – from 40 to 45 degF – at the surface and destroy the colder air-mass in the lower layers. The curves EFGH and E'GH also typify the development which takes place over the ocean when two air-masses move or stagnate under conditions favouring assimilation.

The cases demonstrated in figure 16.15 are similar in nature to the events which led to the interruption of the subsidence surface B over Downham Market, Norfolk, in figure 16.10. It is worth noticing, therefore, that the interruption of the subsidence surface on the occasion referred to was to some extent a localized phenomenon, confined to places where the diurnal

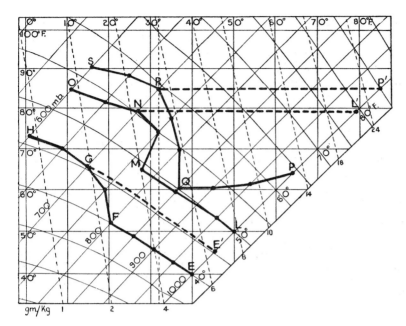

Figure 16.15 Typical tephigrams of ascents over North Africa, Lisbon and Stornoway after the passage of a cold front.

heating was strong enough, and, for that reason, it was possible for the continuity of the subsidence surface to be restored the following night. Similarly, we should guard against assuming too readily that the first day's heating from the ground has completely destroyed a front, even over the African desert. Sections of the front over the Atlas Mountains, for instance, may have been much less weakened than over the open desert further south, particularly in winter if the shallow cold air has filled the shaded valleys. Activity on the front may continue for a while, or may become patchy, and the air-mass difference may take some days to disappear.

Subsidence assists frontolysis in so far as it creates the conditions of clear sky in which solar heating of the ground will be most effective. Downgliding over the frontal surface destroys the frontal cloud-sheets, and turbulence cloud decks will be largely cleared away. Only moderate subsidence accompanied by strong heating from the ground should be most effective.

We may discern from figure 16.15 the conditions which favour or retard the conversion of shallow cold airmasses and disappearance of the fronts and frontal surfaces which mark their limits. Great heating is needed to obliterate a cold airmass by establishing a dry adiabatic lapse rate from the surface up, but this heating is often available in Africa and, in summer, over the Iberian peninsula. Assimilation of shallow cold air-masses then takes place. The shallower the cold airmass and the less pronounced the transition layer above it, the more readily and the sooner it is assimilated. Relics of the front or frontal surface may remain here and there where the cold air-mass was relatively deep or where the heating was not strong enough to obliterate it. Less heating is required to establish, from the surface up, the saturated adiabatic conditions which correspond to total disappearance of a shallow cold air-mass over the ocean. Assimilation occurs in this manner under suitable conditions over the wide oceans and also over large inland seas such as the Mediterranean.

The sequence of events following invasion of the Trade Wind belt over the North Atlantic in the region of the north-west African coast and the Canary Islands by a shallow cold air-mass from high latitudes is interesting. Although the surface air layers show the steep lapse rate of temperature associated with strong heating from below, the existence of a stable, frontal transition layer at no great height aloft, and sometimes as low as 2000 ft, means that the normal Trade Wind cumulus clouds disappear and give place to clear skies or to a sheet of stratocumulus, if the invading air-mass is deep enough to allow condensation within it. The normal appearance of the sky, with cumulus cloud predominating, is restored after a day or two when surface heating has had time to assimilate the shallow cold air-mass to tropical temperatures.

The same course of events may be followed further west over the ocean when there are sufficient observations; but weak cold fronts which have been carried around the Azores anticyclone are still liable to be noticed at Bermuda. As in other cases, the deeper the cold air-mass, the longer the conversion to tropical air takes.

Shallow Arctic air masses are converted to similarity with the air of middle latitudes in exactly the same way in the course of long transit or stagnation over the ocean west and south-west of the British Isles, traces of the former airmass boundary being usually discernible for two or three days.

In high latitudes, air-masses may stagnate over a cold surface and develop surface temperatures approximating to those of the Earth's surface. Stability and absence of turbulence will, however, maintain discontinuities aloft. One factor working against this and tending to establish adiabatic lapse rates is the radiation cooling direct from the cloud top, frequently proceeding from completely overcast layers. It seems, however, that air-mass assimilation in high latitudes proceeds only slowly, except where the air-masses are in transit over open water of considerably higher temperature. Apart from this, frontolysis is a slow process in the polar regions, and old weak frontal lines with residual cloud-sheets and narrow belts of fog and snow maintain their existence over long periods.

Decay of a front due to unfavourable geographical factors (case (d)) usually comes about through the supply of air of progressively higher temperature beneath the frontal surface. For instance, when a warm front, bringing a sector of maritime polar or maritime tropical air from the Atlantic Ocean, moves south-east across Europe in winter, there comes a point at which the continental cold air ahead of the front is replaced by air modified by having been over the warm Mediterranean Sea and, ultimately, perhaps, by air that has been over Africa; the warm front is not likely to survive much of this and the air-mass difference will soon be obliterated. Similar occurrences take place with equal regularity in other parts of the world, as when warm sectors – usually partly occluded – move south-east or east-south-east across the Rockies and encounter air from the Mexican Gulf or from the warm plains of Texas, and when old continental warm sectors reach the Atlantic seaboard of the USA in winter and meet the warm humid air brought northwards over the ocean by the Azores–Bermuda high.

A special type of frontolysis peculiar to occlusions has been described by Bergeron. At the completion of the occluding process, horizontal convergence seems to set in within the uplifted warm air, accompanied by subsidence in the lower levels and lifting of the tropopause. This occurs in the former central regions of the depression where the tropopause had been sucked down. Rising pressure sets in, centred on the occlusion. The occlusion, shown in figures 16.16–16.17, in the rear of the Skagerrak

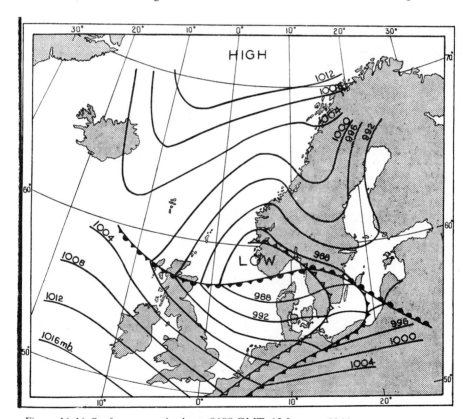

Figure 16.16 Surface synoptic chart, 0600 GMT, 12 January 1946.

depression moved south over the British Isles on 12–13 January 1946, disappearing gradually under the persistent frontolytic effect of strongly rising pressures and clearing skies which helped the influence of the under surface, working alike on the two air-masses moving over it. The charts represent the situation at 0600 on the 12th, soon after the pressure rises near the occlusion first became noticeable. The previously very sharp trough near the Shetland Isles had already shortened remarkably with a backing of the surface winds from E to NNE. The association of the greatest rise of pressure (7.4 mb in 3 h) with the occlusion is clearly seen in figure 16.17. The contours of the tropopause shown in figure 16.18 reveal a broad low area over the British Isles and the North Sea associated with the occluded depression centred just further east, while the line of the upper front of the occlusion (trough-line of the lifted warm sector) – placed on the surface map on the basis of pressure-tendency changes and distribution of snowfall over Norway – corresponds to a distinct ridge on the tropopause. The front disappeared in 36 h. A very similar synoptic situation on 25 August 1945, led to much more rapid dissolution of the occlusion moving south over Britain

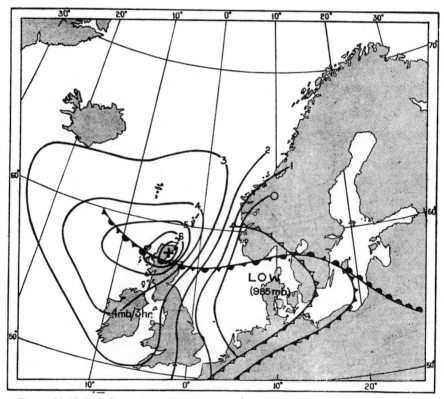

Figure 16.17 Isallobars, 0600 GMT, 12 January 1946.

and the North Sea, probably owing to the much more effective heating influence of the ground in summer assimilating the air-masses, although the isallobaric (pressure-change) effect was a good deal weaker.

Examples are not uncommon over southern Canada in which a long occlusion trailing from a depression over Newfoundland dissolves in the section between the depression centre and the Great Lakes, while the western end of the same front is maintained over Alberta as a sharp boundary between the Canadian and Pacific polar air-masses. Examples have also been described elsewhere.

Conclusion

To sum up our general picture of the life-history of a front: a major frontal system is not easily formed, and once formed, does not easily die away. In some circumstances, both events do take place rapidly. More usually the frontal system travels far across the Earth between its birth and decay and undergoes successive frontogenetic and frontolytic influences, which

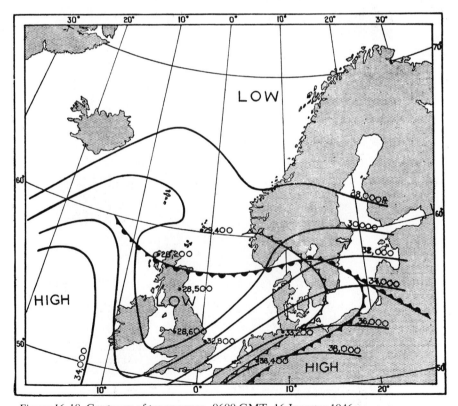

Figure 16.18 Contours of tropopause, 0600 GMT, 16 January 1946.

METEOSAT 1978 MONTH 2 DAY 22 TIME 1155 GMT (NORTH) CH. VIS 2
NOMINAL SCAN/PREPROCESSED SLOT 24 CATALOGUE 1000420076

METEOSAT 1978 MONTH 2 DAY 23 TIME 1125 GMT (NORTH) CH. VIS 1/2
NOMINAL SCAN/PREPROCESSED SLOT 23 CATALOGUE 1000510030

METEOSAT 1978 MONTH 2 DAY 24 TIME 1125 GMT (NORTH) CH. VIS 1/2
NOMINAL SCAN/PREPROCESSED SLOT 23 CATALOGUE 1000510088

Figure 16.19 Satellite cloud scanning pictures for 22, 23 and 24 February 1978.
  A cold front cloud belt over the eastern North Atlantic on the 22nd can be seen advancing across North Africa to merge, together with the remnants of another cloud-band ahead of it, into the equatorial (intertropical) cloud system on the 24th over the Atlantic. The advance towards Africa of another well-marked cold front cloud system from a South Atlantic cyclone is seen to be delayed by the development of a frontal wave depression over these three days. However, the alignment of the equatorial cloud belt over the Atlantic and more patchily over West Africa and the Gulf of Guinea on the 22nd, curving south-east as if to join up with the western leg of the cloud masses over southern Africa, suggests that it belonged to an older depression which had passed eastwards over the South Atlantic and led to the developments seen on 22–24 February over the south-eastern part of Africa. (I am indebted to *Weather*, and Mr A. V. Smith of the Meteorological Office, for permission to publish these Meteosat pictures, which were originally printed in the July 1985 issue of the magazine.)

alternately strengthen and weaken its activity. It suffers, too, continual distortions and changes of slope, as cyclonic disturbances develop and move along it. As the frontal system makes its way around the world, the nature of the air-masses fed into it on either side is continually changing. The activity of the front changes in consequence, but it does not necessarily lose its identity.

In October 1946, the writer had the experience, while sailing south on a whaling ship from Norway to the Antarctic, of having almost continuously within sight the cloud system associated with a cold front which overtook and passed the ship off Cape Finisterre (north-west Spain) on the 17th, until the ship apparently passed through the same front again in the intertropical convergence zone in latitude 9½°N. It was last seen from latitude 8½°N, off west Africa, on the 26th. The front underwent a quiescent phase when it became more or less stationary for three or four days in the belt of high pressure near and just south of the Canary Islands, where the ship also stopped on 20–22 October. The activity of the front, and its cloud development, increased again as it moved on farther south after the 22nd. The writer has published a detailed observation log of the experience (Lamb 1957). The frontal cloud system, still consisting mainly of a stratocumulus and altocumulus belt from the quiescent phase, but which was reaching greater height than before, drew too far ahead (south) of the ship for continued observation when the ship reached latitudes about 18°N on the 23rd.

The sky was then almost cloudless over the cold upwelling water off the African coast, until later that day high cirrus and cirrostratus was again seen across the south-eastern sky and gradually extended along the southern and western sky near the horizon. During the 24th, this high cloud became general, but there was still no low cloud in the N to NE Trade Wind over the relatively cold water from the upwelling west of Africa. No other cloud was seen until in the night of 25–26 October a line of tremendous cumulonimbus cloud, stretching from horizon to horizon, was seen ahead of the ship. Lightning had been seen to the south for some hours before the cloud was seen in the night. An impressive line squall marked the front as the ship passed to the south. At latitude 8½°N on the 26th the ship lay between two similar, well-marked cloud lines, each orientated about south-east to north-west and probably some 10–15 miles apart, consisting of great cumulus and cumulonimbus clouds with their base about 1500 ft, occasionally lowering to 500 ft, and tops of very great height associated with cirrus and cirrostratus high cloud. The most interesting feature of these two cloud lines was that they were each slightly curved, the northern one in the characteristic curved alignment (slightly concave to the north) of a trailing cold front from the northern hemisphere (presumably the front which had been under observation since the 17th) and the southern one with the curvature (slightly concave to the south) to be expected from a trailing cold front from the southern hemisphere. There seems no reason to doubt that this is what they were.

The occlusions of north-west Europe advancing east and south-east across the continent may eventually appear as cold fronts sweeping across China from the heart of Asia, as has been testified by Chinese investigators (Lu 1945). Other similar systems get steered south over European Russia and the Black Sea, and arrive, usually as cold fronts, in the autumn, winter and spring seasons in the eastern Mediterranean.

We find ourselves constantly referring to these fronts emerging from their transformations as cold fronts, because the major, intercellular fronts maintained between the chief anticyclonic areas normally advance east-wards in successive stages as cold fronts and cold occlusions, the air-masses in their rear being drawn from higher latitudes than the air-masses ahead of them.

The frontal sequences in North America are interesting for the great frequency with which new systems enter that continent on one or other of two clearly defined courses. Occlusions, orientated parallel to the great Rocky Mountains barrier, advance on an eastward track. They represent the last stage of former Pacific Ocean storms, but they begin a new phase of their existence east of the mountains when deformations are induced on them, either by topographical disturbances of the wind distribution or by the influence of other circulations. In winter these are warm occlusions, and are often followed by the föhn wind, 'chinook'. The other dominant feature of the North American sequences is provided by the cold fronts which come in from the Arctic shores of northern Canada and sweep south or south-east across the continent east of the Rockies. These fronts, too, are believed to originate as the occlusions of depressions which have come from elsewhere, sometimes merely on a circuitous track from the Davis Strait west of Greenland around the north of Baffin Land and the Canadian Archipelago but more often on longer tracks across the Arctic Ocean. Sverdrup (1933) noticed that climatological mean pressure charts for the winter months show a col, in the Arctic Ocean north of Alaska, between the Canadian and Siberian anticyclones with the line of frontogenesis running south-east into Hudson Bay. This coincides with (d)(i) in the list of geographical belts of strong thermal gradient favouring frontogenesis (p. 271). Occlusions push-ing northwards through the Bering Strait from the Pacific or rounding the polar basin from the North Atlantic must be sharpened up along some such line as this, and may then strike southwards into Canada as cold fronts.

Finally, we may notice an application to the considerations with which long-range forecasters will have to deal. The magnitude of the thermal gradients associated with each of our main geographical belts favouring frontogenesis, as well as details of their precise position and orientation in any given season, must affect the strength and orientations of the jet-streams arising in association with them, and therewith the index of circulation intensity. In so far as the pattern fits a stationary wave length in the upper westerly wind-stream appropriate to the circulation index, we may expect the pattern to be stable or recurrent during that season or at that particular time of the year. This pattern, in turn, will determine the prevailing locations of axes of high and low pressure on either side of the jet-streams and hence the broad character of the weather over wide areas.

300 Weather, Climate and Human Affairs

References

Bergeron, T. (1928) "Uber die dreidimensional verknüpfende Wetteranalyse', Teil 1, *Geofys. Publ., Oslo*, 5, no. 6.
Bergeron, T. (1930) 'Richtlinien einer dynamischen Klimatologie', *Met. Z., Braunschweig*, 47, 246.
Bergeron, T. (1937a) 'On the physics of fronts', *Bull. Amer. Met. Soc., Worcester, Mass.*, 18, 264.
Bergeron, T. (1937b) 'Hur vädret blir till och hur det förutsäges', *Ymer, Stockholm*, nos 2–3, 199.
Bjerknes, J. (1919) 'On the structure of moving cyclones', *Geofysiske Publikasjoner*, I(2), Oslo.
Bjerknes, J. and Solberg, H. (1922) 'Life cycle of cyclones and polar front theory of atmospheric circulation', *Geofys. Publ.*, III(1).
Bjerknes, V., Bjerknes, J., Solberg, H., and Bergeron, T. (1933) *Physikalische Hydrodynamik mit Anwendung auf die dynamische Meteorologie*, Berlin, Springer.
Carpenter, A. B. (1940) 'Occlusions on the Alaskan coast', *Bull. Am. Met. Soc. Worcester, Mass.*, 21, 327.
Chromow, S. P. (1940) 'Einführung in die synoptische Wetteranalyse', Vienna (German trans. by G. Swoboda of Russian original, Moscow, 2nd edn, 1937).
Douglas, C. K. M. (1934) 'The physical processes of cloud formation', *Quarterly Journal of the Royal Meteorological Society*, 60, 333.
Lamb, H. H. (1957) 'Fronts in the intertropical convergence zone: an observer's log and some reflections thereupon', *Meteorological Magazine*, 86, 76–84, London.
Loomis, E. (1981) 'On the storm which was experienced throughout the United States about the 20th of December, 1836', *Transactions of the American Philosophical Society*, VII, 125–163, Philadelphia, Penn.
Lu, A. (1945) 'The winter frontology of China', *Bull. Am. Met. Soc., Lancaster, Pa.*, 26, 309.
Namias, J. (1934) 'Subsidence within the atmosphere', *Harvard Met. Stud. Cambridge Mass.*, no. 2.
Petterssen, S. (1936) 'Contribution to the theory of frontogenesis', *Geofys. Publ., Oslo*, 11, no. 6.
Petterssen, S. (1940) *Weather Analysis and Forecasting*, New York and London.
Refsdal, A. (1930) 'Zur Theorie der Zyklonen', *Met. Z., Braunschweig*, 47, 294.
Rodewald, M. (1937) 'Das Dreimasseneck als zyklogenetischer Ort', *Met. Z., Braunschweig*, 54, 469.
Sawyer, J. S. 'Formation of secondary depressions in relation to the thickness pattern', *Met. Mag., London*, 79, 1.
Shaw, N. and Lempfert, R. G. K. (1906) 'The life history of surface air currents', *Meteorological Office Publication*, no. 174, London.
Sutcliffe, R. C. (1947) 'A contribution to the problem of development', *Quarterly Journal of the Royal Meteorological Society*, 73, 370–383, London.
Sverdrup, H. U. (1933) 'The Norwegian North Polar Expedition with the *Maud*, 1918–25', scientific results. 2, Meteorology', pt. 1, discussion, Bergen.

17

Volcanoes and climate: an updated assessment

A symposium was held at the Hamburg 1983 Congress of the International Union of Geodesy and Geophysics to mark the hundredth anniversary of the great eruption of Krakatoa in the Sunda Straits in the East Indies (about 6°S 105.5°E) on 26–27 August 1883 and review present knowledge of the effects of volcanic explosions on the atmosphere and on the development of climate.* Much of the island of Krakatoa and its mountains, together with other small nearby islands in the Strait, disappeared into the atmosphere or fell as 'bombs' into the sea. Estimates of the amount of solid matter blown up range from 6–18 km³. The column of finer debris, lumped together as 'dust' or 'ash' – though we now know that the gaseous products, particularly sulphur oxides, which form aerosol in the stratosphere, may be more important as regards climatic effect – rose to an observed height of 27 km. Batavia, now known as Jakarta, 160 km from the volcano, was in darkness for 4–5 h in the middle of the day (and there was much rain there, presumably from thunderclouds formed in the rising air column).

The dust veil and aerosol veil in the stratosphere resulting from the eruption spread around the world (cf. figures 17.1 and 17.2) and made itself known by brilliant coloured sunsets and other optical effects (see figures 17.3 and 17.4), of which many descriptions survive. Pyrheliometric measurements of the direct solar beam – a routine that had just been started (at first only at Montpellier, in the south of France) – soon began to show a decline of the beam's intensity (figure 17.5). By 1884–85, a deficiency of 20 per cent or rather more occurred in the average strengths of the beam in some months, perhaps partly due to the contributions from other smaller eruptions elsewhere in the world.

---

* This chapter reprints the opening address at the symposium. Small additions have been made to bring the material up to date.

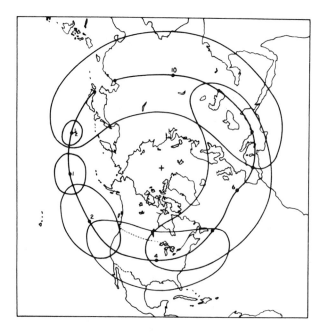

Figure 17.1 Spread of suspended matter in the atmosphere studied by constant (pressure) level balloons released from Japan and floating at about 12 km height; the mean trajectory is shown with numbered points, indicating progress from day to day, while ellipses 'centred' on these points show the region within which 50 per cent of the balloons were still to be found after so many days. (From Angell 1961.)

For 80 years, until the eruption of Agung in Bali, in 1963, this eruption of Krakatoa in 1883 and the atmospheric effects associated with it over the following two or three years remained the most thoroughly reported and investigated case. Certainly no other eruption in all that time approached the same magnitude.* This fact, together with the magnificent report published in 1888 by a special committee of the Royal Society of London, ensured its standing as the classic case appealed to in nearly all writing on volcanic eruptions and climate until after the Agung event. The latter presented the first major opportunity for more comprehensive

* Further eruptions of some importance, in Alaska in October 1883, in Japan in 1888, and in low latitudes south of the equator in 1885, 1886 and 1888, prolonged the observable optical effects and climatic effects attributable to the stratospheric veil which began with the Krakatoa eruption in August 1883. After that the only eruptions prior to 1963 which seem to have affected the northern hemisphere with any observed veil were those of a group of volcanoes in the West Indies and Guatemala in 1902, Ksudatch in Kamchatka in 1907 and Katmai in Alaska in 1912. The status of the 1912 eruption as the only one after 1909 of importance to atmospheric transparency over Europe until 1963 is confirmed by study of the pyrheliometer measurements at Davos in the Swiss Alps from 1909 onwards (Hoyt and Fröhlich 1983). We shall see later, however, that the history of volcanic veils over the southern hemisphere during this period was quite different.

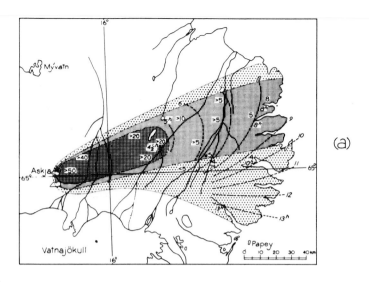

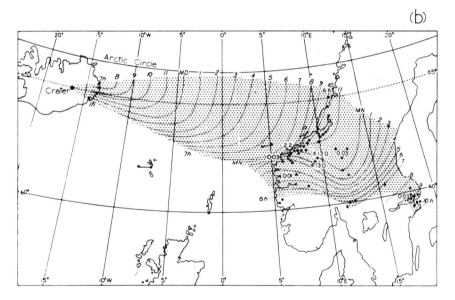

Figure 17.2 Example of the spread of ash from the volcano Askja after its eruption on 29 March 1875. Isopleths indicate depths (in centimetres) of the deposit over eastern Iceland; in the lower diagram depths of the deposit in Norway are given in millimetres (Mohn 1877); the broken lines are isochrones of time GMT of arrival of the dust.

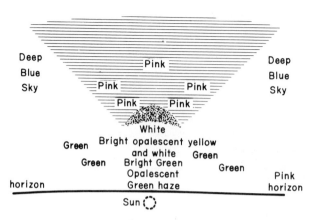

Figure 17.3 Sketch of 'sunset (actually twilight) colours' recorded by F. A. R. Russell at Richmond, Surrey, from 17.00–17.30 hours GMT on 9 November 1883; very similar arrangements of colour have been repeatedly observed after the eruptions of Agung in 1963, Mount St Helens in 1980 and El Chichon in 1982 (cf. Lamb 1970, fig. 1(b)).

measurements with modern techniques, including aircraft sampling of the material in the stratosphere, subsequent particle size measurements, chemical analysis, optical measurements of the veil's density, and so on (e.g. Hofmann and Rosen 1977, 1982; Remsberg and Northam 1975). Some insights into the probable spread of volcanic ash and aerosols in the stratosphere came from observation of the spread of trace elements after the nuclear bomb test explosions in the 1950s (figure 17.6).

1983 was also the 200th anniversary of the great eruption of Laki in Iceland, in May and June 1783, the greatest lava-producing eruption anywhere in historical times (lava volume estimates from 12 to 27 km^3), which also produced dust and ashfalls sufficient to destroy crops in northern Scotland (Caithness).† The dust veil must have been exceptionally dense, because in June of that year the sun was described even in Italy as 'rayless' and could be looked at with the naked eye. About the same date, in southern France, it was noted that the sun was wholly obscured at elevations less than 17° and shone either red or bluish-white most of the day. No stars were seen at night below 40° elevation. The coloured sunsets and the unusual glare of the sky at night, giving light like the full moon even at midnight, were also remarked on. There was another great eruption, of Asama yama in Japan, in August 1783, and the combined dust veils which spread over the Earth seem to have equalled or exceeded the density in the Krakatoa case 100 years later.

---

† In Iceland itself, destruction of the hay crop in this way and poisoning of it by chemicals of volcanic origin has been a repeated problem down through history.

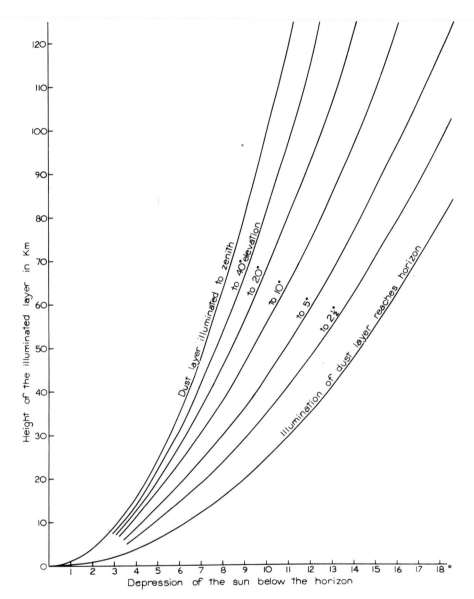

Figure 17.4 Nomogram for finding the height of an illuminated ash or aerosol layer from the depression of the sun below the horizon when illumination to various angles of elevation ceases.

We need not conclude that there is a precise 100-year periodicity in volcanic activity. There was a greater eruption in 1815, of the volcano Tambora in the East Indies, and perhaps another (Coseguina) in Nicaragua in 1835, as well as others in earler times, which do not fit the 100-year spacing. But the 1783 and 1883 eruptions were landmarks in the history of volcanic eruptions and their possible climatic effects as a subject for study. It was the events of 1783, and the widespread reporting of them in that increasingly enlightened age, that led Benjamin Franklin, then living in Paris, as the first diplomatic representative of the newly established United States of America, to write in May 1784:

> During several of the summer months of 1783, when the effects of the sun's rays to heat the Earth in these northern regions should have been the greatest, there existed a constant fog over all Europe and a great part of North America. This fog was of a permanent nature; it was dry and the rays of the sun seemed to have little effect towards dissipating it. . . . They were indeed rendered so faint in passing through it that, when collected in

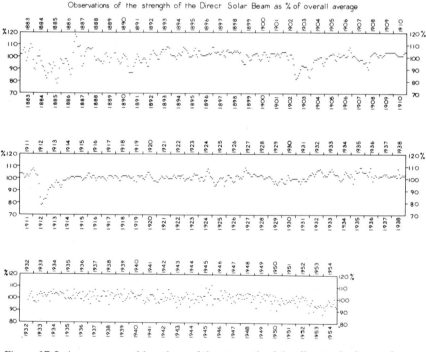

Observations of the strength of the Direct Solar Beam as % of overall average

Figure 17.5 Average monthly values of the strength of the direct solar beam, from observations at mountain observatories in middle latitudes, 1883–1954; the values are shown as percentages of the overall mean (those for the years in the bottom panel are from measurements just at two observatories in southern Japan, between 32° and 37°N.)

the focus of a burning glass, they would scarce kindle brown paper. Of course, their summer effect in heating the Earth was exceedingly diminished.

Hence, the surface was early frozen.

Hence, the first snows remained on it unmelted . . .

Hence, perhaps, the winter of 1783–84 was more severe than any that had happened for many years.

The cause of this universal fog is not yet ascertained . . . whether it was the vast quantity of smoke, long continuing to issue during the summer from Hekla, in Iceland, and from that other volcano which arose out of the sea near the island, which smoke might be spread by various winds over the nothern part of the world . . .

It seems, however, worthy the inquiry whether other hard winters, recorded in history, were preceded by similar permanent and widely extended summer fogs.

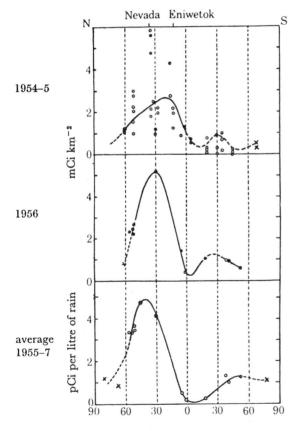

Figure 17.6 Latitude distribution of strontium-90 fall-out after the US nuclear bomb test at Eniwetok (latitude 12°N), in 1954, showing poleward drift.

This is a remarkable statement, glimpsing the necessary global view, for such an early stage in the beginnings of the atmospheric sciences.

We cannot, of course, be sure how far this was even then an altogether novel suggestion. There are hints of similar thinking in various ancient chronicles from Rome, Byzantium and China, as far back as the second and sixth centuries AD. At some times of volcanic activity, we find statements that the sun's light was dimmed and its ray enfeebled for periods up to one year or 18 months 'and the fruits did not ripen' or 'the fruits were killed'.

Little could be done to investigate the suggested coupling of volcanic dust veils and colder climate episodes before Köppen (1873, 1914) provided the first long series of surface temperature values as estimated averages over various areas and, so far as was possible at the time, for the entire northern hemisphere's temperate and tropical zones (figure 17.7). A systematic list of volcanic eruptions was also needed. By far the best early listing – and the only one to attempt an assessment of the magnitude of each eruption in terms of (a) lava and (b) solid matter ejected – was that of Sapper (1917, 1927). By these means, and with the growing series of pyrheliometric

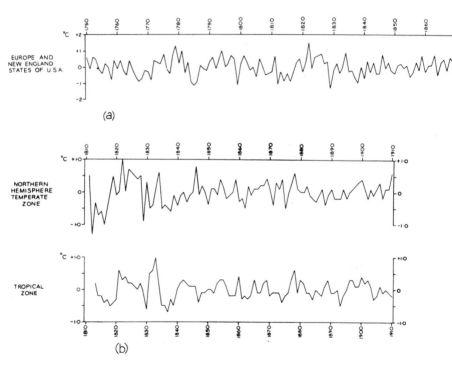

Figure 17.7 Year-to-year variations of the temperature averaged over large areas of the northern hemisphere, as calculated by Köppen (1873, 1914); the values for Europe and New England, 1750–1871 have undergone a ¼ (a + 2b + c) smoothing.

measurements of the sun's radiation, Humphreys (1920, 1940) was able to map out the physics of the most straightforward effects upon climate of the ejecta in the atmosphere after great volcanic explosions. The subject was also treated by Sir Napier Shaw (1930) in his *Manual of Meteorology*, Vol. 3, following Humphreys.

Defant (1924) investigated the possibility of an association between variations of the general atmospheric circulation, using a series of values of the pressure difference between latitudes 30° and 65°N over the North Atlantic Ocean, and volcanic eruptions. He found that in the years between 1881 and 1905, in which the 14 greatest volcanic eruptions of that period took place (including one in latitude 38½°S), the North Atlantic pressure difference averaged 11 per cent less than the overall mean. For the four greatest eruptions, the figure was 18 per cent weakening of the pressure gradient in the eruption year, followed by above-normal values, +9 and +17 per cent, in the next two years. Such a cycle could be explained by slow poleward drift of the dust in the stratosphere and earlier clearance of it over low latitudes, where the tropopause is high and convection vigorous: this should intensify the Equator–pole gradient of effective solar heating of the Earth in the later part of the cycle. Wagner (1940) went on to suggest that the prolonged lull in volcanic activity after the Katmai eruption in 1912, leading to increasing transparency of the atmosphere in the 1920s and 1930s, could

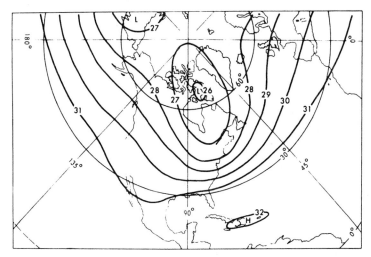

Figure 17.8 Circulation in the middle troposphere defined by mean heights of the 700 mb pressure level in January, as estimated by Wexler (1956) for 20 per cent weakened solar radiation due to volcanic material in the atmosphere; heights are given in hundreds of metres; the trough in the upper westerly wind flow over eastern North America is very much sharpened and extended south in comparison with the situation with full normal insolation.

be the cause of the increased vigour of the general wind circulation and the observed warming of northern climates.

Wexler (1956) indicated how great and prolonged excesses of volcanic dust in the atmoshere could cause an ice age. A circulation pattern with greater southward extension of an upper cold trough over the central part of North America should develop and would steer the Atlantic cyclonic activity northward to initiate formation of the great ice-sheet over Canada. (figure 17.8).

The prevailing 1000–500 mb 'thickness' pattern* at the last glacial maximum, and resulting characteristics of the general wind circulation, as derived by Lamb and Woodroffe (1970) from the geography of ice and surface temperatures, had much in common with Wexler's theoretical proposals of what should result from a 20 per cent reduction of the insolation due to volcanic dust veils. Nevertheless, it is now clear that not volcanic activity, but the regular evolutions of the Earth's orbital variables, account for the timing of the ice ages and interglacials. The effects are enhanced by the high albedo of ice and snow once these are established on the Earth's surface. It may yet be true, of course, that bouts of more frequent volcanic activity world-wide, as well as solar disturbances, recurring on shorter time-scales than the orbital cycles, play a part in the more abrupt phases of glacial onset (cf. Rampino et. al. 1979).

Budyko (1968), arguing that the frequency of great volcanic eruptions within the last 100 years may be taken as a reasonable approximation to their average frequency over a million years or even longer spans of geological time, suggested that ice ages may be no more than the product of random occurrences of clusters of very frequent volcanic explosions. Taking their incidence since about 1870 as 4, and regarding this as the ultra-long-term average, he estimates that:

(i) 40 great eruptions within one century might be expected about once in 10,000 years;
(ii) 130 great eruptions within one century about once in 100,000 years;
(iii) 100 great eruptions within 5 years about once in a million years.

The last-named frequency, he estimated, should reduce the incoming solar radiation by 50 per cent in low latitudes and 80 per cent in high latitudes while the spasm of activity lasted. The extra ice produced should prolong the effect, because when a renewed increase of solar radiation comes, the high albedo of the ice and snow would reflect away most of it.

There must be grave doubts about the relevance of this argument, despite the soundness of the physics. The first half of the recent 100-year period can now be seen as fairly typical of a wave of enhanced volcanic activity world-wide that has lasted 400–700 years: hence the basis of Budyko's

---

* 'Thickness' here means the height difference between the levels where atmospheric pressure is 1000 and where it is 500 millibars. Thickness increases as the air warms and its density decreases.

suggested frequencies is probably too high. Also the recurrence of ice ages within the Quaternary, or last million years, has shown much more regularity than appropriate for a purely random phenomenon. Indeed, the recurrences show prominent periodicities of about 100,000, 40,000 and 20,000 years (Hays *et. al.* 1976; see also Imbrie and Imbrie 1979; Berger 1980), which agreed with those of the Earth's orbit's ellipticity, the obliquity and the precession of the equinoxes. It is clear that the glaciations were not due to any one cause, but the orbital variations and their regularities come into it as well as any effect of varying volcanic activity.

To provide a basis for investigation of the part played by volcanic veils in the atmosphere in the climatic variations of recent centuries, the present writer (Lamb 1970, 1972) published a chronological assessment of known eruptions since AD 1500 in terms of a Dust Veil Index (DVI) calculated according to one or more of three formulae. These derived the index value from either:

(i) the greatest monthly percentage depletion of the incoming direct solar radiation in middle latitudes of the hemisphere in which the eruption occurred;

(ii) the greatest lowering of the year's average temperature over the middle latitudes zone of the same hemisphere, for the eruption year or the next following year;

(iii) the estimated total volume of solid matter dispersed as dust in the atmosphere.

Account was taken of how much of the Earth was likely to be effectively covered by the veil in the stratosphere from eruptions in different latitudes.* In this, modern knowledge of the stratospheric circulation and observations of the drift of trace elements etc. after the nuclear bomb tests in the 1950s and early 1960s was used (figure 17.6). Account was also taken of what was known, or could be deduced as probable, of the duration of the veil in each case, considering particle sizes and heights reached by the ejecta.

The formulae for computation of the Dust Veil Index were designed to give an index value of 1000 for the famous eruption of Krakatoa in 1883. The formulae used are given at the end of this chapter. The resulting list included some great eruptions and dust veils of unidentified origin since AD 1500 of a magnitude (DVI>250) thought likely to have had measurable effects on climate for up to a few years after the eruption. Reporting from the most active volcanic regions in the equatorial zone was established before 1700, and even to some extent in the 1500s; but it is probably not until about 1780 that the list of eruptions becomes substantially complete as regards veils that affected the northern two-thirds of the world. And even in the 1980s one or

---

* Rule-of-thumb calculations were adopted. Explosion products in the stratosphere from eruptions between the equator and latitude 15° spread over both hemispheres about equally, and the DVI was divided equally between them. Eruptions between latitudes 15° and 20° were counted two-thirds to that hemisphere and one-third to the other; dust, etc. from eruptions poleward of latitude 20° was considered effectively confined to the hemisphere concerned.

two veils of matter in the stratosphere have been reported for which no certain origin can be identified. There were 28 great eruptions listed between 1780 and the 1960s. Our knowledge of dust veils over latitudes south of 20° to 40°S cannot, of course, be considered anywhere near complete before some time in the late nineteenth century at the earliest.

This listing nevertheless made possible for the first time a number of statistical studies which, despite the imperfections of the list, were sufficient to confirm the existence of effects on climate. Eruptions with DVI values of >100 over the hemisphere concerned were considered in these studies. In particular, the cycle of changes of strength of the general wind circulation over middle latitudes of the North Atlantic first reported by Defant (1924) – weakening in the first year, followed some time later by enhanced strength – was supported by more cases of great eruptions in low latitudes (figure 17.9(a)) from which the veil spread over the whole Earth. A different sequence was found to follow eruptions in high latitudes (figure 17.9(b)), from which the veil normally never reaches low latitudes to any significant extent. In these cases, not surprisingly, considering the resulting enhanced heating gradient between high and low latitudes, the circulation strength was on average enhanced as soon as any effect attributable to the eruption could be noted. These are, however, statistical generalizations. In the individual cases, there is a good deal of variation, particularly in the time taken by the cycle of circulation changes and their ultimate disappearance.

Another significant association appeared to be a southward displacement of the cyclonic storm sequences, and more frequent northerly winds, near the British Isles in the summers following great eruptions. There were also statistical indications of some lowering of average temperatures in middle latitudes of the northern hemisphere.

Despite the number of statistically significant associations indicated, which may be remarkable in view of the crudity of the observational material, it is clear that volcanic veils are not the cause of all the climatic differences over recent centuries. Figure 17.10(b) compares the course of the 25-year cumulative DVI values for the northern hemisphere with various indicators of temperature. The great frequency of volcanic veils in recent centuries, as compared with the present one, may be regarded as having played some part in the colder – so-called Little Ice Age – climate of those times. But the correspondence appears far from perfect, even if we allow that some descriptive material and the temperatures observed suggest that the Dust Veil Index is underestimated in the 1690s. The maximum of the cumulative DVI values between 1780 and 1840 appears rather as a complication which may have hindered the recovery (globally) from the coldest climate between about 1560 and 1700. Also the twentieth-century minimum of the cumulative DVI values from 1916 onwards and continuing after 1940-62 does not coincide with the period of highest winter temperatures in Britain, Europe and the Arctic in the 1920s and 1930s, nor of the highest average summer temperatures which ran from 1933 to 1949–52.

Schneider and Mass (1975) found that the climatic variation in terms of global temperature over the previous 100 years could be well simulated by an empirical expression taking account of just three variables – volcanic dust, sunspot activity and the man-made increase of carbon dioxide in the atmosphere. Among these, volcanic dust seemed to be the most important, at least up to that time, and the contribution from solar variation was relatively minor.

There have been two important developments in this subject field within the last 20 years or so:

    (i) recurrence of great volcanic outbreaks with a frequency more like that of the recent past centuries;

   (ii) the application of a great range of modern techniques of observation and measurement.

It has become possible, as never before, to study the spread of volcanic materials in the stratosphere, and the heights attained not just by the initial eruption column but by later diffusive processes in the stratosphere, the

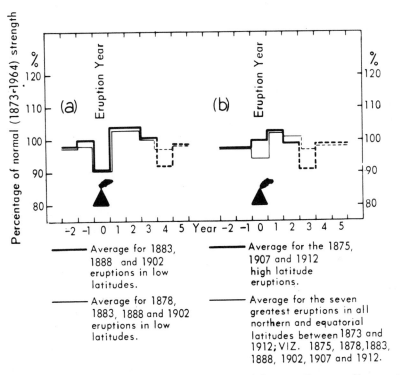

Figure 17.9 Course of the yearly mean values of the overall range of barometric pressure between the 'Azores high' and the 'Iceland low', after great volcanic explosions in low and high latitudes.

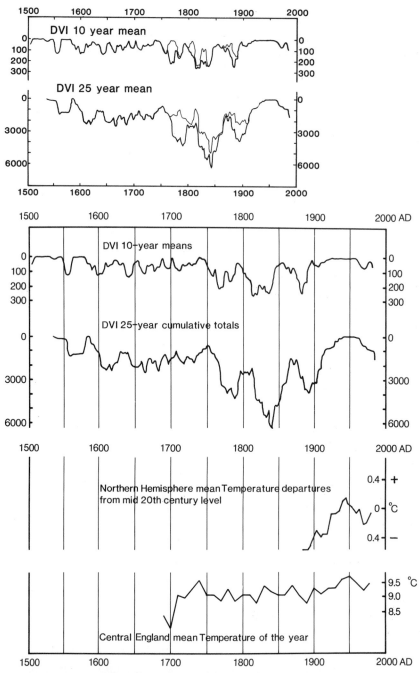

Figure 17.10(a) History of the volcanic 'dust' load in the atmosphere over the northern hemisphere since AD 1500, as indicated by the Dust Veil Index; the thin line indicates the differences that occur in cases where assessment based solely on the evidence of a temperature anomaly are omitted; upper curves 10-year running means of DVI, lower curves 25-year cumulative totals of DVI. (b) Volcanic Dust Veil Index and two indicators of prevailing temperatures in the northern hemisphere compared. DVI is shown as increasingly downward.

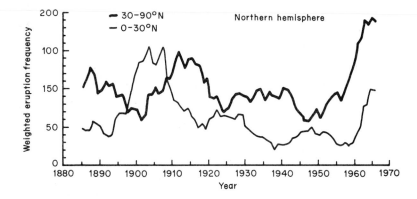

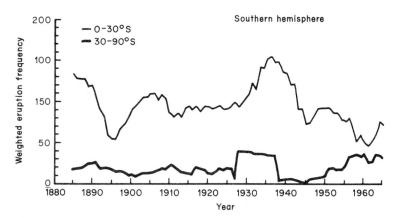

Figure 17.11 History of volcanic eruption frequency in the northern and southern hemispheres, 1885–1965 (after Hirschboeck 1979–80); 10-year running totals, counting each great eruption = 10 moderate eruptions.

sometimes manifold veils, the optical effects, the particle sizes and their chemical composition*, and the subsequent decay of the veils and associated phenomena.

Several other chronologies have also been produced. Bryson and Dittberner (1976), using a chronology compiled at the University of Wisconsin, Madison, believe it possible to show that much weaker eruptions than those considered in Lamb's Dust Veil Index also have their effect. This could be due to the now well-recognized conversion of sulphur dioxide and sulphate ions in the atmosphere into aerosol particles, which are among the most prominent constituents of volcanic veils in the stratosphere. These

* Following the 1982 El Chichon eruption, Hofmann and Rosen (1983) were able to estimate the H_2SO_4 vapour concentration, droplet sizes and vertical distribution in the stratosphere.

workers claimed to simulate the last 100 years' history of northern hemisphere mean temperature better through taking account of the smaller eruptions. K. K. Hirschboeck (1979–80) has described the Madison chronology of eruptions, classified only as minor, moderate or great eruptions, in a study of considerable interest. Over 5000 eruptions in a record starting in 2227 BC are included; but her study concentrated on the 2738 moderate and great eruptions listed between AD 1700 and 1969, and mostly on those between 1880 and 1969. A particularly useful aspect of this chronology is its much bigger sample of eruptions in the southern hemisphere than in Lamb's chronology. (A feature which caused Hirschboeck some doubt, and which possibly distorts her study somewhat, was the decision to weight each great eruption as just ten times more important than a moderate eruption. As she says, the difference could be more like 10,000-fold; and in some cases where more than one cubic kilometre of material is ejected, the latter figure certainly indicates better relative magnitude.) The most interesting results of Hirschboeck's study are the differences in the history of volcanic activity in different latitudes and between the two different hemispheres. The figures confirm a lull in the global total of volcanic activity after the first decade of the present century and a renewal in or about the 1960s and after, but they indicate in the southern hemisphere a considerable maximum of ash-producing activity between 1925 and 1945 (figure 17.11(a) and (b)). Interestingly, this phase coincides with a minimum – even if a very minor minimum – compared with the late nineteenth and early twentieth century in temperatures over the higher latitudes of the southern hemisphere.* Another difference from Lamb's result is that the University of Wisconsin chronology indicates a much greater increase of the global total volcanic dust load in the atmosphere in the 1960s.

Newhall and Self (1982) have produced a chronology (in Simkin et. al. 1981) of over 8000 historical and prehistoric volcanic eruptions, their magnitudes on a scale of 0-8 assessed for their explosive violence, rather as the Richter scale rates earthquake severity. No atmospheric data enter into it, and no adjustment is made for differences of probable dispersion of the eruption products in the atmosphere from volcanoes in different latitudes. The Volcanic Explosivity Index (VEI) is thus designed to quantify volcanic

---

* Other differences have recently been hinted at in studies by Stothers and Rampino (1983a, 1983b). Between 1500 BC and AD 1500, the sequence of great volcanic deposits in the Greenland ice-sheets obtained by Hammer's (1977) acidity measurements corresponds to three known Mediterranean eruptions and two in Iceland out of the nine greatest events, a degree of dominance of activity in these regions which is not repeated in the more recent centuries. The Sicilian volcano Etna is known to have been more active during classical antiquity than recently and seems to have accounted for one-fifth of all the acidity peaks in the Greenland record between 152 BC and AD 43. The greatest known eruptions of Vesuvius were also during the period named, in AD 79 and 469–474.

hazards. It may nevertheless also have some value for climatic studies, but adjustments, at least for latitude of the eruption, should be considered.

Sear and Kelly (1983, and personal communication) have used the VEI in selecting the greatest eruptions (VEI 5 or over) of the period 1881–1980 to look for possible surface air temperature deviations over the northern hemisphere month by month over the following three to four years. The sample was small, and the eruptions were not separated by latitude, apart from considering the two hemispheres separately. Just five northern hemisphere events qualified, including (as was reasonable) the case of May 1902 when there were three VEI 4 events in one month. Statistically significant lowering of average temperatures was observed in all but two of the first 16 months after the eruption, but features unlike all previous studies and reasoning about possible climatic effects also appeared. The strongest response integrated over the northern hemisphere, appeared just 2 months after the eruption. The same applied to temperatures over the Arctic, regardless of the latitude of the eruption; moreover, the response seemed to end sooner over the Arctic, just seven months after the eruptions. These results may be an accident of the very small sample, and it is unfortunate that the eruptions selected as a result of reliance on the VEI rating included one (in Kamchakta, north-eastern Siberia, in 1956) not conclusively indicated as great and omitted the more generally accepted great eruption of Agung, in Bali, in 1963.

Another recent research exploration of the effects of volcanic aerosol veils, by Handler (1984, 1986), has used the VEI chronology to select the largest eruptions (VEI 5 or over) of the period 1869–1983. This time, again accepting the group of VEI 4 events in 1902, ten eruption events or eruption groups were identified for study and were separated into high-latitude (over 50°N), low-latitude (20°N to 20°S) and mid-latitude eruptions (30°–50° latitude in either hemisphere). The sea-surface temperatures in the equatorial Pacific (0°–10°S, 80°–180°W), deviations of which are associated with the Southern Oscillation and El Niño events and have world-wide weather associations (see e.g. Lamb 1972), were then studied in successive three-month seasons, from two seasons before to eight seasons after an eruption. The results showed, with a remarkable degree of consistency, that the equatorial sea temperatures were high in the first five seasons after low-latitude eruptions (highest in the third to fifth seasons) and show (smaller) negative departures in the first four seasons after high-latitude eruptions (greatest in the first season). The five-seasons' duration agrees rather well with Sear and Kelly's finding of 16 months' depression of the average temperature over the northern hemisphere, though their indication of greatest anomaly very soon after the eruption is not supported. (Again, the 1956 Kamchatka eruption was included and behaved partly unlike the rest of the sample, while the much greater eruption of Agung, in 1963, which conformed, was omitted.)

Immediately after the Agung eruption sea-surface temperature at Canton Island rose sharply and continued rising to peak ten months later (almost 2° warmer). Indications were found of a strengthening of the subtropical jet-streams and the Trade Winds around latitude 15° in both hemispheres immediately after low-latitude eruptions. The immediacy of the sea temperature and circulation responses in low latitudes, even after high-latitude eruptions, is remarkable and needs explanation. It points to a surprisingly quick response in the largest scale (global) atmospheric circulation, which once again (see Chapter 11, p. 182) appears as a valuable indicator of the state of world climate.

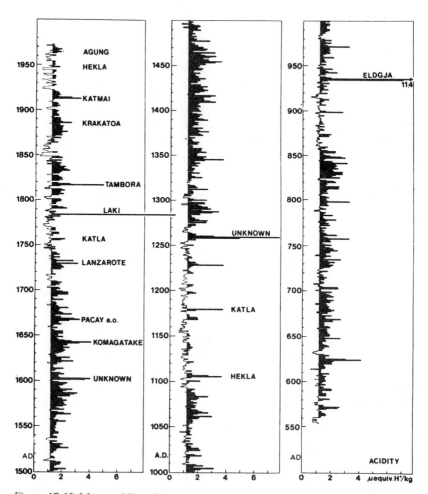

Figure 17.12 Mean acidity of the year-layers from AD 1972 back to 553 in the Greenland ice-sheet at 71°N 37°W, treated as a record of volcanic activity, with known great eruptions named (from Hammer *et. al.* 1980).

More recently, Handler (personal communication, 9 December 1985) has demonstrated a remarkable coherence over the 114 years 1871–1984 between the performance of the Indian south-west monsoon and an updated and corrected list of volcanic explosions. This list included all explosive eruptions other than the small ones which were not followed by any evidence of any significant aerosol in the stratosphere. The theory of this is that veiled (reduced) solar radiation decreases the land–sea temperature contrast between India and the ocean, so weakening the atmospheric circulation and monsoon development (Kutzbach and Otto-Bliesner 1982). Below-average monsoon rainfall was found (chi-square tests show above 99 per cent probability) to be associated with the presence of aerosol veils in the stratosphere over latitudes between 30° and the equator. Absence of aerosol over those latitudes or aerosol veils confined to higher latitudes was a condition associated with above-average monsoons. This result can also explain the very high frequency (77 per cent) of good monsoon years between 1933 and 1962 during an unusually long period of quiescence of the volcanoes. The relationship between the Indian monsoon and the sea-surface temperatures in the equatorial Pacific is such that a decreased monsoon or monsoon failure precedes the warm Pacific (El Niño) event by three months or rather more (Angell 1981), presumably indicating that the effects of the circulation anomaly in the ocean take rather longer to develop, while the sensitivity of the Indian monsoon to low-latitude veils in the stratosphere can be almost immediate. That the relationships found by Handler between volcanic aerosol veils and the Indian monsoon showed themselves without discriminating between the severest (DVI of the order of 1000) and much more moderate (DVI of the order of 100) volcanic explosions speaks for Bryson's (1976) contention that moderate volcanic eruptions are also effective, presumably through conversion of the gases ejected into aerosol in the stratosphere.[*]

Other chronologies have come from new observational techniques applied to ice-cap studies. Hammer (1977) and Hammer, Clausen and Dansgaard (1980) have derived a year-by-year chronology of the volcanic matter reaching the crest of the ice-sheet in central Greenland (71°N 37°W) from AD 553 to the present and a more approximately dated record of deposits from very great eruptions going back 10,000 years. It was not possible to trace silicate particles in the Greenland ice from many of the greatest eruptions in remote parts of the Earth, but study of the amounts of

---

[*] Handler has argued further that the association between volcanic aerosol veils in the stratosphere and deficient monsoons in India is so strong that one might reasonably assume all the poor monsoon years as due to this cause. This assumption would suggest that 10–15 veils due to low-latitude eruptions that were never known about, such as the mystery aerosol veil in early 1982, had indeed occurred. This may then be taken as the number of eruptions missing from the lists during the roughly 100-year period after 1870. Not surprisingly, in the sixteenth and seventeenth centuries, and back to classical times, there were occasional dense veils reported, which visibly weakened the sun's rays but for which no known eruption can be identified as the origin.

volcanic acids (H_2SO_4, SO_2-induced ions, HCl, etc.) by means of electrical conductivity measurements in the year layers yielded a record (figure 17.12) which tallied well with the Lamb Dust Veil Index chronology (Hammer 1977). Moreover, some earlier great volcanic eruptions, known from historical records, e.g. Hekla in AD 1104, could be clearly identified. Finally, these workers compared successive 50-year average values of the volcanic acidity products in the Greenland ice since AD 553 with 50-year averages of such temperature indicators as are available: the central England temperatures derived by the present writer (Lamb 1965) from historical records, the tree-ring growth records at the upper tree-line in California (LaMarche 1972), and the oxygen isotope record derived by Dansgaard and his co-workers from the Greenland ice-sheet itself. The result indicated a long time of increased volcanic activity from about AD 1250 to 1700, coinciding with the development of the cold climate known as the Little Ice Age (figure 17.13). The correspondence even included such details as a pause in the volcanism around AD 1500–50, when the temperatures in Europe and California recovered somewhat, and the recurrence of volcanism around the 1800–50 period, a time of reversion to lower temperatures in all the regions named.

Another chronology of this nature was obtained by Mosley-Thompson and Thompson (1982) from analysis of the solid particle concentration in the year-layers of a 101-metres-long ice-core taken at the South Pole (figure 17.14). (The large particles must be overwhelmingly attributable to volcanic eruption products in the atmosphere owing to absence of other extensive sources of dust.) This record has been analysed back to AD 1056. It reveals a long time of increased deposit from about AD 1450 to 1860. Correspondence with the cold climate period of recent centuries is again suggestive. The timing of the Little Ice Age development was not, however, identical in high

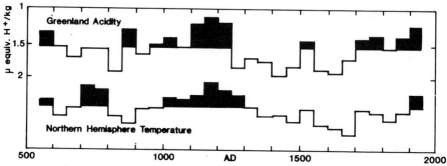

Figure 17.13 Successive 50-year means of the acidity measured in the year-layers of the Greenland ice-sheet at 71°N 37°W, compared with an index of northern hemisphere temperature compiled by Hammer *et. al.* (1980), from published estimates of temperatures in central England, the oxygen isotope record from Greenland and tree growth (mean ring-widths) at the upper tree line in California. The acidity is shown as increasingly downwards, to bring out the parallelism with the temperature changes.

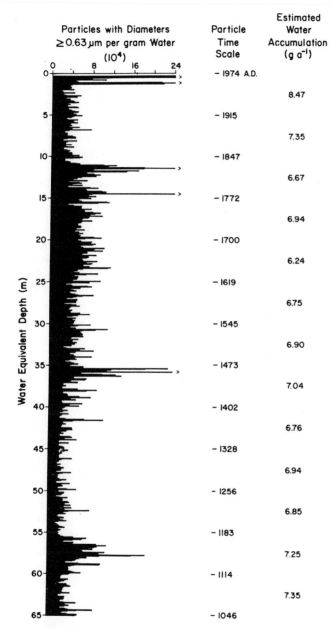

Figure 17.14 Concentration (numbers per thousand microlitres) of large particles in the year-layers of the Antarctic ice-sheet at the South Pole (from Mosley-Thompson and Thompson 1982).

northern and southern latitudes (see, for instance, Lamb 1982, pp. 36, 246): in the Antarctic and sub-Antarctic there is a variety of independent evidence that the coldest phases may have been between about AD 1450 and 1670 and in the later nineteenth century from about 1840 to 1900 or even a little after.

Both these ice-sheet chronologies, which are perhaps our most objective volcanic records, indicate that the Dust Veil Index rating of the Coseguina (Nicaragua) eruption in 1835 in Lamb's chronology may be a considerable overestimate. This could, however, be due to some anomaly in the cross-latitudes spread of the eruption products by the wind circulation on that occasion: there were great anomalies of the general circulation in that decade (Lamb 1963). Lamb believed the Coseguina veil to have been peculiarly long-lasting over middle and lower latitudes, but its poleward spread may well have been impeded.

We turn now to the newest development of volcanic activity. The increased incidence of explosive eruptions that began around 1960 has continued. Figure 17.10 illustrates this by an updating of the Lamb Dust Veil Index series, provisionally completed to the end of 1982. After some quieter years in the 1970s, the great eruptions in 1980–2 seem to have brought the global total DVI yearly values and 25-year cumulative totals back to levels not recorded over either hemisphere since the early years of this century. This is obviously a great opportunity for observational work.

There have been peculiarities in the spread of tephra following each of the greatest eruptions in the latest years, that of Mount St Helens in Washington state in May 1980, and El Chichon in southern Mexico in April 1982, which must have affected their climatic impact. When it became known that in the Mount St Helens eruption possibly 90 per cent of the ejecta were blasted nearly horizontally, assessment of the dust veil in the stratosphere, initially put at 1000 or more, had to be scaled down to 150 or less. In the case of El Chichon, it seems that the initial DVI estimate of 600 based on measured direct insolation deficiency and an estimate of the likely duration of the veil must finally be adjusted to about 800. The total stratospheric burden, with contributions from other volcanoes between January 1982 and spring 1983, was reported to be many times as great as after Mount St Helens. Various types of diffusion and vertical motion were found to have carried the El Chichon aerosol much higher (up to 39 km) in the stratosphere than the heights first observed after the eruption. The final duration of the El Chichon veil has been assessed as about four years. By the end of 1985, however, the remnant of this volcanic veil, at around 18–19 km height, was being supplemented by new material at heights up to 25–28 km attributed to the eruption of the Colombian volcano Ruiz on 13 November 1985.*

The El Chichon stratospheric veil was also unusual in its initially continued concentration over latitudes between about 10° and 40°N for almost nine months after the eruption, although small amounts had much

---

* By late 1986, the persistence of widely observed optical effects makes it appear that the final DVI rating of the Ruiz eruption will hardly be less than 100 to 150.

earlier reached both polar regions. This suggests the likelihood of an unusual consequence that for most of the first year the gradient of solar heating must have been increased over the lower middle latitudes 30° to 50°N. Only in late winter (1982–3) did evidence of substantial poleward spread of the veil over the northern hemisphere suggest conditions for the usual weakening of gradients and wind circulation over those latitudes.

This diversity observed in recent eruptions emphasizes the limitations of what could be learnt from statistical studies of the information on eruptions reported in the historical past, despite the obvious rewards that have come from such studies. From now on the emphasis should surely be on studies directed towards interpretation of atmospheric relationships which study the individual characteristics of each case.

The results obtained by Hammer *et. al.* (1980) from measurements of the sulphuric acid of volcanic origin deposited in the year-layers in the Greenland ice-cap over the last 1500 years, and by Handler (1984) from study of the atmospheric and ocean circulation responses to the volcanic eruptions of the last 115 years, seem to give primacy to the occurrence of veils of volcanic matter in the stratosphere as by far the main cause of the climatic changes observed during these periods. In this, it must be suspected that less than due weight is given to the evidence of anomalies at source in the output of energy from the sun, at least during the Spörer and Maunder minima of solar disturbance between about AD 1410 and 1510 and from 1645 to 1715 respectively (Eddy 1977). Variations in the amount of carbon dioxide in the atmosphere have also occurred during recent centuries, particularly the increasing trend that has accompanied the ever-greater burning of fossil fuels since the Industrial Revolution and the widespread clearing of forests for agriculture since the mid-nineteenth century. The increase of CO_2 since 1850 is estimated at 25–30 per cent (Flohn 1986), an increase that would by many be supposed sufficient to account for the whole warming of climates observed since that time. The various ways by which the natural environment is evidently buffered against the effects of carbon dioxide increase, perhaps partly through the simultaneous increase of water vapour in the atmosphere and increased cloudiness in various latitudes, are apparently not yet fully understood.

That there is an effect from volcanic veils in the stratosphere in weakening, usually slightly, the solar radiation reaching the Earth can hardly be doubted any longer. It seems that it may well have been the most important single influence at work in the climatic variations of the last 200 years, though clearly not the only significant variable. The difficulties of an exact assessment are enormous because of all the other influences that are always varying too. Schneider (1983) has suggested that even the great volcanic veil from the eruption of Tambora (8°S 118°E) in April 1815 in the equatorial Pacific, which was followed by the 'year without a summer' in 1816 in eastern North America and much of Europe, could not by itself account for the frost in July in that year in New England. It is reasonably certain that the jet-stream dipped south in that summer over eastern North America and western Europe, around cold outflows of Arctic air repeatedly affecting just those longitudes. But this seems to be a common feature of the summers in volcanic dust veil years, when a low-latitude eruption occurred at some time between rather over 12 months and three years *before* (allowing time for the veil to have cleared from the low latitudes and drifted over the high latitudes) and also the first one to three summers *after* high latitude eruptions. The 1816 case differs only in the extremity of the anomaly associated with the Arctic air's southward extension in those longitudes and the upper cold troughs which evidently thrust south with the cold air.

It is presumably a testimony to the strength of the association between volcanic veils over the higher latitudes and summer climate in Britain and neighbouring regions that statistical relationships appear to exist between the occurrences of volcanic eruptions in Iceland and human and plant diseases in Europe. K. Matossian reports indications of a significant association between the eruptions in Iceland and the crude death rate in England over recent centuries before about 100 years ago (personal communication, 1 August 1985). And a similar implication lies behind Handler's report of indications of statistical associations between volcanic veils and corn yields in the USA (personal communication, 12 September 1984).

Appendix to Chapter 17: Calculation of the Dust Veil Index

The formulae by which the Dust Veil Index (DVI) is defined and calculated are:

$$DVI = 0.97 R_{Dmax} . E_{max} . t_{mo} \qquad (17.1)$$

where R_{Dmax} is the greatest percentage depletion of the direct solar radiation as measured by the monthly averages in middle latitudes of the hemisphere affected some time after the eruption.

E_{max} stands for the maximum extent of the dust and aerosol veil, rated on a crude scale: whole Earth = 1; extratropical latitudes in one hemisphere plus the whole tropical zone (for eruptions at latitudes between about 20° and 35°) = 0.7; for eruptions between about latitudes 35° and 42° take 0.5; and for eruptions in higher latitudes E_{max} is taken as 0.3.

t_{mo} is the total duration of the dust and aerosol veil until last observed directly or by its effect on the monthly averages of radiation receipt at the observatories in middle latitudes:

$$DVI = 52.5 T_{Dmax} . E_{max} . t_{mo} \qquad (17.2)$$

where T_{Dmax} stands for the estimated lowering of average temperature (degC) for the whole year in the middle-latitude zone of the hemisphere affected in the year most affected:

$$DVI = 4.4q. E_{max}. t_{mo} \qquad (17.3)$$

where q is the estimated total volume of solid matter dispersed from the volcano as dust, measured in cubic kilometres.

When two or more of these formulae can be used, the DVI is given as the round figure nearest to the mean of the values obtained.

Table 17.1   Volcanic eruptions since 1870 known (or reasonably supposed) to have produced moderate to severe dust and aerosol veils in the stratosphere

Date	Volcano	Position	DVI estimate
23 April 1872	Vesuvius	41°N 14°E	70
April 1872	Merapi, Java	7½°S 110°E	(80)
8–10 January 1873	Grimsvötn, Iceland; greatest tephra fall from this volcano in recent times though probably reduced by the Vatnajökull ice-cap overlying the volcano	64½°N 17½°W	?
29 March 1875	Askja, Iceland	65°N 17°W	300
25 June 1877	Cotopaxi, Ecuador	1°S 78°W	50
1878	Ghaie or Raluan	4°S 152°E	?
26–28 August 1883	Krakatoa, Sunda Straits	6°S 105½°E	1000
6 October 1883	St Augustine, Alaska	59½°N 153½°W	20–50?
1885	Falcon Island	20°S 175°W	300
10 June 1886	Tarawera, New Zealand	38½°S 176½°E	(400)
1886	Niafu, Tonga	16°S 175½°E	300
1888	Ritter Island, Bismarck Archipelago	5½°S 148°E	(250)
1888	Bandai San, Japan	38°N 140°E	(250)
February 1890	Bogoslov, Aleutian Islands	54°N 168°W	(50)
7 June 1892	Awu, Great Sangihe	3½°N 125½°E	100
1895 or 1896	Thompson Island; island supposed destroyed entirely by great volcanic eruption	c. 54°S 5°E	(400?)
1898	Una-Una, Celebes	0°S 122°E	140
8 May 1902	Mont Pelée, Martinique	15°N 61°W	(100)
17 May 1902	Soufrìere, St Vincent	13½°N 61°W	(300)
24 October 1902	Santa Maria, Guatemala	14½°N 92°W	(600)
28 March 1907	Ksudatch, Kamchatka	52°N 157½°E	150
6 June 1912	Katmai, Alaska	58°N 155°W	150–200
13 December 1921	Puyehue, Andes	40½°S 72°W	100
4 August–25 September 1928	Paluweh, Indonesia	8°S 121½°E	100?
10 April 1932	Quizopu and others in the Chilean Andes	35½°S 70½°W	30–50

continued

Table 17.1—(continued)

Date	Volcano	Position	DVI estimate
29 March 1947	Hekla, Iceland	64°N 19½°W	20–50
21–24 May 1960	Pontiagudo and others in southern Chile	39°–45°S 72°–73°W	50
19 February–17 March 1963	Agung, Bali	8½°S 115½°E	800
12 August 1966	Awu, Celebes	3½°N 125½°E	200
11–12 June 1968	Fernandina Island, Galapagos	½°S 92°W	200
August–October 1971	Mount Hudson, Chile, and Fuego, Guatemala	45½°S 73½°W 14½°N 91°W	total 100
14 October 1974	Fuego, Guatemala	14½°N 91°W	200
22–25 January 1976	St Augustine, Alaska	59½°N 153½°W	100
13–25 April 1979	Soufrière, St Vincent	13½°N 61°W	total 50?
13–17 November 1979	Chico, Galapagos Islands	1°S 91°W	total 50?
18 May 1980	Mount St Helens, US Rockies	46½°N 122°W	100–150
January 1982	Stratospheric cloud of unknown origin		300
29 March–4 April 1982	El Chichon, Mexico	17½°N 93½°W	800
5 April 1982	Galungung, Java	7½°S 108°E	100–200
13 November 1985	Ruiz, Colombia	5°N 75½°W	?

Note: Brackets are placed around the more uncertain DVI figures.

References

Angell, J. K. (1961) 'Use of constant level balloons in meteorology', *Advances in Geophysics*, 8, 138–219, New York.

Angell, J. K. (1981) 'Comparison of variations in atmospheric quantities with sea temperature variations in the equatorial eastern Pacific', *Monthly Weather Review*, 109, 230–243.

Berger, A. (1980) 'The Milankovitch astronomical theory of palaeoclimates – a modern review', *Vistas in Astronomy*, 24, 103–122, Oxford, Pergamon.

Bryson, R. A. (1982) 'Volcans et climat', *La Recherche*, 7–8(135), 844–853.

Bryson, R. A. and Dittberner, G. J. (1976) 'A non-equilibrium model of hemispheric mean surface temperature', *J. of Atmospheric Sciences*, 33(1), 2094–2106, Lancaster, Pa., American Meteorological Society.

Budyko, M. I. (1968) 'On the causes of climatic variations', *Sveriges Meteorologiska och Hydrografiska Institutets Meddelanden*, series B, 28, 6–13, Stockholm.

Defant, A. (1924) 'Die Schwankungen der atmosphärischen Zirkulation über dem nordatlantischen Ozean im 25-jährigen Zeitraum 1881–1905', *Geografiska Annaler*, 6, 13–41, Stockholm.

Eddy, J. A. (1977) 'Climate and the changing sun', *Climatic Change*, 1(2), 173–190, Boulder, Colo.

Flohn, H. (1986) 'Singular events and catastrophes now and in climatic history', *Naturwissenschaften*, 73, 136–149, Berlin and Heidelberg, Springer.

Hammer, C. U. (1977) 'Past volcanism revealed by Greenland ice sheet impurities', *Nature*, 270(5637), 482–486, London.

Hammer, C. U., Clausen, H. B., and Dansgaard, W. (1980) 'Greenland icesheet evidence of post-glacial volcanism and its climatic impact', *Nature*, 288(5788), 230–235, London.

Handler, P. (1984) 'Possible association of stratospheric aerosols and El Niño type events', *Geophysical Research Letters*, 11(11), 1121–1124, Baltimore, Md, American Geophysical Union.

Handler, P. (1986) 'Possible association between the climatic effects of stratospheric aerosols and the surface temperatures in the eastern tropical Pacific Ocean', *J. Climatology*, 6, 31–41, London, R. Meteorol. Soc.

Hays, J. D., Imbrie, J., and Shackleton, N. J. (1976) 'Variations of the Earth's orbit: pacemaker of the ice-ages', *Science*, 194, 1121–1132, Washington, DC.

Hirschboeck, K. K. (1979–80) 'A new world-wide chronology of volcanic eruptions', *Palaeogeography, Palaeoclimatology, Palaeoecology*, 29, 223–241, Amsterdam, Elsevier.

Hofmann, D. J. and Rosen, J. M. (1977) 'Balloon observations of the time development of the stratospheric aerosol event of 1974–1975, *J. Geophys. Res.*, 82(9), 1435–1440, Baltimore, Md, American Geophysical Union.

Hofmann, D. J. and Rosen, J. M. (1982) 'Balloon-borne observations of stratospheric aerosol and condensation nuclei during the year following the Mt St Helens eruption', *J. Geophys. Res.*, 87(C12), 11039–11061, Baltimore, Md, American Geophysical Union.

Hofmann, D. J. and Rosen, J. M. (1983) 'Stratospheric sulfuric acid fraction and mass estimate for the 1982 eruption of El Chichon'.

Hoyt, D. V. and Fröhlich, C. (1983) 'Atmospheric transmission at Davos, Switzerland, 1909–79', *Climatic Change*, 5(1), 61–71, Boulder, Colo.

Humphreys, W. J. (1920) *Physics of the Air*, New York, McGraw-Hill; 2nd edn, 1940.

Imbrie, J. and Imbrie, K. P. (1979) *Ice Ages: Solving the Mystery*, London, Macmillan.

Kelly, P. M. and Sear, C. B. (1983) 'The short term impact of volcanic activity on climate', ms. University of East Anglia, Norwich.

Köppen, W. (1873) 'Über mehrjährige Perioden der Witterung, insbesondere über die 11-jährige Periode der Temperatur', *Zeitschrift für Meteorologie*, 8, 241–248, 257–267, Vienna.

Köppen, W. (1914) 'Lufttemperaturen, Sonnenflecke und Vulkanausbrüche', *Meteorologische Zeitschrift*, 31, 305–328,Braunschweig.

Kutzbach, J. E. and Otto-Bliesner, B. L. (1982) 'The sensitivity of the African-Asian monsoonal climate to orbital parameter changes for 9000 years BP in a low-resolution general circulation model', *Almos. Sciences*, 39(6), 1177–1188.

LaMarche, V. C. (1972) 'Climatic history since 5100 BC from tree-line fluctuations, White Mountains, east-central California', *Geological Society of America Abstracts with Programs*, 4(3), 189.

Lamb, H. H. (1963) 'On the nature of certain climatic epochs which differed from the modern (1900–39) normal', in *Proceedings of the WMO/UNESCO Rome 1961 Symposium on Changes of Climate*, Paris, UNESCO, Arid Zone Research series, No. XX, 125–150.

Lamb, H. H. (1965) 'The early medieval warm epoch and its sequel', *Palaeogeography, Palaeoclimatology, Palaeoecology*, 1, 13–37, Amsterdam, Elsevier.

Lamb, H. H. (1970) 'Volcanic dust in the atmosphere; with a chronology and assessment of its meteorological significance', *Phil. Trans. Roy. Soc. London*, series A, 266(1170), 425–533.

Lamb, H. H. (1972) *Climate: Present, Past and Future. Vol. 1, Fundamentals and Climate Now*, London, Methuen.

Lamb, H. H. and Woodroffe, A. (1970) 'Atmospheric circulation during the last ice age', *Quaternary Research*, 1(1), 29–58, New York, Academic Press.

Mohn, H. (1877) 'Askeregnen den 29de–30te marts 1875', *Videnskàbs Selskabets Forhandlinger*, No. 10, Cristiania (Oslo).

Mosley-Thompson, E. and Thompson, L. G. (1982) 'Nine centuries of microparticle deposition at the South Pole', *Quaternary Research*, 17, 1–13, New York, Academic Press.

Newhall, C. G. and Self, S. (1982) 'The volcanic explosivity index (VEI): an estimate of explosive magnitude for historical volcanism', *J. Geophys. Research*, 87(C2), 1231–1238, Baltimore, Md, American Geophysical Union.

Rampino, M. R., Self, S., and Fairbridge, R. W. (1979) 'Can rapid climatic change cause volcanic eruptions?', *Science*, 206, 826–829, Washington, DC.

Remsberg, E. E. and Northam, G. B. (1975) A comparison of dust-sonde and lidar measurements of stratospheric aerosols', *Proceedings of the Fourth Conference on the Climatic Impact Assessment Program, Rep. DOT-TSC-OST-75-138*, Springfield, Va., Nat. Tech. Info. Service.

Royal Society (1888) *The Eruption of Krakatoa and Subsequent Phenomena: Report of the Krakatoa Committee of the Royal Society*, G. J. Symons (ed.), London, Harrison & Trübner.

Sapper, K. (1917) 'Beiträge zur Geographie der tätigen Vulkane', *Zeitschift der Vulkankunde*, 3, 65–197, Berlin.

Sapper, K. (1927) *Vulkankunde*, Stuttgart, Engelhorn Verlag.

Schneider, S. H. (1983) 'Volcanic dust veils and climate: how clear is the connection?' (editorial), *Climatic Change*, 5(2), 111–113, Boulder, Colorado.

Schneider, S. H. and Mass, C. (1975) 'Volcanic dust, sunspots and temperature trends', *Science*, 190, 741–746, Washington, DC.

Sear, C. B. and Kelly, P. M. (1983) 'The climatic significance of El Chichon', *Climate Monitor*, 11, 134–139, Norwich, University of East Anglia.

Shaw, N. (1930) *Manual of Meteorology. Vol. 3, The Physical Processes of Weather*, Cambridge, Cambridge University Press.

Simkin, T., Seibert, L., McClelland, L., Melson, W. G., Bridge, D., Newhall, C. G., and Latter, J. (1981) *Volcanoes of the World*, Stroudsberg, Pa., Hutchinson Ross.

Stothers, R. and Rampino, M. R. (1981) 'The major volcanic eruptions, 500 BC to AD 630', ms. Goddard Institute for Space Studies, NASA, New York.

Stothers, R. B. and Rampino, M. R. (1983a) 'Historic volcanism, European dry fogs, and Greenland acid precipitation, 1500 BC to AD 1500', *Science*, 222, 411–413, Washington, DC.

Stothers, R. and Rampino, M. R. (1983b) 'Volcanic eruptions in the Mediterranean before AD 630 from written and archeological sources', *J. Geophysical Research*, 88(B8), 6357–6371, Baltimore, Md, American Geophysical Union.

Wagner, A. (1940) *Klima-änderungen und Klimaschwankungen*, Braunschweig, Vieweg.

Wexler, H. (1956) 'Variations in insolation, general circulation and climate', *Tellus*, 8, 480–494, Stockholm.

Part III

Conclusions: what of the future?

18

The future of the Earth: greenhouse or refrigerator?

This chapter is based on the opening survey given by the writer as Vice-President of Section A (Physics) of the British Association for the Advancement of Science at its Norwich Meeting on 11 September 1984; it is now suitably updated. A shortened version of the address was published in Dr G. T. Meaden's Journal of Meteorology, *vol. 9, no. 92, 1984, pp. 237–42.*

Any forecast of future weather, or of the longer-term climate, must, if it is to be realistic, depend on knowledge and at least some understanding of the physical processes at work. With a public increasingly aware of how science contributes to our competence, this is much more widely appreciated than it used to be. Yet there are still many who do not distinguish between a well-grounded forecast and the most ill-informed guesswork of an amateur. Sometimes, sad to say, such guesswork seems to be featured very publicly on the media – maybe just for the sake of its 'joke value' – though this is really unkind to the amateurs themselves, who are thereby exposed, and to others among them who have a better understanding. Predictions based on mere extension of a recent trend, or on supposed cycles which cannot be related to any known physical process, are worthless.

We have seen that many different changes in the natural environment influence the weather and climate over shorter or longer periods. The immediate cause of a change may often be put down to changes in the prevailing patterns of wind and ocean currents: though the cause of these lies somewhere else, they at least determine where the changes of temperature and weather will be most marked and where they will be milder or even in the reverse direction to the more general experience.

The basis of any climate forecast: causes of changes foreseen

Changes in the extent and location of persistent ice and snow surfaces influence the patterns of the atmospheric circulation while the position lasts (Lamb 1955). The extreme case is an ice-age situation, but the differences from one year to another can be important. Changes in the extent and position of cold and warm water surfaces in the world's oceans, though themselves largely due to previous anomalies in the winds, are often even more important: this has been found to be particularly significant in the tropics, wherever there is a supply of cold water upwelling from the lower layers of the sea at coasts where the Trade Winds blow out to sea and which spread the cold water from there far across the ocean surface. The importance of this is not only in its effect on the large-scale temperature distribution and the positioning, farther north or farther south, of the thermal gradients associated with jet-stream winds. It also makes a very big difference to the uptake of moisture into the atmosphere from the ocean, to supply the formation of clouds and precipitation somewhere along the path of the wind. The latent heat of condensation thereby released provides a major source of energy for the development of storms (and weaker weather systems) over other latitudes. The prime example of the effectiveness of a cold ocean surface supplied by upwelled water is in the equatorial Pacific, seen in the more or less world-wide patterns of anomalous weather that occur when the supply fails. When that supply weakens in the course of the Southern Oscillation, or is cut off in the more extreme El Niño phenomenon, substituting as warm a surface as in the rest of the equatorial zone, there is an enormous increase in the moisture uptake, witnessed commonly by the rainfall in Peru; and the zones of strong thermal gradient shift to the lower-middle latitudes. Also effective, though on a rather less extensive scale, are the cold water surfaces developed by upwelling in the ocean in regions occupied for some long time by low-pressure systems, with cyclonic rotation of the winds about their centre. When such situations develop – as they are liable to do – over the Iceland or Newfoundland regions of the North Atlantic and in parts of the northern Pacific, certain preferred distributions of anomalous winds and weather (as we now know) fairly regularly occur and can be traced right around the hemisphere (Bjerknes 1969; Lamb and Ratcliffe 1972; Lamb 1972, pp. 389–410; Murray and Ratcliffe 1965; Namias 1969).

For a more fundamental view towards what causes the anomalies in the atmosphere and ocean circulations, we may look to external influences on the energy supply. It is not necessarily the case that every change has to have an identifiable external cause. Lorenz (1968, 1970) has argued that the atmospheric and ocean circulation regime may be, or may often be, liable to switch over to another mode of balancing the heat economy which will be, in its turn, equally stable or almost stable. The set of equations that describe the behaviour of the climate system has many solutions, each presumably corresponding to a possible regime. In this indifferent – or 'intransitive' – situation, it may take very little to cause a change of regime. But some changes of climate seem to correspond regularly, and in an intelligible manner, to known changes in the external conditions. And when this can be sufficiently demonstrated, we can speak of causes and forecasting becomes possible, though always subject to the proviso that no other, stronger influence supervenes within the forecast period.

We have mentioned how the possibilities of fluctuation or change in the energy output of the sun, or in the transparency of the atmosphere to incoming and outgoing radiation (from the sun and from the Earth respectively), may have been involved in climatic changes in the past and may be involved today. On longer time-scales, changes of climate have certainly occurred due to the regular cyclic variations in the Earth's orbital arrangements: (i) in the eccentricity, and ellipticity, of the orbit affecting the seasonal range of our planet's distance from the sun; (ii) in the obliquity, or tilt, of the Earth's polar axis, the variations of which change the latitudes of the tropical and polar circles; and (iii) the precession, whereby the times of year at which the Earth is nearest to, and farthest from, the sun gradually change. Cyclical variations found (Berger *et. al.* 1984) in the oxygen-18 content of plankton skeletons in ocean-bed deposits over millions of years past show just the periodicities associated with these orbital variations: (i) about 100,000 years, (ii) 41,000 years and (iii) 19,000 and 23,000 years. The variations over time of ^{18}O in the ocean waters, which these measurements reflect, are associated with changes of global temperature and the volume of ice locked up in ice-sheets on land. Over much longer ranges of the geological past, the climate changed due to drifting of the continents, very slowly changing the extent of land and ocean, and their positions and latitude distribution, and the heaving up and later erosion of mountain ranges.

We have to consider the whole complex of influences under which climate continually varies. Obviously, too, a major problem from now on, on the scales which concern us, with the increasing scope of human activities, is the balance at any time between the effects of environmental changes attributable to human activity and those of natural origin. If we are to gain some perspective over all this, we must simplify. But we should never forget that the full tally of influences affecting the climate is a long one and that many interactions and feedbacks occur.

We are all aware of how greatly humanity has altered the face of the Earth, notably by destroying the natural forest belt of middle latitudes over the last 5000 years (but most of it probably in the last 100–150 years) to make way for agriculture. And now the equatorial rain-forest is fast going the same way. Such human actions causing a precious natural resource to dwindle almost to nothing are beginning to be seen as irresponsible. But, as a species, we are impressed with our power. And it may be that in the realm of climate these things lead us to suppose that our influence is greater than it really is.

The Earth's climate is produced by energy exchanges going on all the time on the hugest scale, between the tropics and the poles. Moreover, natural causes on their own are certainly capable of producing climatic changes of a magnitude that can only be called devastating. And there is evidence that, despite long periods of seemingly stable climate, some of the great changes of the past have taken place surprisingly quickly. The pollens deposited in what is now a peat-bog in the Vosges Mountains in north-eastern France indicate that in the closing stages of the last warm interglacial period the temperate fir and spruce forest in the region, which was mixed with some oak and hornbeam, was replaced by typical pine–birch–spruce forest as in Scandinavia today perhaps within about 150 years (Woillard 1979). And within that time, the really crucial climatic shift to a colder regime may have been even more rapid. Similarly, in the ending of the warmest postglacial times only 3000–4000 years ago, all across Canada the forest retreated rapidly some 200–400 km from its northern limit and, helped along by lightning and fires among the dead wood, was replaced by tundra within about a century (Nichols 1980). Beetle faunas in Britain give evidence of some equally rapid warmings; in one case, in the middle of the last glaciation some 42,000 years ago, an era with summers as warm as today's was quite quickly introduced and then it was all over again within about 1000 years.

This is, perhaps, the right point at which to stress an aspect of especial interest to Britain and Ireland and near-lying countries which has been mentioned in this book. All climatic anomalies and changes have some geographical pattern. We are used to seeing smaller temperature changes in these islands and most of Europe, both seasonally and in the longer term, than in the heart of the great continents. At the other extreme, the variations from year to year are peculiarly big near the edge of the Arctic ice surface, which itself shifts. However, the stability of the climate we know in Britain and western and northern Europe depends not just on the nearness of the great ocean, but on the currents in that ocean. When and if the currents change – in response to some change in the prevailing winds – our situation changes, sometimes drastically. In the great glaciations of the past, the ocean surface, which today is dominated by the warm, saline water of Gulf Stream origin as far north as the British Isles, Iceland and the Norwegian Sea to 72° or 73°N, was radically altered by polar water from east Greenland extending as far south as Portugal (McIntyre *et. al.* 1972). Hence, the temperature change in these islands was greater than almost anywhere else in the world. And that was not just a remote wonder of the geological past. Occasionally, even in these days, a tongue of the polar water reaches the Faeroe Islands, lowering the sea-surface temperature there occasionally for some weeks by 4–5 degC, and in the late 1600s it seems to have dominated the sea surface between there and Iceland for 30 years, even coming close to Shetland for much of the year 1695 (Lamb 1979, 1982a). There were examples of this in the spring of 1968 and 1969 and more briefly in one or two later years.

From analysis of the accumulated deposits on the ocean beds (Hays *et. al.* 1976; Imbrie and Imbrie 1979), major glaciations seem to have come at about 100,000-year intervals throughout the last million years. And within each glaciation, there were important variations of climate, on various shorter time-scales with substantial changes in the extent of ice and of the polar cold water on the surface of the Atlantic Ocean. We now live in an interglacial period. For something like 90 per cent of the last million years climates have been colder than now and with much more extensive ice than now exists. Analysis of the timing of these changes through the last several glaciations (Hays *et. al.* and Imbrie *et. al.*) has by now made it fairly clear that the trigger lies in the astronomical variations concerning the shape, or ellipticity, of the Earth's orbit, the tilt of its polar axis and the gradual progression of the time of year at which the Earth is nearest to the sun. These items, which were first precisely calculated by Milankovitch (1930), vary the radiation available at each latitude and in each month of the year. The changes in the extent of snow and ice, as the Earth receives less or more of the solar radiation at crucial times of the year, plainly amplify the effect by changing the proportion of the radiation that is actually absorbed. And there

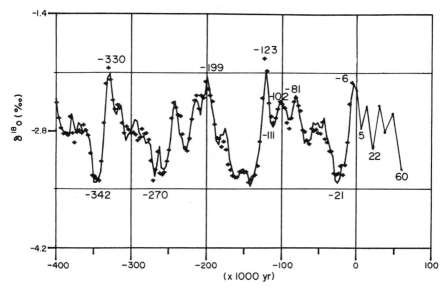

Figure 18.1 Variations of world climate (essentially temperature) over (a) 400,000 years past, and (b) 60,000 years into the future: (a) dots – actual values derived from oxygen isotope ($^{18}O/^{16}O$ ratio) measurements on the material in deep ocean-bed deposits; bold line – a reconstruction of the solar radiation available at the apparently critical latitudes for seasonal melting or net accumulation of snow and ice, based on astronomical measurements (Earth's orbital variables). (b) continuation of radiation curve into future based on same calculations. (From Berger 1980.)

is clearly some phenomenom – presumably a resonance – that makes the 100,000-year time-scale more prominent than it should be on account of the varying eccentricity and ellipticity of the Earth's orbit alone. But the time-scales of these astronomical variations together clearly control the main features of the history of glaciations with the inevitability that night follows day. Hence, the bold pattern of the future can be calculated (figure 18.1). Such calculations, based on the most precise astronomical observations now available (Berger 1980), indicate that the first climax of the next glaciation lies somewhere between 3000 and 7000 years from now into the future. This is probably the most certain element of any climatic forecast that could at present be made.

Causes of change liable to be operative in the shorter term

There are manifestly other causes of climatic change superposed on this long-term framework. Some of these are of importance on various shorter time-scales. They probably account for some, or all, of the most rapid climatic changes. Among these causes are probably fluctuations in the behaviour of the sun and events in the heat economy and circulation of the oceans and their exchanges with the atmosphere. But probably most important are the pollution of the atmosphere by either volcanic eruptions, or from now on (who knows?) by global (especially nuclear) wars and by the increase of carbon dioxide from human activities. The items mentioned are, of course, not the only pollutants of the atmosphere which may have some influence – nitrogen oxides originating in artificial fertilizers, methane from various sources and the chlorofluorocarbons (so-called 'freons') from aerosol sprays may, according to Professor Flohn, together add up to 50 per cent to the carbon dioxide effect – but the items named seem to be the main ones.

Our poisoning of the terrestrial environment by acid rain (through the sulphur oxides in the atmosphere from industrial and domestic burning of coal and oil) and the dumping of nuclear wastes is, of course, quite another cause for concern, in connections that are by no means solely climatic. Winds and water distribute the results. The potential for this is demonstrated in ways that range from the airborne spread of diseases from human viruses and the foot and mouth disease of cattle to potato blight. Even more serious is the spread over hundreds and thousands of miles of many and various sorts of pollution, including the deadly products of nuclear accidents – witness the accident in 1986 at Chernobyl in the Ukraine, the effects of which ruined sheep on the hills of Wales – and the products of nuclear war, which the winds may bring back to visit upon the aggressor as well as the indicated victim.

There is now renewed anxiety about damage to the ozone layer in the stratosphere possibly caused by chemically active agents released by human action, even from such things as the household use of aerosol sprays, as well as the more direct contamination from the exhausts of jet-propelled aircraft and spacecraft. What is in question is the ozone which is formed in the stratosphere at heights around 50 km above the Earth by the action of solar ultra-violet rays upon the oxyen atoms at that height. It protects living organisms, including humanity, on the Earth against lethal concentrations of the ultra-violet light from the sun. Lately, substantial depletion of the ozone has been noted over both the Antarctic and the Arctic. But there is a strong seasonal change in the distribution of the ozone, which is continually transported towards the poles, where it reaches its greatest concentration in late winter to spring and is least in the late summer, despite the amount of local production of ozone which goes on over the polar regions in the long sunlight of summer. Over high latitudes the ozone leaks down out of the stratosphere into the lower atmosphere near the jet-stream and gets destroyed. It is not yet clear whether the recent reports of serious depletion over Antarctica and Svalbard (Spitsbergen), popularly dubbed 'holes in the ozone layer', can be explained within the variability of the seasonal process.

Volcanoes

Of the effectiveness of the sometimes world-wide veils of volcanic dust and aerosol that linger in the stratosphere for several years after great eruptions, there is no doubt. Various chronologies with systematic numerical assessments of the magnitudes of the eruptions of the last several hundred years have been published (Lamb 1970; Newhall and Self 1982; Sapper 1927; Simkin et. al. 1981) and correlations with climate – temperature lowering and atmospheric circulation changes – demonstrated. A lowering of temperature over much of the Earth typically of the order of 0.3–0.5 degC or thereabouts (and occasionally more) over middle latitudes of the northern hemisphere for the first year after a very great eruption has been indicated in some cases. Widespread cooling by 0.1 degC or rather more has been evidenced in many cases. Temperatures in the stratosphere particularly in the layers where the aerosol is most concentrated are found to rise. Two recent studies of the sequels to the greatest eruptions of the past 120 years have brought out further points. A study by Kelly and Sear (1984) has shown that hemisphere-wide effects set in much more quickly than had been thought possible: within just two months, i.e. long before the eruption products have had time to spread out into a uniform or hemisphere-wide veil. Temperature depression was found to be significant over more than a year and sometimes detectable for two to three years after the eruption. Handler (1984, 1986) has discovered changes in the ocean, and specifically in the tropical Pacific, setting in equally quickly: that region is, moreover, one where ocean temperature anomalies are known to be associated with world-wide anomalies of the global wind circulation and weather patterns. It seems clear that the Indian monsoon, in particular, is affected. The mechanism which produces such rapid response is far from clear. Since some of the effects on a wide scale are found to be strongest in the first two to seven months after an eruption, there must be a strong effect from the density of the veil in the early stages, particularly when this is over tropical latitudes, but perhaps also when the radiation received in high latitudes, where the sun is low, is affected by a dust veil over there.

The effects on climate to be expected from the dust and smoke of nuclear war should be of similar nature to those of volcanic explosions but (it has to be expected) of more massive scale because of the numbers of nuclear weapons likely to be fired off in any initial exchange of nuclear hostilities. This has been studied by both American and Soviet scientists, involving the National Center for Atmospheric Research at Boulder, Colorado, and the USSR Academy of Sciences, Moscow, and it has been the subject of a conference (Aleksandrov and Stenchikov 1983; Turco *et. al.* 1983) in Cambridge, Massachusetts, in 1983. It was found that 'for many simulated exchanges of several thousand megatons . . . average daylight levels can be reduced (by the dust and smoke) to a few per cent . . . and land temperatures can reach -15 to -25 degC'. Even in summer, and with only about 100 megatons detonated over just a few cities, temperatures should be reduced to below freezing for months. The prospects for food-growing need hardly be elaborated upon.

Reactions to the unfolding of this understanding in recent years have unfortunately been quite various, as is always the case when humanity is confronted with a new shock to conventional ways of thinking. Among them have been suggestions that military contingency planning for the use of nuclear weapons should be designed to fall just short of precipitating climatic disaster. This, of course, implies an unjustifiable confidence in the precision with which such a borderline could be defined beforehand. It also surely runs against nearly all lessons of experience with attempts to limit the level of weaponry used in war, particularly when a losing nation becomes desperate. Happily, some scientific institutions and institutions of international collaboration in science, as well as some specially convened informal meetings of scientists from East and West, have been prominent in taking a much more responsible line. A statement issued by the Council of the American Meteorological Society on 30 September 1983 recognized 'the inevitable, widespread, devastating consequences of nuclear war by direct explosive effects, and by effects propagated through the atmosphere to the entire globe that could cause the destruction of the biological base that sustains life . . . calls on the nations of the world to take whatever steps are necessary, such as the adoption of appropriate treaties, to prevent the use of nuclear weapons and avoid nuclear war'. And the uncertainties involved have been well expressed in some widely published scientific writings; thus, an article in *Nature* in 1984 by three leading American scientists (Covey *et. al.* 1984) stressed that the materials injected into the lower atmosphere over middle latitudes by large-scale nuclear war, and particularly the global effects of the smoke, 'could cause the temperatures in middle latitudes to drop well below freezing in a matter of days, regardless of season'. Moreover, the wind circulation changes 'could spread the aerosols well beyond the altitude and latitude zones in which the smoke was originally generated'. Needless to say, different approaches, some of them using mathematical models of the general atmospheric circulation, produce slight variations in the severity of the nuclear winter. And it is quite clear that one must expect geographical variations of the effect (Aleksandrov and Stenchikov 1983; Elsom 1985).

Carbon dioxide

The effects of increasing the carbon dioxide in the atmosphere – through our burning of coal, gas, oil, wood and the waste straw and grasses of agriculture – seem at first sight to be as straightforward and grounded in well-understood physics, as those of volcanic dust and aerosol and the products of nuclear war. It is a question of the balance between incoming radiation from the sun and outgoing radiation from the Earth being altered. The Stefan–Boltzmann Law, that a body radiates energy in proportion to the fourth power of its absolute temperature, indicates that a body at the Earth's distance from the sun should have a mean surface temperature around 245–250°K (or about −25 degC). This may indeed be the temperature of the Earth's effective radiating surface, if that is in reality at some height up in the upper atmosphere. But, as we know, the Earth's mean surface temperature is around 288°K (+15 degC). The difference of 40 degC must be accounted for by a 'blanketing' effect in the atmosphere, which has long been attributed to the water vapour and clouds and to the carbon dioxide that absorbs radiation of the long-wave lengths which constitute most of the Earth's emission. This has become widely known as the 'greenhouse effect'. The precise effect of the carbon dioxide, and particularly of increasing the carbon dioxide, in the atmosphere cannot be fully resolved by theoretical modelling because of the complexity of the exchanges between the atmosphere, ocean and ocean-bed, and the biosphere, and the secondary effects of changes in cloudiness. It is agreed that the last-named may be very important, not least because of the possible latitude distribution of the changes. The very high temperatures prevailing on Venus, which have now been measured, are also higher than they should be at the planet's distance from the sun and are normally regarded as establishing the effect of the predominance of carbon dioxide in its atmosphere. However, Newell (1984) has recently argued that the planet Venus may be so much younger than the Earth, and its volcanic activity so much greater, that its internal heat may be the main explanation of its nearly uniformly very high surface temperatures. The situation may be like that on the Earth 3 billion years ago. It seems, therefore, that increasing the carbon dioxide in the Earth's atmosphere – the end-product of all our burning of fossil fuels and of vegetation – should raise the world's temperature, although the effect may be smaller, perhaps very much smaller, than is usually supposed. Much effort in recent years has gone into estimating the rise to be expected (e.g. Manabe and Wetherald 1975; National Academy of Sciences 1977; Smith 1978). The concentration of carbon dioxide in the atmosphere has gone up from around 260 or 270 parts per million (by volume) in the mid-nineteenth century to 340 today and is expected to double (possibly much more than double) during the twenty-first century. Most estimates of the global temperature rise to be expected by AD 2100 are around 2–3 degC (such figures being based on a doubling of the carbon dioxide) but estimates range from near zero to 11 degrees. The figures presently favoured would shift the agricultural belts by many degrees of latitude, upsetting rainfall patterns also, and seem likely to prove beyond the capacity of the world economy and present-day nationalisms to adjust to. Very serious restrictive decisions affecting energy consumption and prevailing practices in most departments of life would have to be faced, starting soon.

There is some disturbing evidence of much greater variations than hitherto supposed in the amount of carbon dioxide in the atmosphere over past centuries before the Industrial Revolution. The record put forward by Stuiver and co-workers, covers the last 1800 years (see Bojkov 1983). What association, if any, the variations may have with changes of world temperature is not clear. There may be a positive correlation between the indicated CO_2 amount and higher temperatures prevailing over the later centuries, but if that were to hold over the whole record, ideas about the variations of average temperature over the world before about AD 700 would need amending. The atmospheric CO_2 variations indicated range from about 280 parts per million around AD 300 to a high of 300–310 p.p.m. between AD 600 and 700, falling sharply to as low as 260 p.p.m. in the later 700s and to about 240 p.p.m. in the 900s. Further minima are indicated, with concentrations around 240 p.p.m. about AD 1250 and 260 p.p.m. between about AD 1560 and 1700. Sharp maxima of about 300 p.p.m. are shown at and soon after AD 1100 and almost as high again more briefly in the early 1500s. The magnitude of the variations indicated before the industrial era is surprising, and the reasons for the rather numerous, and sharp, reversals of trend need explaining.

All evaluations of the effect of increasing carbon dioxide in the atmosphere, even by those such as Idso (1984) who believe the effect may be in the direction of cooling, agree that the effect is likely to be strongest in the highest latitudes, particularly in high northern latitudes. This special sensitivity of the Arctic is because the extent of ice and snow is most easily lessened or increased there, owing to the amount of land and shallow sea between the Arctic circle and latitudes, 75°–80°N.

The argument used by Idso for a cooling effect of carbon dioxide increase is derived from observation of the coincidence between the global, or at least northern hemisphere, cooling observed from 1946 to about 1980 or after with the most rapid increase of carbon dioxide. This argument may not be valid if other variables were responsible, more than counteracting at that time the widely expected CO_2 warming effect. The evidence of recent cooling, noted in Chapter 11, is quite strong.

On the other hand, there are also signs that the eager search in many laboratories for verification of the expected carbon dioxide caused warming of the climates may have been pressed too hard. There is danger in the situation where scientific research is richly funded by interested parties – parties, for instance, whose own support may be affected by the outome of the coal vs nuclear fuel debate and all those other debates surrounding the future of nuclear technologies. The most dispassionate among the informed commentators agree that it is not possible to show any conclusive evidence yet for carbon dioxide warming. Indeed, the growth of the Alpine glaciers, the predominance of cold over warm extremes (see Chapter 11 and, for example, Agee 1982) in North America and Europe during the years of remarkable variability of climate since 1975, the occurrences of snow and killing frosts in low latitudes both north and south of the equator, the continued colder regime in Iceland and shorter frost-free seasons in Canada and northern Europe, and the continuance of temperatures in England averaging below the earlier twentieth-century level, all seem to be preferentially overlooked.

There is unfortunately room for error and misjudgement in the preparation of temperature series which attempt to present averages and trends for the whole Earth, a whole hemisphere or even a whole latitude zone. A large number of temperature-observing stations on land are increasingly affected by artificial sources of heat, particularly from urban growth, as first surveyed by Dronia (1967) and again drawn to attention in work drafted by Professor K. E. F. Watt of the University of California at Davis (personal communication, 13 February 1985). Similarly, many sea-temperature measurements are now made in the engine intakes of ever bigger ships. And temperatures derived by sensing from satellites have their margins of error. Efforts have, of course, been made to adjust for and eliminate these sources of error. But this introduces an arbitrary element, and it may still be open to question how far such adjustments have been correct. Some of the resulting temperature series are hard to reconcile with the indications we have noted of a continuing colder climate than that before 1960, at least in the northern hemisphere. There is no doubt, however, about the warming over the last 50 years in the Antarctic, and since about 1950 in New Zealand. These changes seem to be comparable with the warming of the Arctic between about 1910 and 1945.

The balance, including solar variation

Schneider and Mass (1975) showed that the changes of global, or at least northern hemisphere (of which we are surer), surface temperature from 1880 to 1970 could be rather well explained by indices representing the changes of volcanic matter and carbon dioxide in the atmosphere (counteracting each other), together with a much smaller effect of varying solar activity. Studies by Gilliland (1982), attempting to sort out by several different proposed model constructions what combinations of any two of these variables may best fit the observed changes of northern hemisphere temperature between 1881 and 1975, are also interesting. They support the conclusion that all three variables play a part, and suggest that perhaps also the lunar nodal cycle of 18.6 years, which affects the tidal force, may be important. The solar radiation record was constructed on the basis of relationship to the 11-year sunspot cycle and an assumed correlation of solar luminosity with the recently discovered substantial variation of the sun's radius: this has decreased over the last three centuries, a change believed related to longer-term changes of the sunspot number. This argument makes it possible to construct a supposed record of the sun's radiation output back to the Little Ice Age. The well-known 22-year double sunspot cycle, or Hale cycle, produced by reversals of the sun's magnetic polarity, which has some climatic correlations, does not however come into this – it is not apparent in the solar radius variations. Schönwiese (1983, 1984), using estimates of year-by-year mean northern hemisphere temperature by Borzenkova *et. al.* (1976) for 1880 to 1975, and its extension from 1579 to 1880 partly on the basis of measurements in the Greenland ice (Groveman and Landsberg 1979), found evidence of signfiicant effects of the variations of volcanic material and the increase of carbon dioxide in the atmosphere, together with suggestions of a solar effect besides, though the last-named was more uncertain. Within the whole period since 1579 the increase of carbon dioxide, which may be taken as starting about 1860, appeared as the strongest item, probably accounting for a warming of about 0.6 degC. But the effect of varying volcanic activity was significant over the whole period since 1579 and came out as the strongest one from 1881 to 1980. The CO_2 and volcanic effects were essentially of the same order of magnitude thus far. The suggested effect of solar variation, based on Gilliland's record, Schönwiese found to be of almost half this magnitude.

Budyko (1982) sought to demonstrate that the main course of the Earth's climate through geological ages can be explained by the variations in the carbon dioxide amount in the atmosphere. But not all geologists are agreed that the record fits so well. It seems to be established from analysis of the gas in bubbles trapped in the ice of the Greenland ice-cap that the atmospheric carbon dioxide content was much reduced during the last ice age, probably to about 190 to 200 parts per million. This change must have been at least partly a consequence of the cold climate, because of the great solubility of CO_2 in the colder seas of that time.

It is clear beyond any doubt that there are other influences which also cause the climate to change and which were deeply involved in the last ice age in particular, and in the earlier glaciations in the current (Quaternary) geological period. The Earth's orbital variations can be named with confidence in this connection through the ice ages and interglacials of the last 400,000 years. So one must be more than suspicious of all attempts to explain all the major turns of the Earth's temperature history in terms of just one variable, whether carbon dioxide or volcanic veils or any other, or to present it as overriding all other effects. Our discussion above makes that very clear in the case of the last 500 years.

Nevertheless, attempts have been made to present the whole rise of world temperature since the Industrial Revolution, from the eighteenth to the twentieth century, as attributable to the increase of carbon dioxide, and this is still often implied in current discussions. Schönwiese's analysis shows that this is unwarranted.

If we take the long record of temperatures prevailing in central England since 1659 (Manley 1974) as reasonably representative of the changes that have affected much of Europe (see Chapter 4 in this book), and probably a much wider area of the northern hemisphere (see Chapter 11), a great oscillation of the general temperature level amounting to almost 1.5 degC in the 10-year averages is seen between about 1659 and 1730 (figure 18.2). This

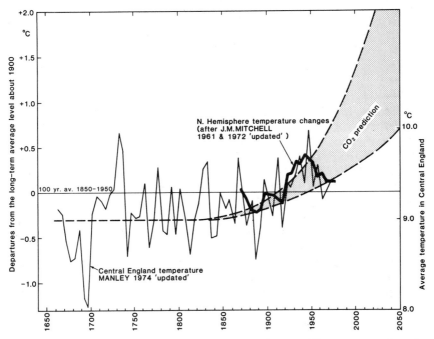

Figure 18.2 The range of 'predictions' of past and future world temperature, based solely on the commonly estimated effect of differences in the amount of carbon dioxide in the atmosphere, and compared with observations in England: Bold curve – observed changes of northern hemisphere average temperatures 1870–1974 (5-year averages after J. M. Mitchell); thin curve – observed changes of prevailing temperatures in central England (5-year averages from Manley 1974).

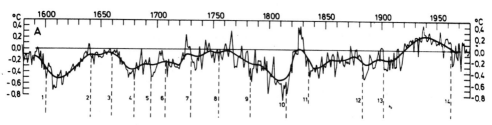

Figure 18.3 Average temperature over the northern hemisphere since AD 1579: before 1880, reconstructed by Landsberg and Groveman (1979) from the indications of oxygen isotope measurements in the Greenland ice-cap; from 1880, actual thermometer measurements averaged over the hemisphere.

is by far the greatest fluctuation in the course of the instrumental record. It is seen also – though rather less sharply – in the northern hemisphere temperatures reconstructed from the Greenland ice record (Groveman and Landsberg 1979) here reproduced as figure 18.3, but in Greenland it seems to have been surpassed by the change (warming) from 1810–16 to the late 1820s and the preceding cooling from the 1760s to 1810–16. These rapid changes certainly cannot be accounted for by carbon dioxide variations or any intrusion originating in the human economy. The 1690s–1730 warming spans the end of the Maunder minimum of solar disturbance. And the 1760–1830 period saw very big variations in the amount of volcanic material in the atmosphere as well as some of the biggest sunspot activity variations in the record. There is an acknowledged possibility that there was also much greater volcanic activity in and around the 1690s than our chronologies reveal, owing to possible underestimation of a succession of great volcanic eruptions known to have occurred in low latitudes in the East Indies, as well as the 1693 eruption of Hekla in Iceland. So it seems likely that both these great fluctuations in the temperature record were caused by a combination of short-term, but drastic, swings in the volcanic aerosol production and in the solar energy itself.

   Of the external influences on climate which seem to have been significant in the record of past climate, these are the two – volcanic activity and fluctuations of the sun – which are liable to the most rapid changes. It seems reasonable to suppose that it is their changes, superimposed on the slower, progressive changes in the radiation available because of the Earth's orbital cycles (and the still slower effects of drifting continents changing the geography), that have caused most of the rapid changes of climate traceable in the past. Some may, however, have occurred for no greater reason than that some critical point had been reached in the slow change of radiation budget in the orbital cycles or in the geographical development, as when rising sea-level after the last ice age led to a sudden removal of permanent ice – removal of the ice dome which was the largest relic of the former North American ice-sheet – from Hudson's Bay within a few centuries around 7000 years ago. Flohn has examined (1986) the evidence of a number of catastrophic climatic events – rapid changes – in ancient and modern times. Other cases may have been brought about through build-up during relatively warm moist interglacial periods of the great ice-sheet on Antarctica until it surged far out over the ocean. Another may have happened through the sudden collapse of a great ice dome based on the bed of the Barents and Kara Seas at a certain stage around 10,800 years ago in the postglacial warming. This has been suggested as the cause of a 600-year reversion of the climate in Europe, which brought back glaciers to the Lake District in north-western England.

Reverting to consideration of our present situation, we must recognize that the world's climate is subject to changing influences from various sources, and at least the influences of the sun, volcanoes and human pollution of the atmosphere can have drastic effects. It is indeed doubtful whether the present-day international economy could be adjusted to the effects of the bigger and more rapid changes that have occurred in the past much better than the more primitive economies of the past 1000 years. In attempting to foresee the likely changes of climate, no one of the three main variable external influences named can be neglected.

The future increase of CO_2 expected and volcanic variations

The atmospheric carbon dioxide is widely expected to double during the course of the twenty-first century, though the time taken would be affected by possible moderation of the rate at which humanity burns coal and oil. Expectations of what the climatic effect should be vary widely, though a resulting rise of world temperature by 1.5–3 degC – much more than this in the polar regions – is probably favoured by most investigators. This would be enough to move the various grain belts – and the desert zone – poleward by several degrees of latitude and thereby change or disrupt the economies of many nations.

The behaviour of the volcanoes cannot be predicted on this time-scale. Any substantial change of prevailing air temperatures such as that looked for from carbon dioxide increase, and the resulting more gradual changes in the extent of land-based ice and of the sea-level and coastal positions, could have some effect in instigating volcanic activity through changes in the loading of different portions of the Earth's crust. The activity of the world's volcanoes has increased since the long lull in the northern hemisphere between about 1912 and 1960, and this seems to have contributed to the cooling of climates after about the 1940s.

Solar changes

Solar changes may also have played a part in this recent cooling. The idea that sunspots, and perhaps other variations affecting the energy output from the sun (or some parts of that output), might be implicated in the variations of weather and climate was popular in the early part of this century until the conspicuous failure of an all-too-simplistic forecast of changes in the level of the great African Lake Victoria at the 1933 minimum of the 11-year sunspot cycle (see Lamb 1972, p. 441). The supposed strong and simple statistical relationship between sunspots and the lake level had been based on less than two complete cycles. The subject fell into disrepute, and any supposition about sun–weather relationships was widely regarded as cranky, particularly among British meteorologists, for some decades afterwards. Willett (1949a, 1949b, 1950, 1951) was the first to revive interest in the topic, and set the research on a new basis, with a remarkable exposition of relationships between the behaviour of the large-scale wind circulation and climate over the globe and the longer sunspot cycles of about 22 years (the Hale cycle) and 80 or 100 years' length (the Gleissberg or secular cycle). This account was based on research which he began in the late 1930s on the general circulation of the atmosphere over the northern hemisphere, with C.-G. Rossby and J. Namias and other colleagues at the Massachusetts Institute of Technology, in connection with the development in the 1940s of a 5-day (and later longer ahead) forecasting technique for the then US Weather Bureau. From about 1945 onwards Willett began to notice the associations with long-term solar variations which led him to conclude that these were important and were the only extra-terrestrial influence of importance to the sequence of climatic behaviour. His views still deserve serious attention because of the success of the long-term forecasts which they led him to make. Much more recent work by J. A. Eddy (1975, 1977, 1979) has highlighted the climatic associations with the Spörer and Maunder extreme minima of solar activity in the fifteenth and seventeenth centuries AD. Eddy also drew attention to the evidence for contraction of the sun's radius after about AD 1700 to very recent years.

It has not been possible to establish statistical significance of the associations appearing between climate and the secular and longer-term solar cycles, because of the shortness of the firm and detailed climatic record in comparison. But, as Gilliland has argued, 'it would be a grievous mistake to ignore a physically reasonable term (i.e. the process, solar forcing), 'merely because a statistically well defined study is not possible' (Gilliland 1984).

Willett divided the secular solar cycle of 80 or 100 years into four quarters, each of the long cycles starting with a sudden break from very active 11-year sunspot cycles (maximum sunspot numbers well over 100) to periods of much more subdued activity, with the later 11-year cycles then gradually building up again to a new active period. During the first quarter of the secular cycle (which may be more sharply pronounced in alternate cycles at 170- to 200-year intervals), Willett found the circumpolar vortex and its polar core to be expanded: the general wind circulation over the hemisphere showed a prevalence of zonal (west to east) orientation with the mainstream of the flow in a rather low latitude. Low temperatures, wetness and glacier growth prevailed in middle latitudes. Over the next two quarters of the secular cycle, the average position of the mainstream of the still predominantly zonal wind circulation shifted, at first slowly and then faster, towards higher latitudes. In the final quarter of the cycle, the wind circulation tended to break down into meridional (northerly, southerly and blocking) patterns, with subpolar anticyclones over the continental sectors of the hemisphere giving warm dry summers and cold dry winters. This analysis has much similarity with the later findings of other investigators (Lamb 1972, p. 450).

On this basis, Willett predicted in 1951 a trend towards cooler and wetter conditions in middle latitudes of the northern hemisphere starting in the 1950s, and that this trend would moderate somewhat in the early 1970s before continuing to further extremes in the 1980s and 1990s. Until 1975 this forecast must be regarded as having been a noteworthy success, especially since it was almost the first scientifically based forecast of climate over such a period ever published. Nevertheless, it became clear that some aspects of the solar behaviour were not properly foreseen, when the highest sunspot maximum was in 1957, not the 1947 maximum as had been supposed (see Lamb 1977, p. 700). The misunderstanding may have been partly due to the apparent nature of this sun–climate relationship, that the warmest phase in middle latitudes tends to come about 20 years before the sharp decline of solar activity – a point on which there is now more evidence, from sunspot and climatic records extending back at least to the sixteenth century (e.g. in the appendices in Lamb 1972, 1977).

It may be worth recalling in passing that the great rise of northern hemisphere temperatures between about 1900 and 1940, which accompanied the increasing sunspot activity up to the 1957 cycle, has been frequently attributed to the increasing carbon dioxide in the atmosphere. The CO_2 increase accelerated after 1950–60, just when the northern hemisphere temperatures were falling. The quiescence of the volcanoes from 1912 to about 1960 was also a feature of the marked warming period. The warming should, perhaps, be attributed more to these features of the period than to the increasing carbon dioxide.

Recent observations of the sun

After 1975, changes have been observed in the course of solar behaviour for which the science was not prepared (Willett 1980). These emphasize how our ability to predict the sun at present limits the possibilities of forecasting the climate when changes in the sun, as well as intense volcanic activity on the Earth, may supervene and change the outcome. The discrepancy between the expected nature of the 11-year sunspot cycles in the latter end of the twentieth century, as predicted up to the mid-1970s, and the 1979–80 sunspot maximum is well seen in figure 18.4. Another change in these latest years is that measurements made regularly since 1979 show the sun expanding again and its luminosity fading, although it is not yet known whether this is a short-term cycle or a long-term trend towards the situation as it was in the Little Ice Age in the seventeenth century. The rate is so far only about 0.02 per cent a year, but the effects may be important especially if the trend were to continue. Figure 18.4 also shows the correspondence

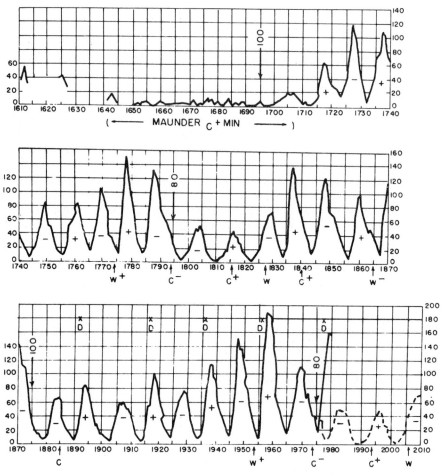

Figure 18.4 Sunspot numbers (vertical scale) and the climate record of northern hemisphere middle latitudes, from AD 1610 to 2010, as presented by H. C. Willett (1980); the plus and minus signs in each solar cycle refer to the predominating polarity of the sun's magnetic field, reversing from one '11-year' sunspot cycle to the next; the W and C indices below the diagram refer to predominating warmth or cold of the climate in middle latitudes, more intense in each case if a plus is added, weaker if a minus is shown; the diagram is designed to show the discrepancy between the very active sunspot maximum around 1980 and the weaker (and longer than 11-year) cycle expected.

between the seventeenth-century cold climate period and the very low sunspot activity of the Maunder minimum.

There may be less reason for surprise than at first appears in the occurrence of one more highly active sunspot cycle with the 1979 maximum. (As is typical of high activity, it was also a short cycle, only ten years long, after the much weaker 12-year cycle that produced the weak 1969 maximum.) Eddy (1977) had noted evidence of at least five other periods of prolonged inactivity of the sun in the last 7000 years, resembling the Spörer and Maunder minima. The latest occurrence of the kind, and the shortest to be considered in the same class, was the period of low activity (and cold climate) between 1795 and 1823, and this was preceded by some erratically active cycles. On the basis of studies linking such solar variations to disturbances caused in the solar system by the regular long-term excursions of the big planets Jupiter and Saturn from the centre of mass of the system (and allowing some influence of the same sort from Neptune and Uranus), Fairbridge and Shirley (1987), supported by Landscheidt (1983), predict another period of quiescent sun, resembling the Maunder minimum from about the present time until around AD 2030 or after.

Conclusions: the forecasting position and implications

The great outburst of research activity in climatology since about 1960–1970 has given us a much more specific knowledge of the history of climate and of the things that cause climate to change, as well as revealing some of the impacts on the economy, on society and on the international order. But science is not yet in a position to make reliable climatic forecasts over many years ahead, apart from the near-certainty of development of the first stages of the next ice age around 3000–7000 years from now. Our understanding makes it fairly clear that the world-wide warmth of the twentieth century in comparison with the immediately preceding centuries, should be attributed partly to carbon dioxide increase; but it seems that the roughly 50-years-long lull in volcanic activity in the northern hemisphere, which led to a clearer, more transparent atmosphere, also played an important part – it may have been the main part. Also the intensity of the sun's beam, apparently in the course of the secular cycle and some still longer cyclical evolution, seems to have been greater during much of the first half of this century than in earlier years (or than it has become in the years 1979–85).

With further increase of the atmospheric carbon dioxide and with new trends in the behaviour of the sun and in volcanic activity, there are grounds for supposing that climate's behaviour may become more erratic in the years ahead. There have indeed been signs of increased variability and incidence of many kinds of extreme since about 1960, as listed elsewhere by the writer (Lamb 1982a, pp. 256–8) and noted in this book. This would also be one implication of the type of forecast indicated in Chapter 11 on the basis of the 200-year (supposedly solar-influenced) cycle seen in the long history of the southwesterly wind in England. That forecast is in fact much in line with the outlook that would be based on the solar forecast referred to above. Over the next half-century or more, the expected relatively low frequency of the general westerly-wind type situation in this part of the world would imply a high frequency of blocking and climatic influences bearing at least some resemblance to those in the periods 1580–1650 and 1780–1850.

Another way of looking at the climatic events of recent years is to see them in the framework of the longer-term glacial–interglacial cycles. The fact that calculation indicates the first climax of the next glaciation, determined by the Earth's orbital variables, should be only 3000–7000 years from now implies that the climatic changes which must before then alter much of the European and temperate North American environment to part of the boreal forest (birch–pine–spruce) zone could be imminent. In this connection, the very long pollen-analytical record obtained by the late Dr Geneviève Woillard (1979) of Louvain la Neuve from a peat bog at Grand Pile in the Vosges Mountains in north-east France, covering the last 140,000 years, is interesting. The earliest part of the record presents the vegetation of the beginning of the last interglacial warm period, next the warmest part of the interglacial and then – after only 11,000 years of that warmth – the decline. The change from temperate spruce and fir forest, mixed with oak and alder and other trees, like today's forest in the area, to the northern birch and pine forest seemed to have taken only about 150 years according to her diagnosis. The warmth-demanding oaks and firs disappeared even more quickly. Woillard's findings have since been called in question by a newly identified break in the Grand Pile deposit record. Frenzel and Bludau (1987) now believe that this vegetation took at least 1500 years, and 3000 to 4000 years to the beginning of glaciation. The implications of such a transition starting about now would, however, be quite similar to a conclusion by J. R. Bray in 1971. On grounds of the recurrent 2000- to 2600-year cycle apparent in solar/auroral and climatic records, Bray suggested that a climate similar to, but more severe than, the Little Ice Age of recent centuries should be expected about 2000 to 2300 years from now – possibly severe enough to start the development of glaciation on the continents in middle latitudes.

Analogy with the duration of previous interglacial warm climates, such as the present, has suggested to Woillard and others, especially botanists, that the Earth may now be at or close to the beginning of the final, cool-temperate stage of the present interglacial – a natural change perhaps sufficient to 'swamp' any climatic warming tendency from the products of human activity, whether over the next 200 or 2000 years.

As the end of the last interglacial approached, a drastic stage may have been introduced by just one very dry, hot summer, analogous to 1976, which killed many of the firs at Grand Pile. This is an interpretation of our present situation which some suggest may have as much to do with the widely reported death of trees in Europe and North America in recent years as acid rain. It is also quite reasonable to suppose, from the evidence of past climates, that the succession of extreme summers and winters in North America and Europe since 1975 may be a signal of serious climatic change.

Climatic fluctuations that menace the stability of the world economy could be before us, particularly in relation to droughts in Africa and elsewhere in the Third World. Our models serve to warn us also of the drastic changes that could come from further increase of carbon dioxide and other effluents originating in human activities. This side of concern for the environment has to be heeded along with the more obvious warnings about gases which acidify the rain and the dangers of emissions from accidents in the nuclear industry and from nuclear-active wastes. At the time of writing, it seems doubtful whether our politicians are as aware as the general public of the dangers and responsibilities implied.

It must be understood in connection with future climate that there is no necessary contradiction between forecasts of: (i) continued or renewed cooling over the next few decades due either to volcanic activity or solar output changes (or both); (ii) a rather strong warming, lasting some centuries, due to increase of carbon dioxide and other pollution from human activities; and (iii) a new ice age developing quite strongly some 3000–7000 years from now and continuing with ups and downs for tens of thousands of years, due to the Earth's orbital variations.

The knowledge that we now have as a result of climatic studies contributes strongly to the dawning appreciation of our responsibility to care for the future of the environment which the Earth provides for mankind, for the animal kingdom and for the plant world.

References

Agee, E. M. (1982) 'A diagnosis of twentieth century temperature records at West Lafayette, Indiana', *Climatic Change*, 4(4), 399–418.

Aleksandrov, V. V. and Stenchikov, G. L. (1983) 'On the modelling of the climatic consequences of the nuclear war', Moscow (Computing Centre of the Academy of Sciences of the USSR).

Berger, A. (1980) 'The Milankovitch theory of paleoclimates – a modern review', *Vistas in Astronomy*, 24, 103–122, Oxford, Pergamon.

Berger, A., Imbrie, J., Hays, J., Kukla, G., and Saltzman, B. (eds) (1984) *Milankovitch and Climate*, Vol. 1, NATO ASI. series, Dordrecht, Reidel.

Bjerknes, J. (1969) 'Atmospheric teleconnections from the equatorial Pacific', *Monthly Weather Review*, 97(3), 163–172, Washington, DC.

Bojkov, R. D. (1983) 'Report of the WMO (CAS) meeting of experts on the CO_2 concentration from pre-industrial times to the IGY', *WMO Project Research Monitor, Atmospheric CO_2 Report No. 10*, Geneva, World Meteorological Organization, 1–34.

Borzenkova, I. I., Vinnikov, K. Ya., Spirina, L. P., and Stekhnovskiy, D. L. (1976) 'Change in the air temperature of the northern hemisphere for the period 1881–1975', *Meteorologia i Gidrologiya*, 7, 27–35, Leningrad.

Bray, J. R. (1971) 'Solar-climatic relationships in the post-Pleistocene', *Science*, 171, 1242–3.

Broecker, W. S. (1975) 'Climatic change: are we on the brink of a pronounced global warming', *Science*, 189, 460–461.

Budyko, M. I. (1974): *Climate and Life*, New York and London, Academic Press.

Budyko, M. I. (1982) *The Earth's Climate: Past and Future*, New York and London, Academic Press.

Covey, C., Schneider, S. H., and Thompson, S. L. (1984) 'Global atmospheric effect of massive smoke injections from a nuclear war: results from general circulation model simulations', *Nature*, 308, 21, London.

Dronia, H. (1967) 'Der Stadteinfluss auf den weltweiten Temperaturtrend', *Met. Abhandlungen*, 74(4). Berlin, Inst. f. Met., Geophys. der Freien Universität.

Eddy, J. A. (1975) *A New Look at Solar–Terrestrial Relationships*, Boulder, Colorado, High Altitude Observatory, National Oceanic and Atmospheric Administration.

Eddy, J. A. (1977) 'Climate and the changing sun', *Climatic Change*, 1(2), 173–190.

Eddy, J. A. and Boornarzian, A. A. (1979) In *Bulletin of the American Astronomical Society*, 11, 437.

Elsom, D. M. (1985) 'Climatological effects of a large-scale nuclear exchange: a review', in *The Geography of Peace and War*, D. M. Pepper and A. Jenkins (eds), Oxford, Blackwell.

Fairbridge, R. W. and Shirley, J. H. (1987) 'Prolonged minima and the 179-year cycle of the solar inertial motion', *Solar Physics*, 110, 191–220.

Flohn, H. (1986) 'Singular events and catastrophes now and in climatic history', *Naturwissenschaften*, 73, 136–149, Berlin and Heidelberg, Springer.

Frenzel, B. and Bludau, W. (1987) 'On the duration of the interglacial to glacial transition at the end of the Eemian Interglacial (Deep Sea stage 5E): botanical and sedimentological evidence.' In *Abrupt Climatic Change* (W. H. Berger and L. D. Labeyrie (eds), pp. 151–62. Reidel, Dordrecht.

Gilliland, R. L. (1982) 'Solar, volcanic and CO_2 forcing of recent climatic changes', *Climatic Change*, 4, 111–131.

Gilliland, R. L. (1984) 'Reply to comments on "Solar, volcanic and CO_2 forcing of recent climatic changes"', *Climatic Change*, 6, 407–408.

Groveman, B. S. and Landsberg, H. E. (1979) 'Reconstruction of the northern hemisphere temperature 1579–1880', *Publication No. 79 181/182*, University of Maryland, Department of Meteorology.

Handler, P. (1984) 'Possible association of stratospheric aerosols and El Niño-type events', *Geophysical Research Letters*, 11(11), 1121–1124, Baltimore, Md, Amer. Geophys. Union.

Handler, P. (1986) 'Possible association between the climatic effects of stratospheric aerosols and sea surface temperatures in the eastern tropical Pacific Ocean', *J. Climatology*, 6, 31–41, London, Roy. Soc. Meteorol. Soc.

Hays, J. D., Imbrie, J., and Shackleton, N. J. (1976) 'Variations in the Earth's orbit: pacemaker of the ice ages', *Science*, 194, 1121–1132.

Idso, S. B. (1984) 'An empirical evaluation of Earth's surface air temperature response to radioactive forcing, including feedback, as applied to the CO_2-climate problem', *Archiv für Meteorologie, Geophysik und Bioklimatologie*, series B, 34, 1.

Imbrie, J. and Imbrie, K. P. (1979) *Ice Ages: Solving the Mystery*, London, Macmillan.

Kelly, P. M. and Sear, C. B. (1984) 'The climatic impacts of explosive volcanic eruptions', *Nature*, 311(5988), 740–743, London, 25 October.

Lamb, H. H. (1955) 'Two-way relationships between the snow or ice limit and 1000–500 mb thickness in the overlying atmosphere, *Quart. J., Royal Meteorol. Soc.*, 81, 172–189, London.

Lamb, H. H. (1970) 'Volcanic dust in the atmosphere; with a chronology and assessment of its meterological significance', *Phil. Trans. Roy. Soc. London*, series A, 266(1170) 425–533.

Lamb, H. H. (1972) *Climate: Present, Past and Future. Vol. 1, Fundamentals and Climate Now*, London, Methuen.

Lamb, H. H. (1977) *Climate: Present, Past and Future. Vol. 2, Climate, History and the Future*, London, Methuen.

Lamb, H. H. (1979) 'Climatic variation and changes in the wind and ocean circulation: the Little Ice Age in the north-east Atlantic', *Quaternary Research*, 11, 1–20.

Lamb, H. H. (1981) 'Climatic changes and food production: observations and outlook in the modern world', *Geo-Journal*, 5.2, 101–112.

Lamb, H. H. (1982a) *Climate, History and the Modern World*, London, Methuen.

Lamb, H. H. (1982b) 'Climatic changes in our own times and future threats', *Geography*, 67(3), 203–220.

Lamb, H. H. and Ratcliffe, R. A. S. (1972) 'On the magnitude of climatic anomalies in the oceans and some related observations of atmospheric circulation behaviour', *Rapports et Procès-Verbaux*, 162, 120–132. Copenhagen, Conseil International pour l'Exploration de la Mer.

Landscheidt, T. (1983) 'Solar oscillations, sunspot cycles, and climatic change, in *Weather and Climate Response to Solar Variations*, B. M. McCormack (ed.), Boulder, Colo., Colorado Associated University Press, 293–308.

Lorenz, E. N. (1968) 'Climatic determinism', *Meteorological Monographs*, 5, 1–3, Boston, Mass., American Meteorological Society.

Lorenz, E. N. (1970) 'Climatic change as a mathematical problem', *J. Applied Meteorol.*, 9, 325–329, Lancaster, Pa., American Meteorological Society.

McIntyre, A. Ruddiman, W. F., and Jantzen, R. (1972) 'Southward penetrations of the North Atlantic polar front: faunal and floral evidence of large-scale surface water mass movements over the last 225,000 years', *Deep-sea Research*, 19, 61–77.

Manabe, S. and Wetherald, R. T. (1975) 'The effects of doubling the CO_2 concentration on the climate of a general circulation model', *Journal of Atmospheric Science*, 32(1), 3–15.

Manley, G. (1974) 'Central England temperatures: monthly means 1659 to 1973, *Quart. J., Royal Meteorological Society*, 100, 385–405.

Milankovitch, M. (1930) 'Mathemathische Klimalehre und astronomische Theorie der Klimaschwankungen', *Handbuch der Klimatologie* (Köppen and Geiger), I, Teil A. Berlin.

Mitchell, J. M. (1961) 'Recent secular changes of global temperature', *Ann. New York Acad. Sciences*, 95, 235–250.

Murray, R. and Ratcliffe, R. A. S. (1969) 'The summer weather of 1968: related atmospheric and sea temperature patterns', *Meteorol. Mag.*, 98, 201–219.

Namias, J. (1969) 'Seasonal interactions between the North Pacific Ocean and the atmosphere during the 1960s', *Monthly Weather Review*, 97(3), 173–192, Washington, DC.

National Academy of Sciences (1977) *Energy and Climate*, monograph, *Studies in Geophysics* series, Washington, DC.

Newell, R. E. (1984) 'Is Venus younger than Earth?', *Speculations in Science and Technology*, 7(1), 51–57.

Newhall, C. G. and Self, S. (1982) 'The volcanic explosivity index (VEI): an estimate of explosive magnitude for historical volcanism', *J. Geophys. Research*, 87(C2), 1231–1238.

Nichols, H. (1980) Personal communication, Institute of Arctic and Alpine Research, University of Colorado, Boulder, 25 July.

Sapper, K. (1927) *Vulkankunde*, Stuttgart, Engelhorn.

Schneider, S. H. and Mass, C. (1975) 'Volcanic dust, sunspots and temperature trends: climatic theories in search of verification', *Science*, 190, 741–746, Washington DC.

Schönwiese, C.-D. (1983) 'Northern hemisphere temperature statistics and forcing. Pt A, AD 1881–1980', *Archiv für Met., Geophys., Biokl.*, series B, 32, 337–360, Vienna, Springen.

Schönwiese, C.-D. (1984) 'Northern hemisphere temperature statistics and forcing. Pt B, AD 1579–1980', *Archiv für Met., Geophys., Biokl.*, series B, 35, 155–178.

Simkin, T. Seibert, L., McClelland, L., Nelson, W. G., Bridge, D., Newhall, C. G., and Latter, J. (1981) *Volcanoes of the World*, Stroudsberg, Pa., Hutchinson Press.

Smith, I. (1978) 'Carbon dioxide and the greenhouse effect', Report ICTIS/ER 01, London, International Energy Agency Coal Research.

Turco, R. P., Toon, O. B., Ackerman, J., Pollack, J. B., and Sagan, C. (1983) 'Nuclear winter: a global consequence of multiple nuclear explosions', *Science*, 222, 1283.

Watt, K. E. F. (1985) 'Tree mortality, acid rain, carbon dioxide and the greenhouse effect', unpublished memorandum, University of California, Davis, Department of Zoology.

Willett, H. C. (1949a) 'Long-period variations of the general circulation of the atmosphere', *J. Meteorol.*, 6(1) 34–50, Lancaster, Pa., Amer. Meteorol. Soc.

Willett, H. C. (1949b) 'Solar variability as a factor in fluctuations of climate during geological time', *Geografiska Annaler* (issue on Glaciers and climate), 36(1–2), 295–315, Stockholm.

Willett, H. C. (1950) 'Temperature trends of the past century', *Centennial Proc. Royal Meteorol. Soc.*, 195, London.

Willett, H. C. (1951) 'Extrapolation of sunspot–climate relationships', *J. Meteorol.*, 8(1), 1–6, Lancaster, Pa.

Willett, H. C. (1980) 'Solar prediction of climatic change', *Physical Geography*, 1(2), 95–107.

Woillard, G. (1979) 'Abrupt end of the last interglacial in north-east France', *Nature*, 281(5732), 558–562, London, 18 October.

INDEX